Proceedings of the 21st Lake Louise Winter Institute

Fundamental Interactions

Proceedings of the 21st Lake Louise Winter Institute

Fundamental Interactions

Lake Louise, Alberta, Canada 17 – 23 February 2006

editors

Alan Astbury
University of Victoria, Canada

Faqir Khanna
University of Alberta, Canada

Roger Moore
University of Alberta, Canada

World Scientific

NEW JERSEY • LONDON • SINGAPORE • BEIJING • SHANGHAI • HONG KONG • TAIPEI • CHENNAI

Published by

World Scientific Publishing Co. Pte. Ltd.

5 Toh Tuck Link, Singapore 596224

USA office: 27 Warren Street, Suite 401-402, Hackensack, NJ 07601

UK office: 57 Shelton Street, Covent Garden, London WC2H 9HE

British Library Cataloguing-in-Publication Data
A catalogue record for this book is available from the British Library.

FUNDAMENTAL INTERACTIONS
Proceedings of the 21st Lake Louise Winter Institute 2006

ISBN-13 978-981-270-367-5
ISBN-10 981-270-367-5

Printed in Singapore by World Scientific Printers (S) Pte Ltd

PREFACE

The twenty-first Lake Louise Winter Institute, entitled "Fundamental Interactions", was held from February 17-23, 2006, at the Chateau Lake Louise, situated in the scenic Canadian Rockies.

The pedagogical talks focussed on Precision measurement at Hadron colliders, evidence of Quark-Gluon plasma, Neutrino Physics, Nuclear Astrophysics, Low Energy tests of the Standard model and Physics beyond the standard model. These main talks were supplemented by contributed talks from all the collider facilities and from laboratories considering physics with non-accelerator experiments. The combination provided a variety of physics with the experiments providing the latest details of the new results.

We wish to thank Lee Grimard for a phenomenal job of organising various details of the Winter Institute, in particular providing a quiet and peaceful interface with the hotel staff. In all the support of the staff at Chateau Lake Louise is greatly appreciated.

Finally, we wish to thank the Dean of Science, and the University Conference fund at University of Alberta for financial support. We thank TRIUMF and Institute of Particle Physics for providing funds generously for a continuing operation of the Winter Institute. It is a pleasure to thank Theoretical Physics Institute and Physics department for infrastructure support that makes our task for organising the Winter Institute much easier.

Organizing Committee
A Astbury
F.C. Khanna
R. Moore

CONTENTS

Preface v

Contents vii

I. Making Precision Measurements at Hadron Colliders: Two Lectures
HJ Frisch 1

II. Evidence for a Quark-Gluon Plasma at RHC
J. Harris 33

III. Fundamental Experiments at Low Energies
H. Jurgen Kluge 52

IV. Neutrino Physics: A Selective Overview
S Oser 63

V. Low-Energy Tests of the Standard Model
M Pospelov 93

Charged Particle Multiplicities in Ultra-Relativistic AU+AU and CU+CU Collisions
BB Back, et al 111

Standard Model Physics at CMS
S Beaucheron 116

Recent Results in Diffractive *ep* Scattering at HERA
M Beckingham 121

High-P_T Suppression in Heavy Ion Collisions From the Brahms Experiment at RHIC
S Bekele 126

The Suppression of High P_T Non-Photonic Electrons in AU+AU Collisions at GEV at $\sqrt{S_{NN}} = 200$ RHIC
J Bielcik 131

Direct CP Violation Results in $K^{\pm} \rightarrow 3\pi^{\pm}$ Decays from NA48/2 Experiment at CERN
C Bino 136

Neutrino Astronomy at the South Pole
DJ Boersma 146

Direct CP Violation in B Decays at BELLE
M Bracko 151

Standard Model Physics With The Early Data
P Bruckman De Renstrom 156

Leptonic B Decays
M-C Chang 161

Searches for the Higgs Boson in CMS
Georgios Daskalaskis 166

New Physics at EP Collisions
Jerome De Favreau 171

Recent Electroweak Results From DØ
JD Degenhardt 177

Search for Exotic Physics with Atlas
PA Delsart 182

Getting Ready for Physics at the LHC with the CMS Detector
V Drollinger 187

W and Z Cross Section Measurement at CDF
I Fedorko 193

New Resonances and Spectroscopy at BELLE
B Golob 198

Cosmic Neutrinos Beyond the Standard Model
U Harbach 204

Lepton Flavor Violating τ Decays at *BABAR*
C Hast 210

Measurement of sin $2\phi_1$ In b $\rightarrow$ *s* Decays at BELLE
NC Hastings 215

DVCS Measurements with Nuclear Targets at HERMES
M Hoek 220

Measurement of Foreward-Backward Asymmetry in B $\rightarrow$ K* $\ell^+\ell^-$ at BELLE
A Ishikawa 225

The Latest from Minos
H Kang 230

Black Holes and Quasistable Remnants at the LHC
B Koch 235

Noncommutative Geometry and the Particle Content of the Universe
T Kopf 240

A Study of Mixing in the $B_s^0-B_s^0$ System Using the DØ Detector
D Krop 244

Expected Physics Performance of the LHCB Experiment
O Leroy 249

Isoscalar Extraction of ΔS in the Nucleon at HERMES from Semi-Inclusive DIS
LA Linden Levy 254

Electroweak Physics at HERA
J List 260

Recent Results on Radiative Penguin and Leptonic *B* Decays at *BABAR*
TB Moore 266

The Cobra Double Beta Decay Experiment
B Morgan 271

Proton Structure from HERA
K Nagano 276

The LHCB Tracking System and Its Performance
J Nardulli 281

New States at *BABAR*
M Negrini 287

Measurement of the CKM Angle γ at *BABAR*: Status and Prospects
N Neri 292

Search for New Phenomena at the DØ Experiment
C Noeding 298

Charm Physics at *BABAR*
A Oyanguren 303

Mass-Dependent O(α_S^2) Corrections to Semileptonic B-Quark Decay at $M(\ell\nu_\ell)=0$
Alexey Pak 308

Top Physics at CDF
E Palencia 313

DI-Electron Widths of the Y(1S), Y(2S), and Y(3S) from the CLEO-III Detector
J Pivarski 318

Electron Reconstruction and Calibration with Single Z and W Production in CMS at the LHC
C Rovelli 323

Searching for Dark Matter Annihilation in Z=-1 Cosmic Rays with AMS
G Rybka 328

Atlas Construction: A Status Report
A Sfyrla 333

Double Longitudinal Spin Asymmetry in Inclusive Jet Production in Polarized P+P Collisions at $\sqrt{s} = 200$ GEV
F Simon 338

New Phenomena Searches at CDF
A Soha 343

Electroweak Results at LEP2
P Spagnolo 348

Picasso: Direct Dark Matter Detection Using The Superheated Droplet Technique
CE Storey 356

LHCB Particle Identification and Performance
P Szczypka 367

B_c at CDF
W Wester 372

Transversity Measurements at HERMES
B Zihlmann 377

List of Participants 383

MAKING PRECISION MEASUREMENTS AT HADRON COLLIDERS: TWO LECTURES

HENRY J. FRISCH

Enrico Fermi Institute
University of Chicago
5640 S. Ellis Ave.
Chicago, Il. 60637

These two lectures are purely pedagogical. My intent is to enable non-experts to get something out of the individual presentations on collider physics that will follow- measurements of the W,Z, top, searches for SUSY, LED's, the Higgs, etc. We often forget that we are talking about instruments and the quantities they actually measure. The surprise is how precise the detectors themselves are; the challenge will be to exploit that precision in the regime where statistics is no longer a problem, and everything is dominated by the performance of the detector ('systematics'). Precision is necessary not only for measuring numbers such as masses, mixing angles, and cross-sections, but also for searches for new physics, comparing to the Standard Model.

Lecture I: The Electroweak Scale: Top, the W and Z, and the Higgs via M_W and M_{top}

1. Introduction and Purpose

My intent in these pedagogical lectures is to enable non-experts to get something out of the detailed individual presentations on collider physics that will follow. We are presented with so much detail that one often forgets that we are talking about instruments and the basic quantities they actually measure. The surprise is how precise the detectors themselves are; the challenge will be to exploit that precision in the regime where statistics is no longer a problem, and everything is dominated by the performance of the detector ('systematics').

This challenge also extends to the theoretical community- to look for something new we will need to understand the non-new, i.e. the SM predictions, at an unprecedented level of precision. Some amount of this can be done with control samples- it is always best to use data rather than Monte

Carlo, but it's not always possible. The detectors are already better than the theoretical predictions.

2. Problems in Making Precision Measurements

The emphasis here will be more on problems to be addressed than on new results. I have used mostly CDF plots just because I know the details better- no slight to DØ or the LHC experiments is intended. The problems however are general. I have cut some corners in places and been a little provocative in others, as teachers will.

I have intentionally used older public results from CDF and *DØ* instead of the hot-off-the-press results generated for the 2006 'winter conferences' so as not to steal the thunder of the invited speakers who are here to present new results from CDF and D0, and so that you can recognize the evolution of the results as the integrated luminosity grows.

The two one-hour lectures included a very large number of plots; in the interest of space I have included only a small fraction here- many (updated with more luminosity) are included in the D0 and CDF invited talks in this volume.

3. Some History and Cultural Background

3.1. *Luminosity History: Orders of Magnitude*

A history of luminosity, starting with the $SP\bar{P}S$ and the discovery of the W and Z^0, followed by the race between CERN and Fermilab to discover the top, is shown in Figure 1.

Figure 2 shows the luminosity 'delivered' and 'to tape' from the current Run II, in inverse femtobarns (right), and from the 1987 run, in inverse nanobarns (left). As a reminder, the $\mathrm{W}^{\pm} \rightarrow e^{\pm}\nu$ cross-section times BR is about 2.2 nb at 1.8 TeV, so 30 nb^{-1}means that $\approx$ 66 $\mathrm{W}^{\pm} \rightarrow e^{\pm}\nu$ decays were created in the 1987 exposure. The cross-section for a 115 GeV Higgs in $\mathrm{W}^{\pm} \rightarrow e^{\pm}\nu + H$ production is ~20 fb, and so the right-hand plot indicates that if $M_H = 115$, $\approx$ 20 $\mathrm{W}^{\pm} \rightarrow e^{\pm}\nu + H$ events would have been created in the present 1 fb^{-1}at each of DØ and CDF.

3.2. *Hubris: The 50 GeV Top Quark and No Quarkonia*

Figure 3 is an historical reminder both that we should not be over-confident about what we know, and that Nature has a rich menu of surprises. The left-hand page is the 1984 discovery of something that did not exist- a top

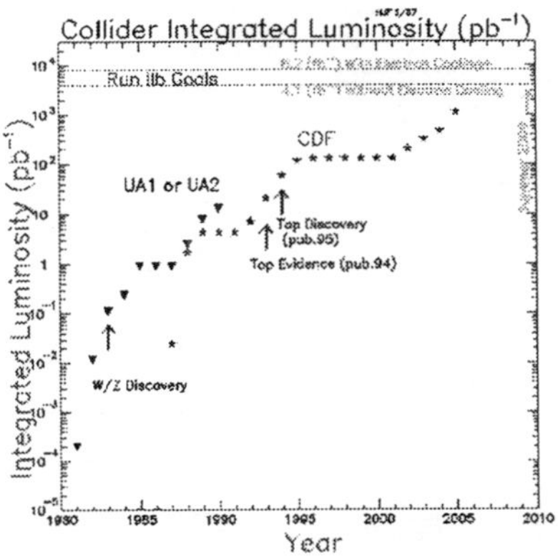

Figure 1. A history of high-energy (no ISR) hadron colliders: integrated luminosity by year.

quark with mass less than 50 GeV (it was largely W+2 jets). The right-hand page is a prediction from 1974 that there are no narrow states with masses between 3 and 10 GeV decaying into lepton pairs.

4. The Tevatron and the LHC

By now everybody should know about the Tevatron and LHC. I will spare you pictures and boilerplate; the main differences that everybody, including mathematical theorists, should know are:

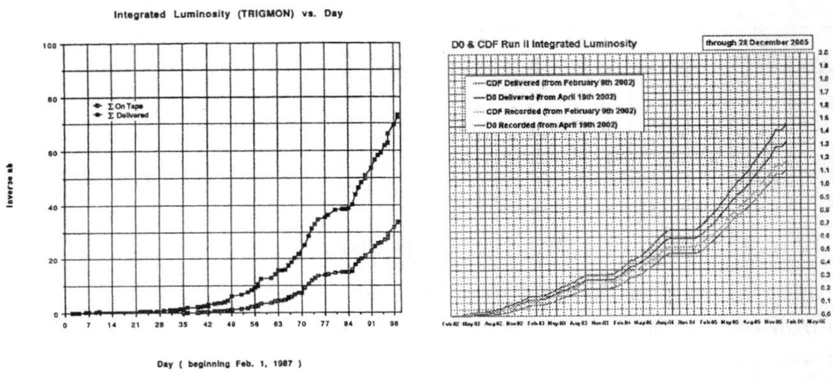

Figure 2. The integrated luminosity in the 1987 Tevatron run (Left), in Inverse Nanobarns, and in Run II (Right), in Inverse Femtobarns. Note that 1 fb^{-1}= 10^6 nb^{-1}. Note also the efficiency to tape has improved substantially.

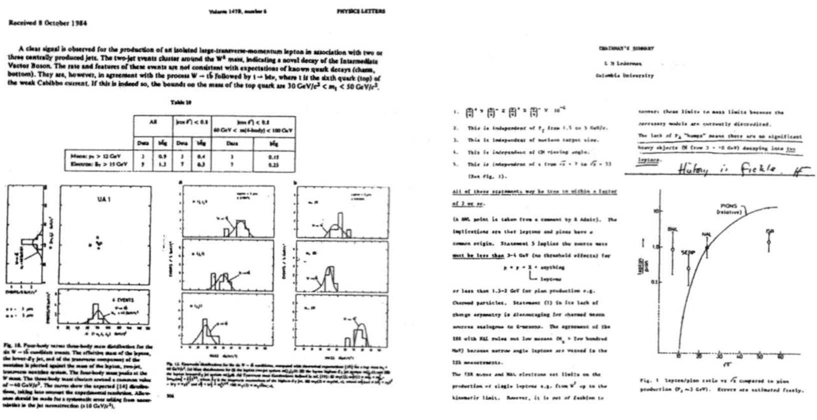

Figure 3. Left: The 1984 Top 'discovery'; Right: The 1974 'no discovery' announcement of the J/ψ and Upsilons.

	Tevatron	LHC
Parton Source	Antiproton-Proton	Proton-proton
Energy (TeV)	1.96 (not 2!)	14
Peak Luminosity ($cm^{-2}s^{-1}$)	2×10^{32}	1×10^{34}
Crossing Spacing (ns)	396	24.95
Peak Interactions/Crossing	5	19
Luminous Line σ (cm)	30	4.5 [1]
Luminosity Lifetime (hours)	3.8/23 [2]	15
$< x >$ at M_W	0.04	0.006
$< x >$ at $2M_T$	0.18	0.025

An LHC upgrade to $2 - 9 \times 10^{34}$ is planned [3].

5. The Anatomy of Detectors at Hadron Collider: Basics

For those moving to the LHC from Cornell, SLAC, or LEP, working at a hadron collider is really different from at an e^+e^- machine– at CDF it took several years for experienced physicists who have worked only at e^+e^- machines to understand 'whatever you ask for in your trigger will you get' (the story of jets at ISR and Fermilab fixed-target as well).

Figure 4 shows a 'cartoon' of the production process for the W and Z, which are the ‘standard candles’ along with the J/ψ and Υ at the Tevatron (top will be another at the LHC).

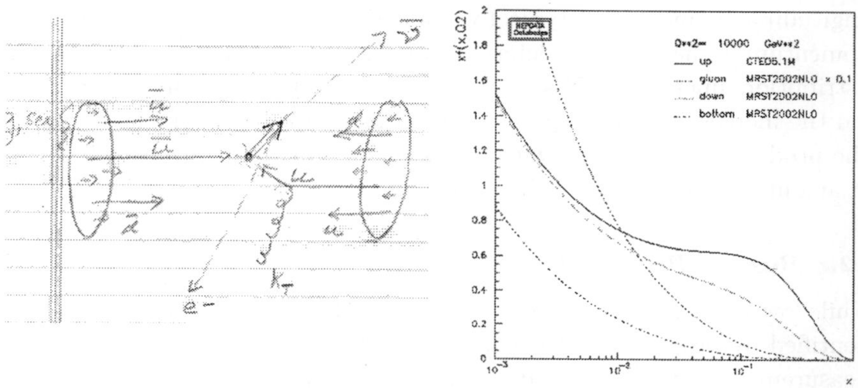

Figure 4. Left: A ‘cartoon’ of the production process for W’s and Z’s. Right: The CTEQ6.1M PDF’s at Q=100 (from Joey Huston).

5.1. *Basics: Kinematics and Coverage:* $\mathbf{p_T}$ *vs* $\boldsymbol{P_{||}}$

The phase space for particle production at a hadron collider is usually described in cylindrical coordinates with the z axis along the beam direction, the radial direction called ‘transverse’, as in ‘Transverse Momentum’ (p_T), and the polar angle expressed as Pseudo-rapidity η, where $\eta \equiv -ln(tan\theta/2))$. Pseudo-rapidity is a substitute for the Lorentz-boost variable, y, where $y \equiv 1/2ln(E+p_z)/(E-p_z) = tanh^{-1}(p_z/E)$. Since in most cases one does not know the mass of a particle produced in a hadron collision (most are light- pions, kaons, baryons,..), we use pseudo-rapidity, which assumes zero mass. (This is a common error when doing kinematics with W’s, Z’s, and top, where the mass truly matters). Note that typical particle production is 4-6 particles per unit-rapidity; in the central region one unit at CDF is about 14 m^2; the density in a min-bias event is very low. Hadron colliders are not intrinsically ‘dirty’- only complex.

Two simple equations contain much of the physics for the production of heavy states at a collider: the mass and longitudinal momentum of the heavy state (e.g. a W, Z, $t\bar{t}$ pair, or WH) are determined by the difference in momentum carried by the interacting partons, and the mass by the product.

$$m^2 = x_1 * x_2 s \qquad\qquad p_z = (x_1 - x_2) p_{beam} \tag{1}$$

Note that a heavy object typically has a velocity $\beta << 1$, even though the longitudinal momentum is typically not small. Note also that the transverse momentum of the system is determined by the competition of falling parton distribution functions (PDFs) as the total invariant mass of the system rises, and the increase in phase space as the momentum of the system increases. The production thus peaks with a total system energy above threshold by an amount characteristic of the slope in $x_1 * x_2$.

5.2. *Basics: Particle Detection*

While low-momentum– typically up to a few GeV– charged particles can be identified by processes that depend on their velocity, β, as a simultaneous measurement of $p = \beta\gamma m$ and β allows extracting the mass, for momenta above a few GeV, pions, kaons, and protons cannot be separated. However electrons, muons, hadrons, and neutrinos interact differently, as shown in Figure 5. The measurement of their energies and/or momenta stem from their different modes of interaction.

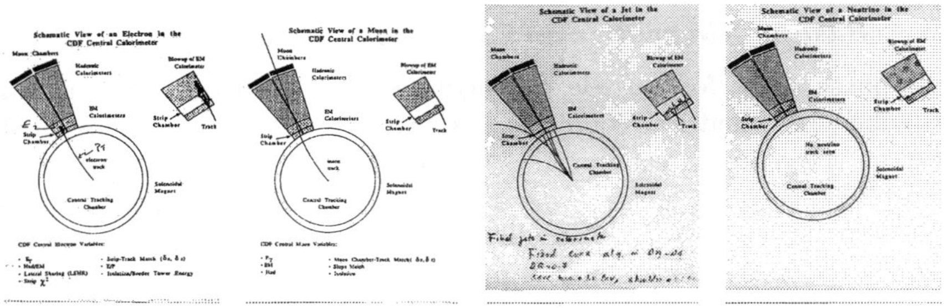

Figure 5. A 'cartoon' of how electrons,muons, jets, and neutrinos are identified in a solenoidal detector (by Sacha Kopp).

6. Calibration Techniques: CDF as an Example

6.1. *Momentum and Energy Scales: E/p*

In contrast to LEP, at the LHC or Tevatron the overall mass (energy) scale is not set by the beam energy- there is a continuum of c.m. energies in the parton-parton collisions. Moreover the hard scattering is not at rest either longitudinally or transversely in the lab system- there is 'intrinsic Kt' as well as 'hard' initial-state radiation (ISR). Finally, the beam spot is a line and not a spot- the vertex point, used to calculate transverse energies, including those of missing energy and photons for which no track is observed, has to be determined from the event.

Dealing first with the issue of setting the scale for momentum, energy, and mass measurements: the current big detectors consist of a solenoidal magnetic spectrometer followed by calorimeters. The magnetic spectrometer uses a precisely measured (NMR) magnetic field and the precise geometry of the tracking chambers to measure the curvature ($\propto 1/P_T$)of the tracks of charged particles. This is an absolute measurement- if perfect one has the momentum scale. One can then use particles with measured momentum as an *in situ* 'test beam' to calibrate the energy scale of the calorimeters.

The momentum scale can be checked by measuring the masses of some calibration 'lines' provided by Mother Nature- the J/Psi and Υ systems, and the Z^0in its $\mathrm{Z}^0 \rightarrow \mu^+\mu^-$ decays ($\mathrm{Z}^0 \rightarrow e^+e^-$ doesn't work for momentum calibration!).

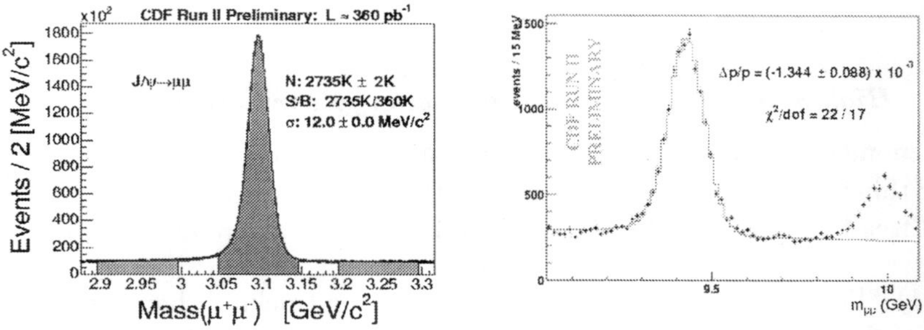

Figure 6. Left: The reconstructed $J\Psi$ invariant mass in dimuons (CDF). Right: The similar plot for the Upsilon system.

However the momentum scale can be incorrect due to mis-alignments in the tracking chamber. The combination of a calorimeter and a magnetic spectrometer allows one to remove the 1st-order errors in both by measuring 'E' (calorimeter energy) over 'p' (spectrometer momentum). With perfect resolution, no energy loss, and no radiation, these two should be equal: $E/p = 1.0$.

The 1st-order error in momentum is due to a 'false-curvature'- that is that a straight line (0 curvature= ∞ momentum) is reconstructed with a finite momentum. The 1st-order error in calorimeter energy is an offset in the energy scale, and does not depend on the sign ($\pm$) of the particle [7]. Expanding both the curvature and calorimeter energies to first order:

$$1/p = 1/p_{true} + 1/p_{false} \qquad (\mu^+) \qquad 1/p = 1/p_{true} - 1/p_{false} \qquad (\mu^-) \quad (2)$$

$$E = E_{true} * (1 + \epsilon) \qquad (e^+) \qquad E = E_{true} * (1 - \epsilon) \qquad (e^-) \qquad (3)$$

The first-order false curvature p_{false} then is derived by measuring E/p for positive and negative electrons with the same E

$$1/p_{false} = ((E/p(e^+) - E/p(e-))/2E \qquad (4)$$

The first-order calibration scale error ϵ then is removed by setting the calorimeter scale for electrons so that E/p agrees with expectations. In CDF, this is done initially to make the calorimeter response uniform in $\phi - \eta$.

$$1/p_{false} = ((E/p(e^+) + E/p(e-))/2 \qquad (5)$$

6.2. *Higher-order momentum and energy corrections*

The momentum and energy calibrations at this point are good enough for everything at present exposures except the W mass measurement. There are three higher-order effects that are taken care of at present:

(1) 'Twist' between the two end-plates of the tracking chamber;
(2) Systematic scale change in the z-measurements in the chamber;
(3) Non-linearity of the calorimeter for electrons that radiate hard photons, due to $e(E/2) + \gamma(E/2) \neq e(E)$

Figure 7 shows the use of the J/Ψ mass to correct for the first two of these effects. What is plotted is the correction to the momentum scale versus the cotan of the difference in polar (from the beam axis) angle of the two muons. There is a linear correction to the curvature of $\delta c = 6 \times 10^{-7} cot(\theta)$ that corrects for the twist between the endplates, and a change in the scale of the z-coordinate by 2 parts in 10^4, $z_{scale} = 0.9998 \pm 0.0001$. This is precision tuning of a large but exceptionally precise instrument!

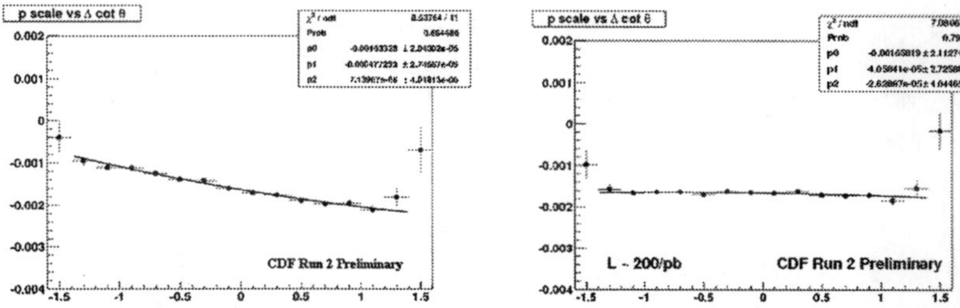

Figure 7. Left: The correction to the momentum scale versus the cotan of the difference in polar angle of the two muons in J/psi decay before corrections: Right: The same after correcting the curvature by $\delta c = 6 \times 10^{-7} cot(\theta)$ the scale of the z-coordinate by 2 parts in 10^4.

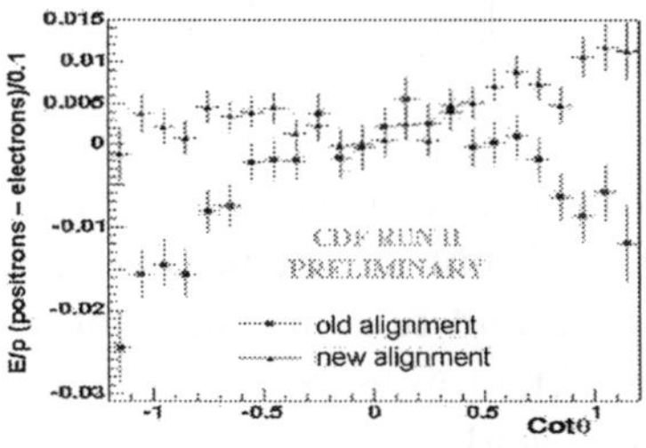

Figure 8. Measuring a higher-order correction to track curvature: the calorimeter to momentum ratio E/p versus $cot\theta$ for e^+ and e^-, before and after the curvature and z-scale corrections.

6.3. *Calibrating the Hadron Calorimeters and the Jet Energy Scale*

Much of the top mass information is encoded in its jets: the b-jets are first-generation daughters of a 2-body decay, one W decays into 2 jets, and the missing-Et of the neutrino is measured in the calorimeter.

There are a number of ways to calibrate the calorimeter response to jets:

(1) In situ calibration by isolated hadrons ('E/p')
(2) Test beam (for higher momenta- not easy to transport though.)
(3) Dijet balancing (D0 uses this cleverly at large η for Et reach.)
(4) γ-jet balancing
(5) Z^0-jet balancing

However with much higher statistics and the goal of precision recon-

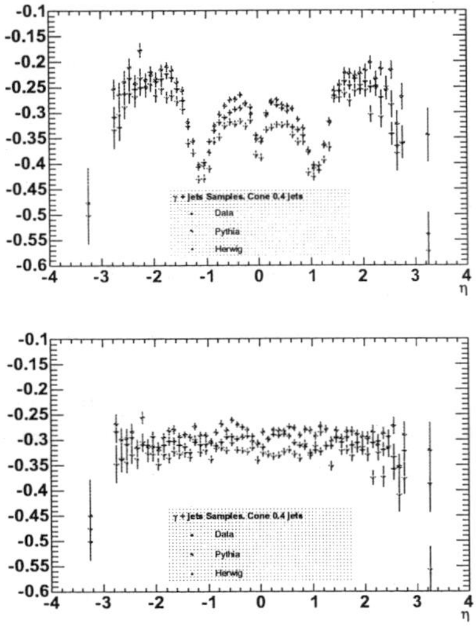

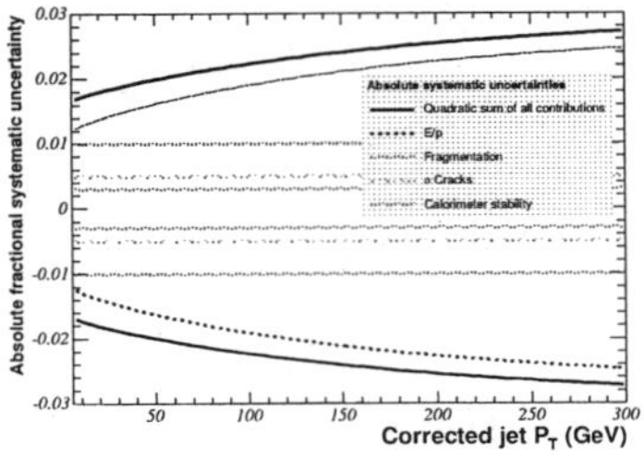

The total (black), and individual contributions (calorimeter response in blue dashes, e.g.,..) systematic uncertainty on the jet energy scale from CDF versus the measured p_T of the jet after corrections.

struction of masses in states such as the Higgs and possible SUSY states we will need a new level of sophistication. For example, we already see that fake rates for photons and taus are different for gluons and quarks, and probably for charm. The top mass analyses already use b-specific algorithms. Areas that need work are:

(1) Energy flow algorithms
(2) Quark/gluon-dependent corrections
(3) Heavy-flavor dependent algorithms
(4) Integrated QED and QCD higher-order corrections

For example, the quark/gluon/flavor content of jets in photon-jet and Z-jet balancing will be different. Another example is in W-decay to jets, where half the jets are a $u\bar{d}$ pair and half a $c\bar{s}$ pair.

7. W and Z^0Production as PDF Probes

W and Z production provide a precise measure of the up and down quark parton distribution functions (PDF's).

Since we measure W's and Z's in their leptonic modes, the kinematics of the decay also matter. Consider the W's: they are polarized, as the u and d quarks are light and couple through V-A so quarks have helicity -1 and antiquarks +1. The W decays also by V-A, so the charged leptons come out opposite to the helicity direction. However, the dominant effect, at least at the Tevatron, is that the W is moving in the rest frame, and since the (valence) u quark momentum is generally higher than the (sea) $\bar{d}$ anti-quark; W^+ go in the proton direction, and W^- in the $\bar{p}$ direction (the LHC, being proton-proton, doesn't have this useful asymmetry).

Figure 9 shows the distribution in the difference of e^+ and e^- versus η (pseudo-rapidity) of the electron ($e^{\pm}$) measured by CDF. The left-hand plot shows the full range as well as the experimental uncertainty band; the right-hand plot shows a comparison with the predictions using the CTEQ6 PDF's. One can see that the PDF's do not fit well.

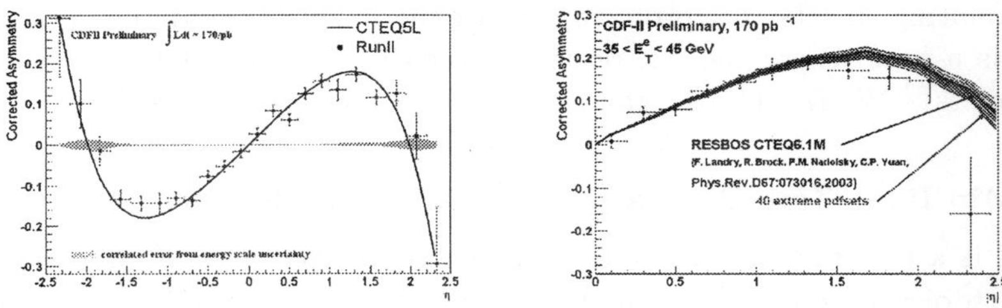

Figure 9. Left: The CDF forward-backward charge asymmetry in $W^{\pm} \rightarrow e^{\pm}\nu$ decays plotted versus pseudo-rapidity. The black error band gives the experimental uncertainty; also shown is the prediction using the CTEQ5L parton distribution functions. Right: The same data, folded around zero in η (remember this is $\bar{p}p$), compared to a prediction using the RESBOS MC generator and the CTEQ6.1M PDF's.

8. 'QCD'- Jet Production, Quark and Gluons, ISR, FSR

The dominant feature in the hadron collider landscape is the production of jets- the hard scattering of partons. Figure 10 reproduces the abstract from a seminal paper in 1971, when the idea of partons was brand new, by Berman, Bjorken, and Kogut [4], pointing out that the existence of partons would lead to point-like scatterings and hence high p_T phenomena, including 'cores' (jets); the paper is amazingly prescient about high-Pt phenomena, including W and Z production and hard-scattering (and there is a wonderful Peyrou plot representation of hard processes- I've omitted these pages due to space limitations.).

The precision needed for QCD measurements of jet production spectra and angular distributions is not as difficult as is needed to reconstruct final states such as top, and, we hope, the Higgs, SUSY states, etc. However

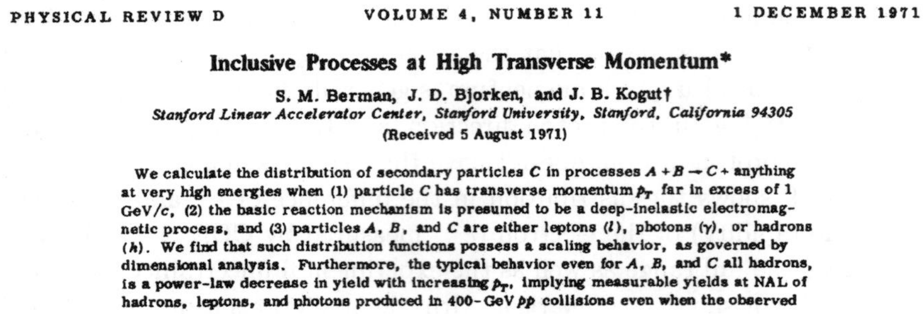

PHYSICAL REVIEW D VOLUME 4, NUMBER 11 1 DECEMBER 1971

Inclusive Processes at High Transverse Momentum*

S. M. Berman, J. D. Bjorken, and J. B. Kogut†
Stanford Linear Accelerator Center, Stanford University, Stanford, California 94305
(Received 5 August 1971)

We calculate the distribution of secondary particles C in processes $A+B \to C+$ anything at very high energies when (1) particle C has transverse momentum p_T far in excess of 1 GeV/c, (2) the basic reaction mechanism is presumed to be a deep-inelastic electromagnetic process, and (3) particles A, B, and C are either leptons (l), photons (γ), or hadrons (h). We find that such distribution functions possess a scaling behavior, as governed by dimensional analysis. Furthermore, the typical behavior even for A, B, and C all hadrons, is a power-law decrease in yield with increasing p_T, implying measurable yields at NAL of hadrons, leptons, and photons produced in 400-GeV pp collisions even when the observed

Figure 10. The start of the seminal paper by Berman, Bjorken, and Kogut.

the counting of jets is a problem, both experimentally and theoretically. This is a case where the use of ratios, such as (W+njets)/(Z+njets), or (W+njets)/(W+(n+1)jets), can cancel systematics.

9. Min Bias, Zero Bias, and the UEV: 'Soft' Physics

There is a lot of interesting physics in the large cross-section interactions. These topics are out-of-favor at the moment, but are rich and intellectually challenging. Among the topics are:

(1) Comparison of particle production in color-singlet (e.g. W,Z) production vs jet production.
(2) Rapidity gaps
(3) Multiple parton collisions
(4) Fragmentation functions, including the $p_T^{hadron}/p_T^{parton} = 1$ limit

In addition to their intellectual merit in their own right these topics need to be understood for high-p_T physics as well. For example, the 'fake rates' for jets passing photon or tau identification criteria are measured to be different for gluon jets and quark jets. The probability for a jet to appear as an isolated π^0 or charged track thus needs to be understood.

10. The $M_{Top} - M_W$ Plane and the Higgs Mass

10.1. *Motivation*

The top quark is remarkable for its physics and useful as a tool for calibration. It may also be a window into the world of heavy weakly-interacting particles (such as a Higgs of one sort or another) in that it is produced strongly in pairs, but due to its strongly-conserved flavor quantum number (top-ness), has to decay electroweakly. Due to radiative corrections, the masses of the W, Z, Higgs, and top quark are related in the SM; precise measurements of the W and top quark masses thus predict the SM Higgs mass.

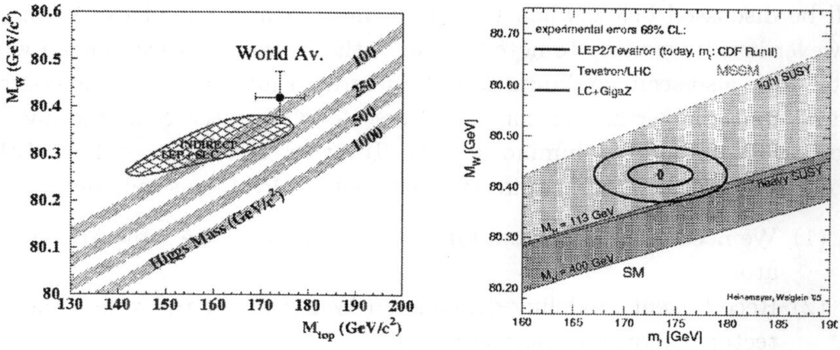

Figure 11. Left: The M_W vs M_T plane as of March 1998. Right: The M_W vs M_T plane as of the summer of 2005. Note the difference in the scales of the abscissas.

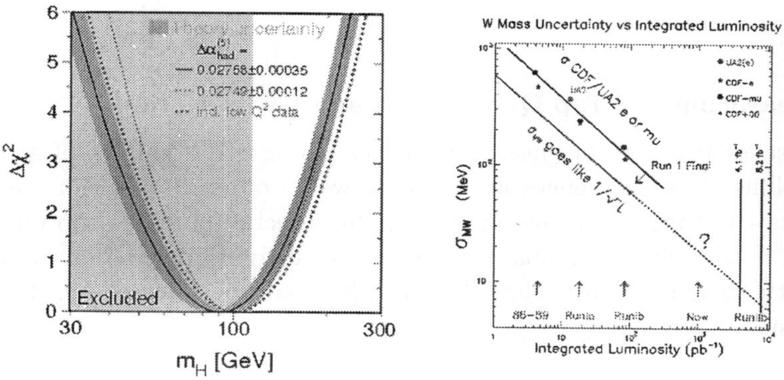

Figure 12. Left: The LEP EWKWG fit for the mass of the SM Higgs, showing the region excluded at 68% C.L.; Right:The total uncertainty on the W mass as measured at the Tevatron, versus integrated luminosity. If the control of systematic uncertainties continues to scale with statistics as the inverse root of the integrated luminosity, the Tevatron can do as well as LHC projections, and with different systematics.

10.2. *What limits the precision on the W mass and the top mass measurements?*

Figure 12 also shows the history of the uncertainty on the W mass as a function of the square-root of luminosity. The statistical uncertainty is expected to scale inversely with the exposure. The systematic uncertainties will be discussed below when we get to the measurement of the W mass; however it is interesting to note that since the systematics are studied with data, they also seem to diminish with integrated luminosity. If the control of systematic uncertainties continues to scale with statistics as the inverse root of the integrated luminosity, the Tevatron can do as well as LHC projections, and with very different systematics. Problems include:

(1) We need NLO QCD and QED incorporated in the same MC generator;
(2) Recoil event modeling depends on W p_T at low p_T, where the detector response is hard to measure;
(3) The underlying event energy is typically 30 MeV/tower/interaction, which implies one has to get the detector response from data; the Z will play a critical role.
(4) Using the Z for calibrating detector response will require Monte Carlos to treat W and Z production with NLO QED and QCD corrections in a consistent manner, and to understand any higher order differences.

11. Measuring the Top Quark Mass and Cross-section

I will discuss two specific measurements as pedagogic examples of some specific difficulties (challenges is the polite word) of doing precision measurements - the measurements of the top cross-section and the top mass. The idea is to make it possible for you to ask really hard questions when you see the beautiful busy plots that we all usually just let go by. First some basics.

11.1. *$t\bar{t}$ Production: Measuring the Top Cross-section Precisely*

The prime motivation for a precise measurement of the top cross-section is that new physics could provide an additional source for the production (leading to a larger cross-section than expected) or additional decay channels (leading to a smaller measured cross-section into Wb) [9]. More prosaically, the cross-section is a well-defined and in-principle easy-to-measure quantity that tests many aspects of QCD and the underlying universe of hadron collider physics- the PDF's, LO, NLO and NNLO calculations, and provides a calibration point for calorimeters and the energy scale. Lastly, and less defensible scientifically, is the uneasy feeling that too high a cross-section (e.g.) means that the top mass is really lighter than we measure, and so relying at 1-σ on the crucial EWK fits and limits on the Higgs mass may be misleading us.

11.2. *Total Cross-section for $t\bar{t}$ Production: Parsing the CDF and DØ Plots*

A brief history of theoretical predictions and a summary of the D0 and CDF measurements in different channels for σ_{top} is tabulated in Figure 13.

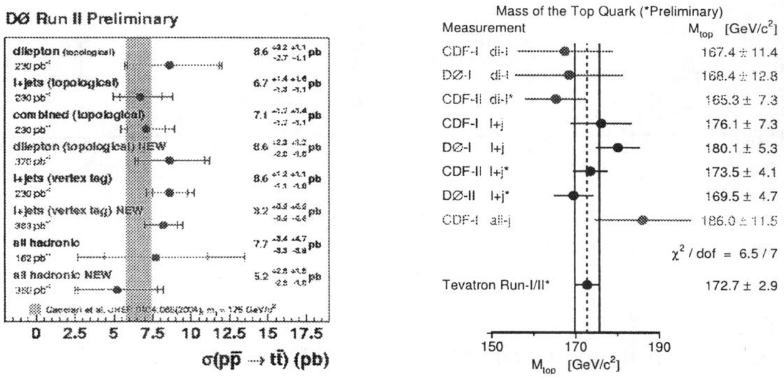

Figure 13. A summary of the CDF and D0 top cross-section measurements (left) and top mass measurements (right) as of Summer 2005.

11.3. *Precision Measurement of the Top Mass*

A summary of top mass measurements as of July 05 is given in Figure 13. Note the dilepton measurements tend to be systematically lower. Specific techniques are discussed below. (A comment- the D0 Run I measurement when analyzed by a different technique (same data, same calibration) moved from 173.3±7.8 to 180.1±5.3. We see a somewhat similar effect in the CDF cross-section measurement shown in Fig 14; both raise the question of whether we are correctly including the systematics on data selection and algorithms in precision measurements.)

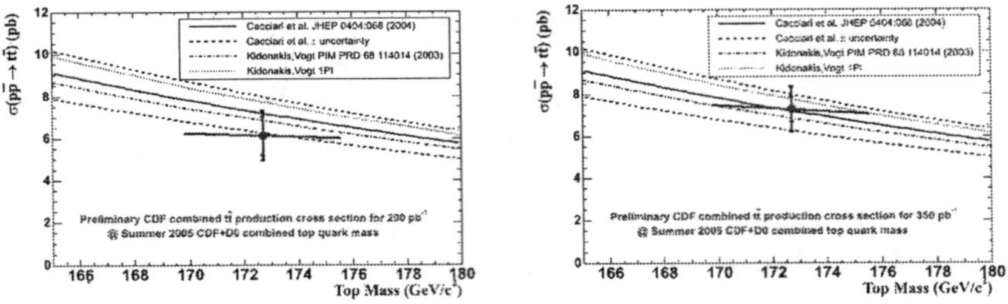

Figure 14. The measured and predicted CDF top cross-sections versus mass with approximately 200 pb^{-1}(Left) and with approximately 350 pb^{-1}(Right).

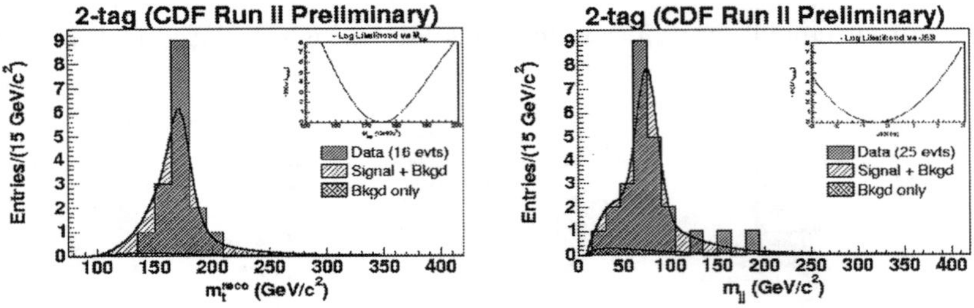

Figure 15. The fits for M_{top} and M_W in the 2-tag sample of the CDF '2D template analysis' as of summer 05.

11.4. *CDF Templates in M_{top} and M_jj (2D) in Lepton+Jets*

Figure 15 shows the fits for M_{top} and M_W in the 2-tag sample of the CDF '2D template analysis' as of summer 05.

11.5. *DØ Matrix Element Likelihood in M_{top} and M_jj (2D) in Lepton+Jets*

Figure 16 shows the results of the DØ 'matrix element' analysis of the top and W mass. However along with the extra statistical power there is a drawback from more complex methods - it is often hard to judge how good it looks. There's a net gain in precision, but one really needs simpler parallel analysis for sanity checks.

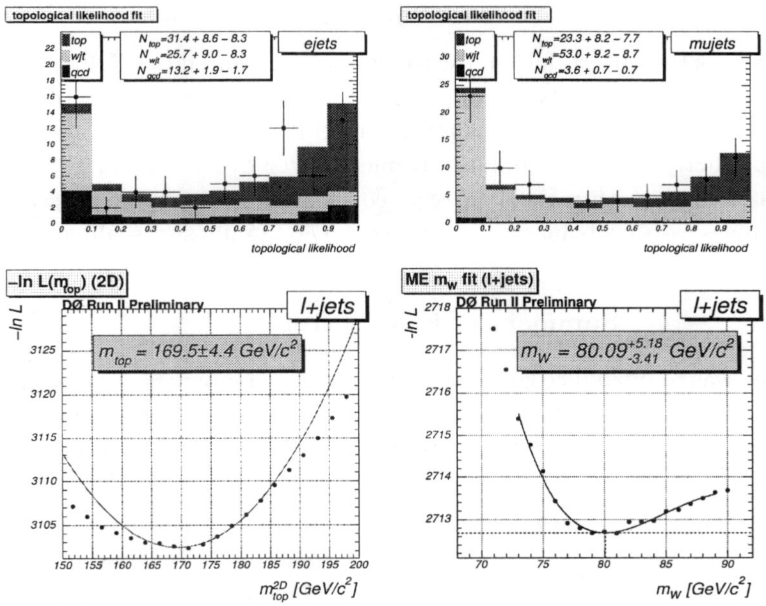

Figure 16. The D0 top and W mass fits in the 2D analysis. Closer to the dilepton number (all with 2 sigma, but...?)

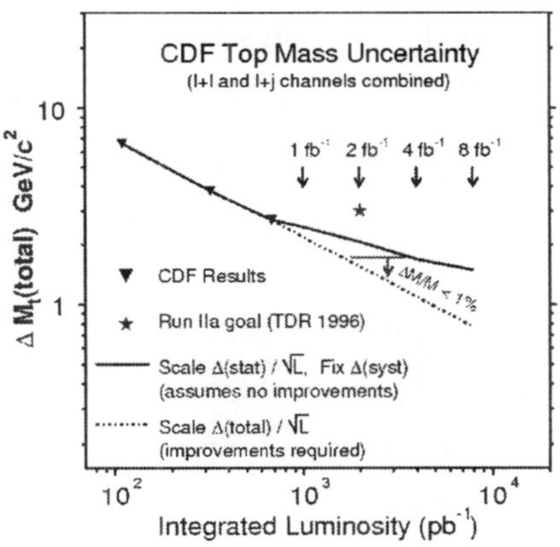

Figure 17. The CDF uncertainty on the top mass vs integrated luminosity. The red star is the projection from the Run II Technical Design Report;

11.6. *Ultimate Precision on the Top Mass Measurement*

Figure 17 shows the uncertainty in the top mass as measured by CDF vs the inverse root of the luminosity. The star shows the uncertainty predicted in 1996 for 2fb^{-1}; the present uncertainty with 0.8 fb^{-1}is significantly better than the prediction. New techniques to measure or evade systematic uncertainties open up with more data (e.g. constraining the jet-energy scale by the W mass, γ-b-jet balancing,..) and so a $1/\sqrt{\text{Luminosity}}$ scaling may be possible.

Summary of First Lecture

- Idea was to introduce key measurements and numbers from previous data so you can look at detailed presentations with a critical eye.
- Things to watch for in the following talks on Top, and Electroweak Topics:

  (1) $M_{top} - M_W$ off in (upper) left-field? What is the top mass?
  (2) σ_{top} and m_{top} consistent with predicted cross-section?
  (3) Systematics- just entering an era of enough data to measure systematics better - new methods, new ideas,...
  (4) Transparency- can we show more 'under the hood'? (less black box)
  (5) Transparency- can CDF and DØ (and soon Atlas and CMS) work harder on making comparisons- e.g. making the same plots with the same axes and scales!

Lecture II: Searching for Physics Beyond the SM, and Some Challenges for the Audience

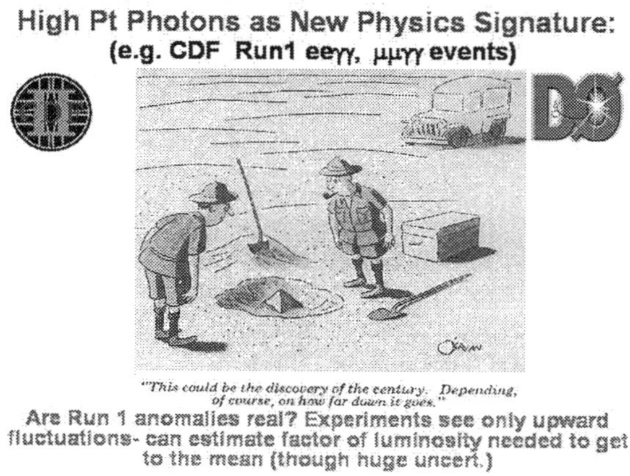

Lecture II emphasizes the problems that high statistics will bring for Beyond-the-Standard Model searches- first at the Tevatron, and then (in spades) at the LHC. Many of these problems are theoretical- in almost all cases we need precise Standard Model predictions in order to find new things (exceptions being new bumps- e.g. a Z-prime, KK excitations of the Z, etc.) Our parochial hope at the Tevatron is, of course, that we find something new before the LHC. We had hints of new things in Run I:

Some Run I oddities (none significant):

(1) The top dilepton sample looked odd (too many e-mu events, e-mu close in ϕ, a few with kinematics that didn't fit the top, and the 'tri-lepton' event.)
(2) Top mass in dileptons was consistently lower than in lepton+jets.
(3) Top mass resolution seemed too good- luck?
(4) The $ee\gamma\gamma \not{E}_t$ event and the 2.7σ excess in $\ell + \gamma + \not{E}_t$.
(5) High-Pt $Z + \gamma$ event, and a $W + 2\gamma$ event at start of Run II.

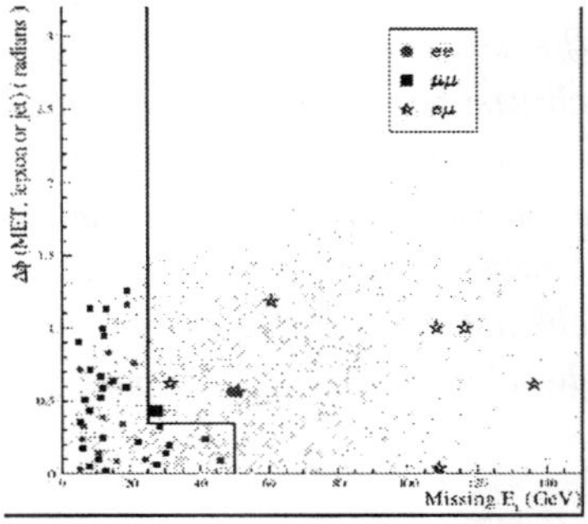

The Run-1 Dilepton events in the $\not{E}_t - \Delta\Phi$ plane. The yellow dots are SM expectations for $t\bar{t}$.

11.7. *Strategies: Signature-Based vs Model-Directed, Blind vs A Priori*

None of the above effects was significant statistically- but made one want more data. We now have 10-times the data, and are scheduled for a factor of 4-10 beyond that. What to do? There are two major kinds of direct searches, and in each three kinds of strategies have been followed (all this categorization is arguable):

(1) Model-Directed: Optimize sensitivity for points in theory parameter space;
(2) Signature-Based: Look broadly under selected lamp-posts

Avoiding biases is important (see next slide)- two solid strategies:

(1) 'A Priori'- use the same cuts as published in Run I, or in the 1st

1/3rd of the data; then run on rest of data without changing anything;

(2) 'Blind'- this is heavily used now-very useful and appropriate in some cases such as precision measurements: W mass, B lifetimes and masses, and classic well-defined searches: e.g. $B \to \mu\mu$.

A brief anecdote about a blind analysis around 1900:

There was a controversy over two conflicting measurements of a line in the solar spectrum. The famous spectroscopist at Princeton asked his machinist to rule a grating at a non-standard (blind!) lines/inch, and to put the value in a sealed envelope. The Prof. then measured the line in terms of an unknown dispersion, wrote a Phys Rev with an accompanying letter that said 'under separate cover you will receive the grating spacing from my machinist, Mr. Smith; take this number, multiply it by my number, put it in the blank space in the paper, and publish it'. Now, that's blind.

12. Theoretical Motivation and Experimental Caution

As in the search for the W and Z, there is a defining energy scale for the new physics beyond the SM. In the case of the W, Fermi's effective field theory of a 4-fermion interaction predicted that $\nu_e + e^- \to \nu_e + e^-$ scattering violated S-wave unitarity at a c.m. energy $\gtrsim 300$ GeV. For the SM, it's more complicated (see, Gunion et al. in the Higgs Hunter's Guide, e.g.), but the conclusion is the same- there must be something new at that scale. We experimentalists are consequently primed to find something new at the Tevatron and/or LHC. New means comparing data to precise predictions of the SM. Figure 18 shows what can happen when eagerness combines with insufficiently understood SM predictions.

12.1. *Lepton+Gamma+X: The $\ell\gamma\not{E}_t$ and $\ell\ell\gamma$ Signatures*

One of the anomalies of Run I was the famous CDF $ee\gamma\gamma\not{E}_t$ event. This spawned the advent of 'signature-based' searches at the Tevatron. In particular there were two follow-ups: $\gamma\gamma + X$ (Toback) and $\ell\gamma + X$ (Berryhill). The $\ell\gamma + X$ search resulted in a 2.7σ excess over SM expectations.

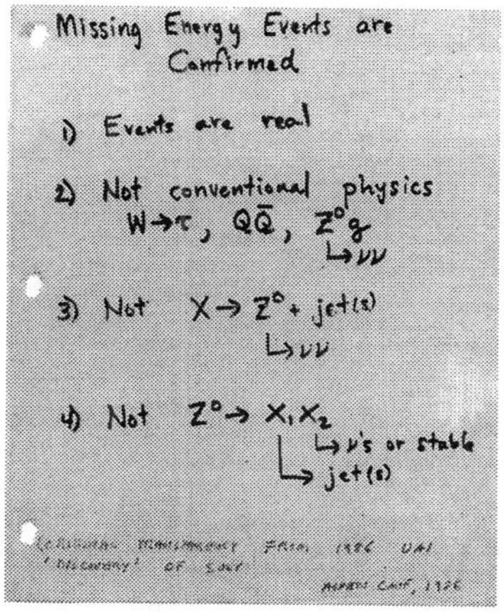

Figure 18. An example of why the careful calculation of SM predictions is so crucial: the announcement of the 'discovery' of SUSY at the 1986 Aspen Conference. The right explanation turned out to be a cocktail of SM processes.

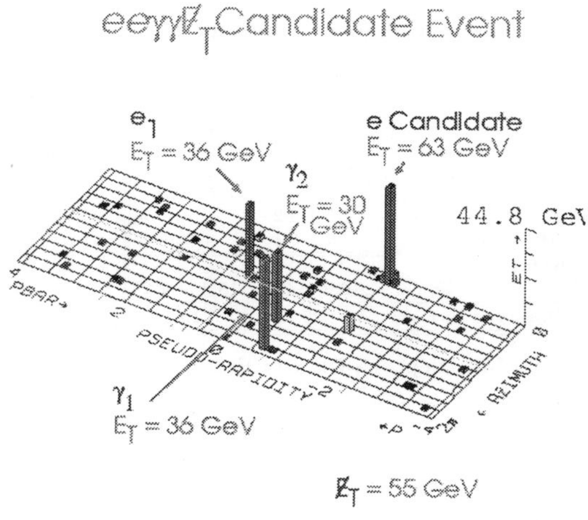

The analysis is being repeated with exactly the same kinematic cuts so this time it is a priori- (i.e. not self-selected to be interesting). Figures 19-21 show the CDF Run II results on the 2 signatures $\ell\gamma\not{E}_t + X$ and $\ell\ell\gamma + X$. This is a repeat of the Run I search- the $ee\gamma\gamma\not{E}_t$ event would show up in both, and so would an excess in $\ell\gamma\not{E}_t$. No more $\ell\ell\gamma\gamma\not{E}_t$ events have been seen with > 3 times the data (305 pb^{-1}/86 pb^{-1}) and higher energy (30% increase in $t\bar{t}$ crossection, e.g.).

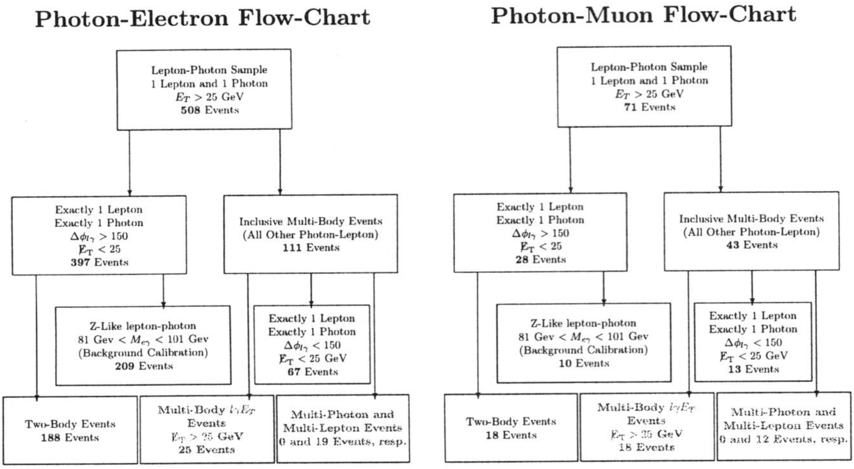

Figure 19. Left: The flow of the $\ell+\gamma+X$ signature based search in electrons. Right:The flow in muons.

Lepton+Photon+$\not{E}_T$ Predicted Events

SM Source	$e\gamma\not{E}_T$	$\mu\gamma\not{E}_T$	$(e+\mu)\gamma\not{E}_T$
$W^\pm\gamma$	11.9 ± 2.0	9.0 ± 1.4	20.9 ± 2.8
$Z^0/\gamma+\gamma$	1.2 ± 0.3	4.2 ± 0.7	5.4 ± 1.0
$W^\pm\gamma\gamma, Z^0/\gamma+\gamma\gamma$	0.14 ± 0.02	0.18 ± 0.02	0.32 ± 0.04
$(W^\pm\gamma$ or $W^\pm) \to \tau\gamma$	0.7 ± 0.2	0.3 ± 0.1	1.0 ± 0.2
Jet faking γ	2.8 ± 2.8	1.6 ± 1.6	4.4 ± 4.4
$Z^0/\gamma \to e^+e^-, e \to \gamma$	2.5 ± 0.2	-	2.5 ± 0.2
Jets faking $\ell+\not{E}_T$	0.6 ± 0.1	< 0.1	0.6 ± 0.1
Total SM Prediction	**19.8 ± 3.2**	**15.3 ± 2.2**	**35.1 ± 5.3**
Observed in Data	**25**	**18**	**43**

Multi-Lepton+Photon Predicted Events

SM Source	$ee\gamma$	$\mu\mu\gamma$	$ll\gamma$
$Z^0/\gamma+\gamma$	12.5 ± 2.3	7.3 ± 1.7	19.8 ± 4.0
$Z^0/\gamma+\gamma\gamma$	0.24 ± 0.03	0.12 ± 0.02	0.36 ± 0.04
$Z^0/\gamma+$ Jet faking γ	0.3 ± 0.3	0.2 ± 0.2	0.5 ± 0.5
Jets faking $\ell+\not{E}_T$	0.5 ± 0.1	< 0.1	0.5 ± 0.1
Total SM Prediction	**13.6 ± 2.3**	**7.6 ± 1.7**	**21.2 ± 4.0**
Observed in Data	**19**	**12**	**31**

Figure 20. The SM predictions and observations in the CDF $\ell\gamma\not{E}_t$ and $\ell\ell\gamma$ signatures.

However, this has proved another educational example of MC predictions being the limiting factor. We do not have a control sample, but depend on SM predictions, largely $W\gamma$ and $Z\gamma$. We have 2 MC generators-

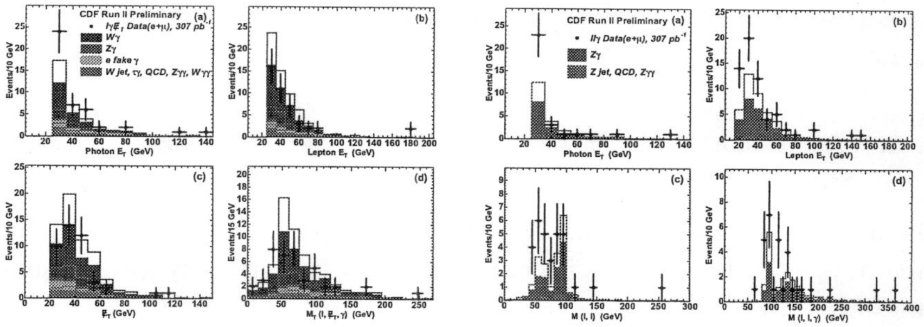

Figure 21. The distributions in photon and lepton E_T, $\not{E}_T$, and 3-body transverse mass from the CDF $\ell\gamma\not{E}_t$ search in 305 pb^{-1}.

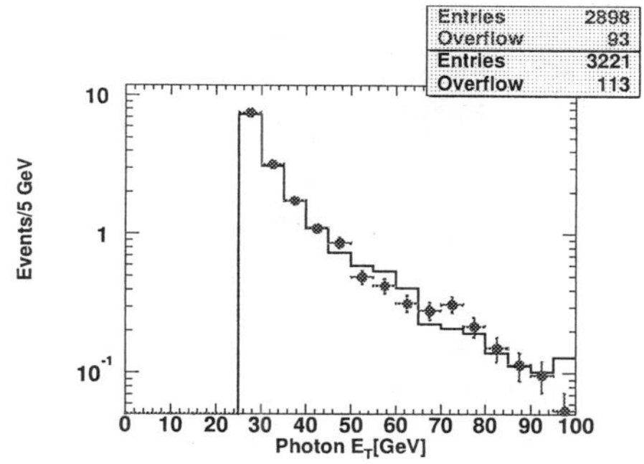

Figure 22. Comparison of the E_T spectrum of isolated photons in Drell-Yan+γ from MadGraph (red) and Baur (black) MC generators. There was disagreement *after* fragmentation and ISR with Pythia-now understood.

MadGraph and a program from Uli Baur. They agree beautifully. However after running them through Pythia they disagreed by 15% in yield, including a different identification efficiency for muons (!). Problems were in the interface (diagnosed by Loginov and Tsuno) for *both*- the Les Houches accord format is not precisely enough defined.

Lessons:

(1) Always use 2 MC's- you may find both samples are flawed;
(2) Both MC generators can be ok and you still can get it wrong;
(3) CDF has lost huge amounts of time to the generator interfacing-needs re-examination by the theoretical community.

There is a problem coming up- we do not yet have the SM event generators with integrated higher-order QED and QCD at a precision comparable to the coming statistics. We can normalize to data at low E_T^γ, but we need the next step up in prediction sophistication.

12.2. *Inclusive High Pt W's and Z's: A Weak Boson Signature*

The idea: many models of new physics- Extra Dimensions, Z-primes, Excited Top, $t' \to Wb$, SUSY, Right-handed Quarks- naturally give a signature of a high-Pt EWK boson- W, Z, or photon. This is natural in the strong production of pairs followed by weak decays: e.g. top.

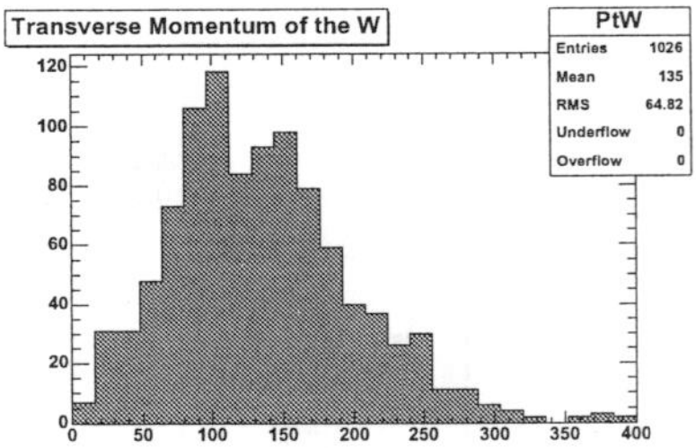

Figure 23. The p_T spectrum for Z's from the decay of a 300 GeV right-handed singlet down quark $Q\bar{Q} \to uWdZ$ in the Bjorken-Pakvasa-Tuan model.

CDF has done a search for anomalous inclusive high-Pt Z production, as would come from the decay of new heavy particles. The analysis selects on dilepton mass $66 < m_{\ell\ell} < 106$ and then compares the p_T spectrum with SM expectations (Figure 24. However the inclusive Z+X is dominated by SM Z+jets- we cannot yet predict this at the level needed, and at present rely on a tuning of the spectrum for $p_T < 20$ GeV.

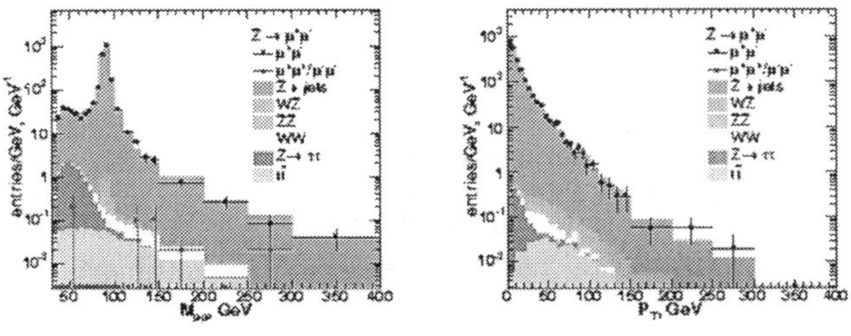

Figure 24. The inclusive search for anomalous high-p_T Z+X production (CDF).

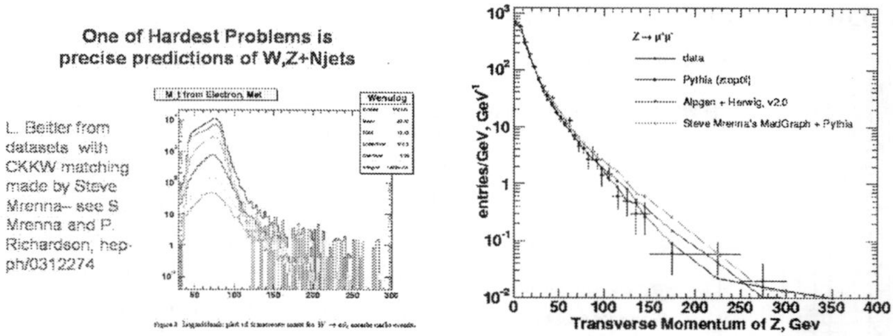

Figure 25. Left: Transverse mass distributions from 'matched' W+ njet samples (Mrenna); Right:Inclusive high p_T Z production and 3 monte-carlo predictions, showing that we cannot yet a priori test the data against the SM.

To increase sensitivity to the decays of new particles, the search adds objects other than light jets (e.g. other leptons, photons, heavy-flavor) to the signature- subsignatures of Z+Njets, $Z+\gamma$, $Z+\ell$,... For example: CDF recorded a Z with 200 GeV Pt balanced by a photon with 200 GeV Pt in Run I (100 pb^{-1}). Figure 26 shows Run II results from 305 pb^{-1}on the number of photons accompanying a Z boson with photon E_T greater than 25, 60, and 120 GeV.

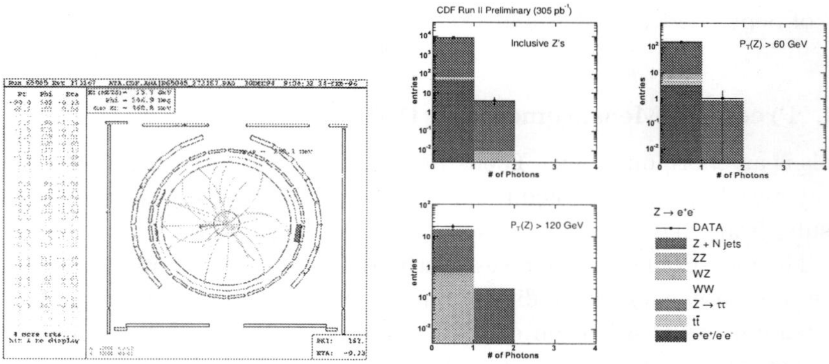

Figure 26. Left: A CDF Run I event with a 200 GeV (E_T)photon balanced by a Z with 200 GeV. Right: The number of photons seen with a Z boson with photon E_T greater than 25, 60, and 120 GeV, reading clockwise.

13. Direct Search for the Higgs

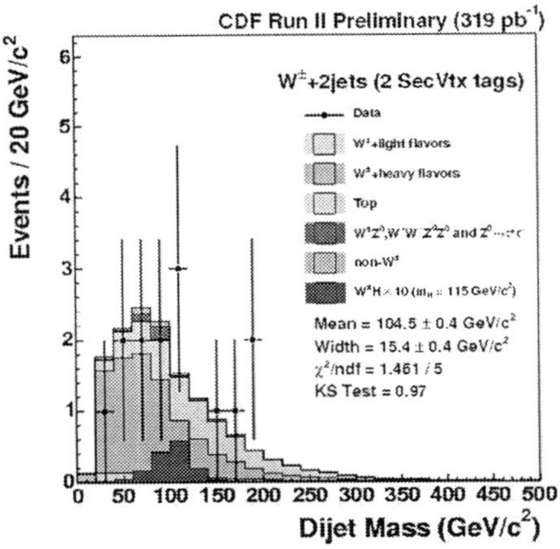

We saw in Lecture I that the EWK precision data favor a light Higgs (too light, even). Although it's not a precision measurement (my title), it is relevant to summarize the current status of direct Higgs searches. We show the dijet mass spectrum for 'W+H' channel from CDF with 319 pb^{-1}above.

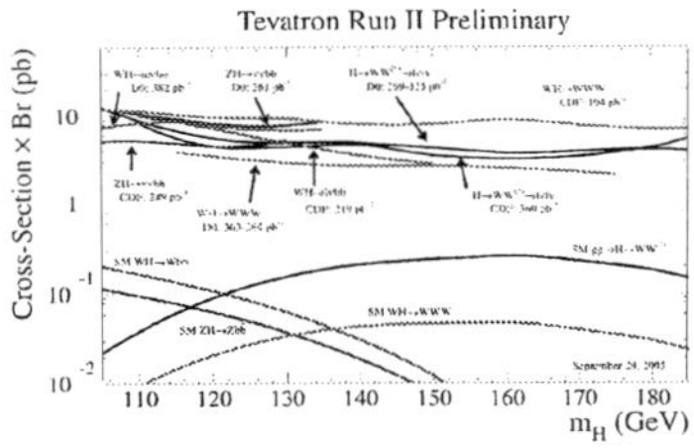

The cross-section limits from direct searches for the Higgs as of Sept., 05 from CDF and D0

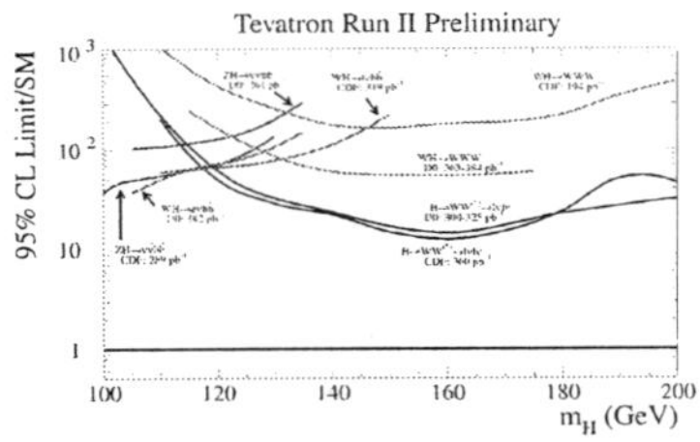

The ratio of cross-section limits from direct searches to SM predictions for the Higgs as of Sept 05, from CDF and D0.

14. Precision Measurements in B Physics

This is an enormous topic. There are many new results in the works. I will limit myself to some thoughts on precision, illustrated by several specific results from CDF.

The measurements of masses, lifetimes, and mixing are inherently precision measurements. They differ from the other standard-bearing precision measurements at a hadron collider, the top mass, the W mass, W and Z and top cross-sections, W and Z decay asymmetries, etc., in that they are usually entirely tracking-based, rather than depending on the calorimeters. Precision thus depends on different quantities: alignment, resolution, tracking and trigger efficiencies, and tracking trigger biases. While difficult, these are more tractable than the calorimeter response to jets and the underlying event (for $\not{E}_t$) in top decay, to pick an example. Consequently with much beautiful hard work by many people in CDF and D0, these measurements are often limited by statistics.

The statistics limitation is not intrinsic: there are plenty of B's. The measured single-b cross-section at the Tevatron for $|y| < 1$ is 30 μb, so at 1.8E32 (present peak luminosity) Fermilab is making more than 5000 b's per second. Realistically one could expect more than 10^{10} per year produced.
Now it seems a given that a detector at a hadron collider must have a silicon vertex detector; in 1981, however, it wasn't at all obvious that one would survive or work [10]. The success of the CDF SVX is a remarkable achievement, as is the subsequent SVT trigger system that allows triggering on displaced vertices.

14.1. *CDF measurement of the Λ_b lifetime:*

Figure 27 shows the mass peak and decay length distribution for the CDF measurement of the Λ_b lifetime.

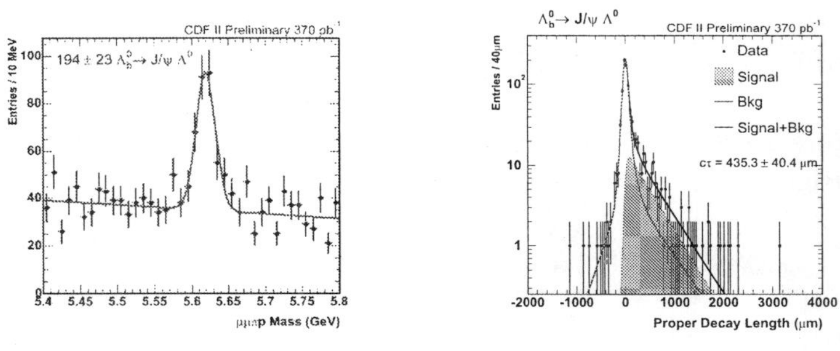

Figure 27. Left:Fully reconstructed $p\pi\mu\mu$ mass; Right: The proper decay length distribution.

The results are: $\tau(\Lambda_b) = 1.45^{0.14}_{-0.13}(\text{stat})\pm 0.02(\text{sys})$ psec and $\tau(\Lambda_b)/\tau(B^0) = 0.944 \pm 0.089$ (CDF).

14.1.1. *CDF measurement of the B_s lifetime:*

Figure 28 shows the fully reconstructed B_S mass in $\Phi\pi\pi$ and the distribution in $c\tau$ in $\Phi\pi\pi\pi\pi$ from CDF. The lifetime is measured to be $\tau(B_s) = 1.381 \pm 0.055(\text{stat})\pm^{0.052}_{0.046}(\text{sys})$ ps (See D. Krop's talk for the DØ number).

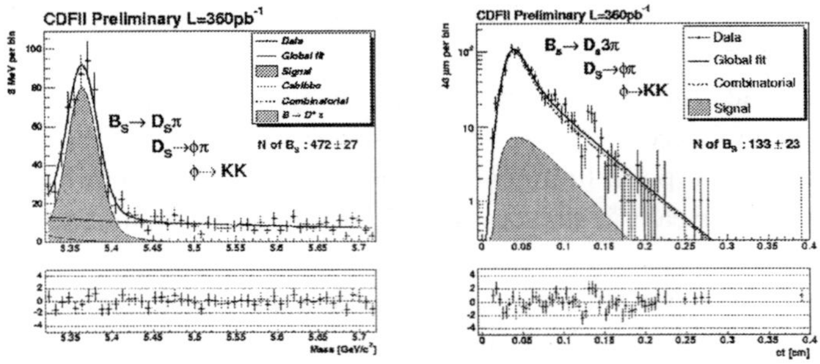

Figure 28. Left: Fully reconstructed B_S mass in $\Phi\pi\pi$; Right: The distribution in $c\tau$ in $\Phi\pi\pi\pi\pi$.

14.1.2. *CDF B_s Mixing:*

The remarkable resolution (~10 microns) achievable with silicon detectors around the small beam pipe of a hadron collider allows the measurement of lifetime oscillations up to $\sim$ 30 ps^{-1}. Both DØ and CDF scan in frequency space for the amplitude A, where:

$$\text{Prob } (B_s^0 \to B_s^0) = 1/2\Gamma e^{-\Gamma t}(1 + Acos(\Delta m_s t)$$

$$\text{Prob } (B_s^0 \to \bar{B_s^0}) = 1/2\Gamma e^{-\Gamma t}(1 - Acos(\Delta m_s t)$$

Figure 29 shows the reconstructed $\Phi\pi\pi$ mass (Left) and the amplitude scan (Right) [11].

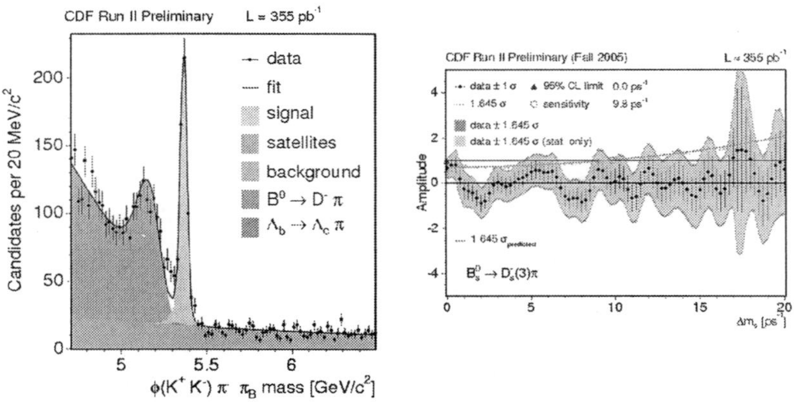

Figure 29. CDF: Exclusion (95%C.L.): Δm_s < 8.6 ps^{-1}

15. Expert Topics: Challenges for Students

I will briefly touch on an list of topics that I think lie ahead of us on the road to exploiting the higher precision inherent in our detectors.

15.1. *B-jet Momentum Scale: Gamma-bjet Balancing*

The response of the calorimeter to the b-quark jets from top decay is critical for the top mass; sharpening the resolution is also critical for discovering the Higgs. One source of b's of known momentum is $Z^0 \rightarrow b\bar{b}$; even at the Tevatron this is very difficult as the rate of 2-jet production prohibits an unprescaled trigger threshold well below $M_Z/2$. At the LHC this will be hopeless, I predict. However the 'Compton' process gluon b $\rightarrow \gamma$ b will give a photon opposite a b-jet. Figure 4 shows the flux of b-quarks versus x at $Q = 100$ GeV (CTEQ6.1M); one can see that at x=0.01 ($p_T = 70$ GeV at the LHC) the b-quark flux is predicted to be only a factor of 3 lower than the gluon flux.

15.2. *Rethinking Luminosity*

To make precision measurements of cross-sections, we need both to measure the numerator and the denominator precisely, where the numerator is the number of events corrected for acceptance and efficiency, and the denominator is traditionally the proton-proton (antiproton) luminosity. However the denominator is harder to measure than the numerator. To improve the precision on crossection measurements, it should be standard to measure the ratio to W and or Z production [6]. A secondary benefit would be in book-keeping- we could (should!) keep each W or Z in every file (small record)- to short-circuit the current nightmare of missing files and cockpit errors.

15.3. *Changing the Paradigm: W/Z ratios, Color Singlet/Color Triplet Ratios, and Other New Precision Tests*

Are there quantities that we can measure more precisely than ones we traditionally have been using? One example - instead of searching in the W+N jets and Z+ N jets for new physics, search in the ratio (W+N)/(Z+N): The cross section corresponding to a 1-sigma uncertainty in the W/Z ratio in 2 fb^{-1}, and in 15 fb^{-1} is shown below. The bins up through N=4 use the cross sections from CDF Run I; the N=5 and higher bins have been

Event and W Properties		W/Z Ratio Method Reach	
N(Jets)	σ_W	σ_{new} 2 fb^{-1}	σ_{new} 15 fb^{-1}
0	1896 pb	20 pb (1.0%)	20 pb (1.0%)
1	370 pb	4.4 pb (1.2%)	3.7 pb (1.0%)
2	83 pb	1.5 pb (1.8%)	0.9 pb (1.1%)
3	15 pb	0.5 pb (3.5%)	240 fb (1.6%)
4	3.1 pb	230 fb (7.5%)	95 fb (2.9%)
5	650 fb	100 fb (16%)	40 fb (6%)
6	140 fb	50 fb (36%)	18 fb (13%)
7	28 fb	20 fb (78%)	8 fb (29%)
8	6 fb	——	4 fb (63%)

extrapolated, Using the dimuon channel one can gain approximately root-2 on these uncertainties.

15.4. *Particle ID: Distinguishing $W \to c\bar{s}$ from $W \to u\bar{d}$, $b\bar{b}$ from b in Top Decays*

We take it for granted that we can only identify hadrons (π, K, and p) up to a few GeV by dE/dx and by conventional TOF. Based on simulations, 1 psec resolution may be eventually possible, extending particle ID to momenta over 10 GeV in a detector the size of CDF. A Japanese group (Ohnema et al.) has recently achieved 5 ps resolution in TOF. This would have a big impact on precision measurements- for example, same-sign tagging in B_s mixing, identifying the b and $\bar{b}$ in the measurement of the top mass, and also separating $c\bar{s}$ from $u\bar{d}$ in top decays.

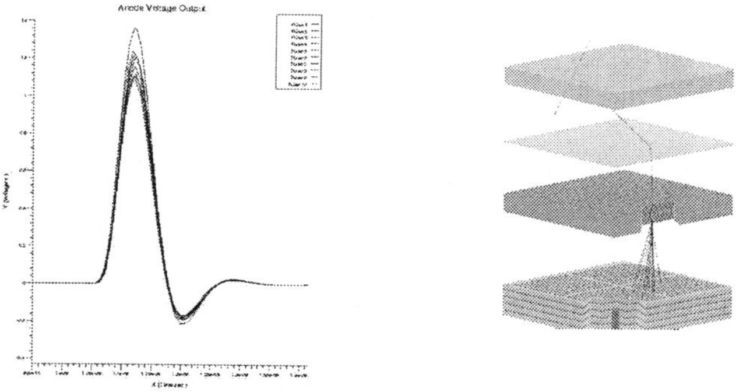

Figure 30.

16. Summary

- The Tevatron is just now moving into the domain where the W, Z, and top have enough statistics so that we are systematics dominated in many analyses. The LHC will turn on and immediately be in the systematics-dominated domain in almost all channels.
- In addition, the statistics is such that the theoretical SM predictions are sensitive to QED as well as QCD higher-order corrections- a new regime.
- Challenge- can we make systematics on top and W masses go down as $1/\sqrt{Luminosity}$?
- B_s mixing is not systematics dominated- it's a trigger problem. Challenge- can we accumulate the statistics for B_s mixing up to the inherent precision of the detector (trigger and DAQ question)?
- Watch the top mass, the W mass, B_s mixing, and for surprises out on the tails of kinematic distributions.
- These detectors are remarkable precision instruments, and are presented with a wealth of measurements. We need not only to exploit them as they are but also to support those folks working on hardware who concentrate on further developing their precision.

17. Acknowledgments

I thank all the CDF and DØ collaborators who have contributed to the topics I discussed. For understanding, wisdom, plots, and discussions I thank in particular: Eric Brubaker, Andrzej Czarnecki, Robin Erbacher, Rick Field, Ivan Furic, Doug Glenzinski, Chris Hays, Matt Herndon, John Hobbs, Joey Huston, Steve Levy, Andrei Loginov, Ashutosh Kotwal, Vaia Papadimitriou, Jon Rosner, Jim Strait, Evelyn Thompson, and Carlos Wagner.

Talks I have found useful and/or taken plots from: Florencia Canelli, Feb. 2005, Tev4LHC; Rick Field, XXXV Symposium on Multiparticle Dynamics, Kromericz; Kenichi Hatakeyama, Top2006, Coimbra, Jan, 2006; Aurelio Juste, Lepton-Photon, July, 2005; Cheng-Ju S. Lin, Aspen, Feb. 2006; Fabio Maltoni, HCP2005, Les Diableret, July 2005 Vaia Papadimitriou, XXXVth Multiparticle Dynamics, Kromericz; F. Ruggiero: http://chep.knu.ac.kr/ICFA-Seminar/upload/9.29/Morning/session1/Ruggiero-ICFA-05.pdf; Evelyn Thompson, Top2006, Coimbra, Jan, 2006; Eric Varnes, Top2006, Coimbra, Jan, 2006; Carlos Wagner, EFI Presentation, February 2006.

Lastly I would like to thank the organizers of the Winter Institute, in particular Faqir Khanna, Lee Grimard, and Roger Moore, for their unfailing hospitality and remarkable organization for what was a wonderful week.

References

1. LHC Design Report CERN-2004-003 (June 2004), Section 2. I have taken the 7.75 cm quoted for the RMS bunch length, multiplied by the geometric luminosity reduction factor of 0.836, and divided by $\sqrt{2}$. I hope this is correct.
2. The initial luminosity has a lifetime of 3.8 hours, which crosses the longer lifetime after 2 hours, at which point the luminosity is half the peak.
3. See the talk by F. Ruggiero at:http://chep.knu.ac.kr/ICFA-Seminar/upload/9.29/ Morning/session1/Ruggiero-ICFA-05.pdf
4. S.M. Berman, J.D. Bjorken, J. B. Kogut, Phys.Rev.D4:3388,1971.
5. I first learned of this method from A. Mukherjee and A. B. Wicklund, who used it in the CDF early precise (at that time) measurement of the Z^0mass.
6. H.Frisch, CDF/Phys/Top/Public/2484; Feb. 1994; M. Dittmar, F. Pauss, D. Zurcher; Phys.Rev.D56:7284-7290,1997
7. J. D. Jackson and R. McCarthy; "Z^3 Corrections to Energy Loss and Range", Phys. Rev. B6,4131 (1972).
8. Fabio Maltoni, *Top Physics: Theoretical Issues and Aims at the Tevatron and LHC*, HCP2005, July 8, Les Diablerets, Switz.;
9. G. L. Kane and S. Mrenna, Phys.Rev.Lett.77:3502-3505,1996.
10. Technical Design Report, CDF Collaboration; Aug. 1981
11. D0 announced the result $17 < \Delta M_S < 21$ ps^{-1} at 90% C.L. at the Moriond EWK conference March 12, 2006.

EVIDENCE FOR A QUARK-GLUON PLASMA AT RHIC

JOHN W. HARRIS
Physics Department, Yale University,
P.O. Box 208124, 272 Whitney Avenue, New Haven CT, U.S.A. 06520-8124
E-mail: John.Harris@Yale.edu

Ultra-relativistic collisions of heavy nuclei at the Relativistic Heavy Ion Collider (RHIC) form an extremely hot system at energy densities greater than 5 GeV/fm^3, where normal hadrons cannot exist. Upon rapid cooling of the system to a temperature T $\sim$ 175 MeV and vanishingly small baryo-chemical potential, hadrons coalesce from quarks at the quark-hadron phase boundary predicted by lattice QCD. A large amount of collective (elliptic) flow at the quark level provides evidence for strong pressure gradients in the initial partonic stage of the collision when the system is dense and highly interacting prior to coalescence into hadrons. The suppression of both light (u, d, s) and heavy (c, b) hadrons at large transverse momenta, that form from fragmentation of hard-scattered partons, and the quenching of di-jets provide evidence for extremely large energy loss of partons as they attempt to propagate through the dense, strongly-coupled, colored medium created at RHIC.

1. Introduction

All matter in the Universe existed in the form of quarks, leptons and the gauge bosons that carry the fundamental forces of Nature just a few micro-seconds after the Big Bang. As the Universe cooled, a quark-hadron phase transition occurred and the nuclear particles formed from quarks and gluons. Quantum Chromodynamics (QCD) on a lattice reveals such a quark-hadron phase transition at a temperature of 1.75 $\times 10^{12}$ Kelvin (175 MeV).[1] Above this temperature lattice QCD calculations predict that hadrons "melt" into a form of hot QCD matter consisting of quarks and gluons, known as the Quark-Gluon Plasma (QGP). Fig. 1 shows a schematic phase diagram of matter as a function of temperature and baryo-chemical potential. Several different phases of QCD matter are indicated. A first order phase transition is expected along a curve up to the critical point as shown in Fig. 1.[1,2] The QGP phase is expected at higher temperatures. The early Universe cooled down from higher temperatures close to the vertical axis, as shown, where the baryo-chemical potential $\mu_{baryon} = 0$. The region of high μ_{baryon} appears rich with structure and is presently an area of intensive theoretical investigation.[3] Understanding the nature of these phase transitions has implications for nuclear physics, astrophysics, cosmology and particle physics.

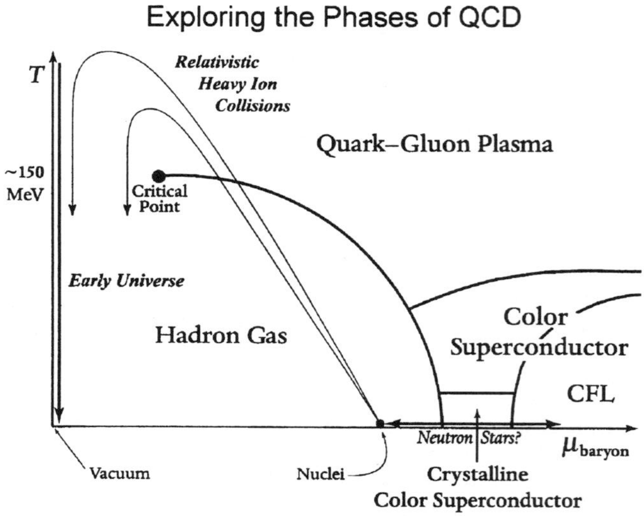

Figure 1. Schematic phase diagram of QCD matter as a function of temperature T and baryo-chemical potential μ_{baryon}.

The equation of state of hot QCD matter and its properties depend critically on the number of flavors and on the quark masses that are used in the lattice calculations.[1,2] Calculations of the energy density on the lattice as a function of temperature are displayed in Fig. 2. A relatively sharp deconfinement transition occurs at a temperature of approximately 175 MeV in 2-flavor QCD and at about 20 MeV lower temperature for 3-flavors (both in the chiral limit). There has also been recent success in implementation of techniques to calculate on the lattice at small but finite baryon density.[4]

The Relativistic Heavy Ion Collider (RHIC) at Brookhaven National Laboratory was constructed to collide nuclei at ultra-relativistic energies to form hot, dense QCD matter, to study its properties and better understand the quark-hadron phase transition. The RHIC facility, which commenced operation for physics in the year 2000, accelerates and collides ions from

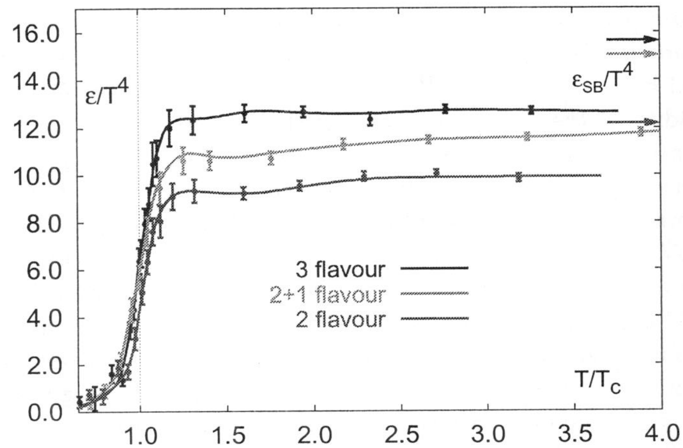

Figure 2. Lattice QCD results for the energy density ϵ/T^4 as a function of the temperature T relative to the critical temperature T_c. The flavor dependence is shown by the three curves, depicting results for 3 light quarks (u,d,s), 2 light quarks (u,d) plus 1 heavier (s) quark, and 2 light quarks (u,d) as indicated. The Stephan-Boltzmann values of ϵ_{SB}/T^4 are depicted by the arrows on the right.

protons to the heaviest nuclei over a range of energies, up to 250 GeV for protons and 100 A-GeV for Au nuclei. In the five production runs for physics, RHIC has collided Au + Au at center-of-mass energies $\sqrt{s_{NN}}$ = 19.6, 62, 130, and 200 GeV, Cu + Cu at center-of-mass energies $\sqrt{s_{NN}}$ = 62 and 200 GeV, d + Au at $\sqrt{s_{NN}} = 200$ GeV, and p + p at $\sqrt{s} = 200$ GeV. RHIC has also begun to collide polarized protons for studies of the spin content of the proton. RHIC and its four experiments (BRAHMS, PHENIX, PHOBOS, and STAR) are described comprehensively in Ref. [5]. The Large Hadron Collider (LHC) heavy ion program will start in a few years and is expected to explore farther into the high energy density regime of QCD.

2. Large Energy and Particle Densities Created at RHIC

An initial objective of RHIC was to determine the energy density in the initial colliding system and to establish whether it surpasses the critical energy density from lattice QCD that is necessary for creating the QGP phase transition. Measurements of the transverse energy per unit pseudorapidity $dE_T/d\eta$,[6,7] and the total particle multiplicity density and mean transverse momentum per particle[8] were used to estimate the energy density assuming a Bjorken longitudinal expansion scenario.[9] The energy density can be estimated by $\epsilon_{Bj} = \frac{1}{\tau_o \pi R^2} \times \frac{dE_T}{dy}$, with dE_T/dy the transverse energy per unit rapidity, R the transverse radius of the system, and τ_o the formation time. Assuming a maximum value for the formation time τ_o = 1 fm/c, a conservative estimate of the minimum energy density for the 5% (2%) most central Au + Au collisions at $\sqrt{s_{NN}}$ = 130 GeV is 4.3 GeV/fm^3 [8] (4.6 GeV/fm^3)[6] and 4.9 GeV/fm^3 [7] for the 5% most central Au + Au collisions at $\sqrt{s_{NN}}$ = 200 GeV. Note that the energy density derived from the RHIC experiments using the Bjorken formlation is a lower limit and may be much larger ($\sim$ ten times), since τ_o at RHIC is expected to be significantly less than 1 fm/c. This lower limit on the energy density at RHIC is approximately twenty-five times normal nuclear matter density ($\epsilon_{n.m.}$ = 0.17 GeV/fm^3) and seven times the critical energy density (ϵ_c = 0.6 GeV/fm^3) predicted by lattice QCD for formation of the QGP.

Hadrons with transverse momentum less than 2 GeV/c are produced abundantly in collisions at RHIC. The measured charged hadron multiplicity density at midrapidity is $dn_{ch}/d\eta \mid_{\eta=0} = 700 \pm 27(syst)$ [10] for the 3% most central collisions of Au + Au at $\sqrt{s_{NN}}$ = 200 GeV. This corresponds to a hadron multiplicity density $dn_{total}/d\eta \mid_{\eta=0} = 1050$ and a total hadron multiplicity in the most central events of $\sim$ 7000.[10,11] In terms of the number of created quarks, consider the case where all observed hadrons in the final state are mesons, each containing a quark and anti-quark. These 7,000 hadrons correspond to 14,000 quarks and anti-quarks. A lower limit for the number of created quarks and anti-quarks can be obtained by subtracting off the valence quarks that originally enter the collision in the incident nuclei. The number of original valence quarks in a head-on collision of two Au nuclei is 2 (Au nuclei) x 197 nucleons/Au x 3 quarks/nucleon $\sim$ 1200. Thus, more than 90% of the more than 14,000 quarks and anti-quarks in the final state are produced in the collision.[a]

3. Observation of Strong Elliptic Flow at RHIC

The observation of an unexpectedly large elliptic flow at RHIC has led to exciting consequences for understanding the dynamical evolution of these collisions. Unlike the case for collisions of elementary particles, nuclei colliding with non-zero impact parameter have an inherent spatial asymmetry

[a]Note that there are also a large number of gluons present that have not been considered in this estimate.

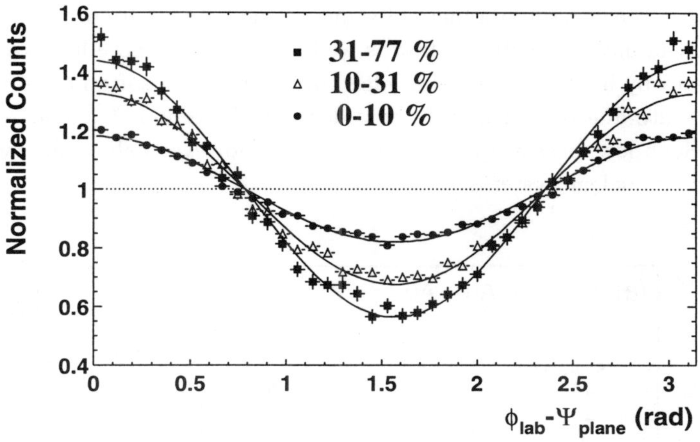

Figure 3. Azimuthal correlations of charged hadrons as a function of the azimuthal angle relative to the reaction plane for three different centrality ranges selected (as denoted in the legend) in 130 GeV/n Au + Au collisions.

associated with the asymmetric region of overlap. The larger the impact parameter, the larger the asymmetry perpendicular to the reaction plane.[b] Displayed in Fig. 3 are the azimuthal angular distributions for collisions over three different impact parameter ranges at RHIC.[12] Hadrons are emitted preferentially in the reaction plane providing evidence for large pressure gradients early in the collision process that generate the elliptic flow. The pressure gradients and elliptic flow in-plane increase with increase of the initial spatial asymmetry out-of-plane. These results indicate that the initial spatial asymmetry is transformed efficiently into the observed momentum-space anisotropy during the brief traversal time of the incident nuclei ($<$ 1 fm/c at these energies). Thus, the system must be dense and highly interacting to accomplish this transformation efficiently in such short time.

To study this azimuthal anisotropy in quantitative detail the second Fourier harmonic component of the azimuthal distribution of particles in momentum space is constructed with respect to the reaction plane, $v_2 = \langle\cos(2\phi)\rangle$ where $\phi = \text{atan}\,(p_y/p_x)$. The v_2 is called the elliptic flow. Displayed in Fig. 4 is v_2 for $\pi^{\pm}$, K_s^0, $\overline{p}$, and $\Lambda + \overline{\Lambda}$ as a function of p_T in $\sqrt{s_{NN}}$

[b]The reaction plane is defined as the plane containing the incident beam and impact parameter vectors.

= 200 GeV Au + Au minimum bias collisions.[13] Also shown are predictions from hydrodynamics. The elliptic flow is well described at these low transverse momenta by hydrodynamical models incorporating a softening of the equation of state due to quark and gluon degrees of freedom[14,15] and zero viscosity.[16] Such low viscosities were not expected nor ever before observed for hadronic or nuclear systems.

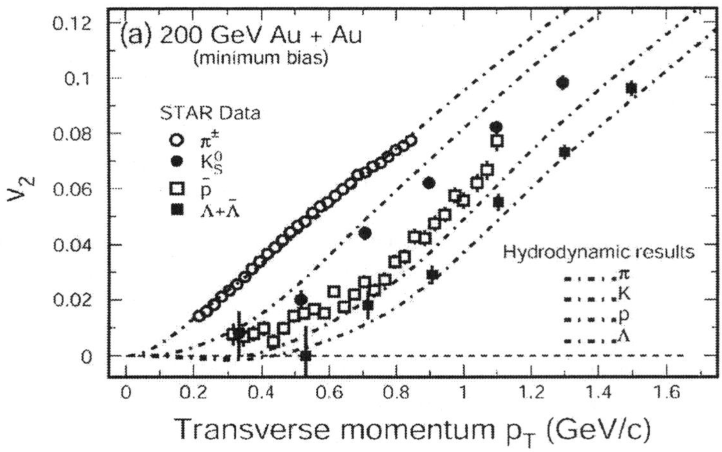

Figure 4. Elliptic flow v_2 for $\sqrt{s_{NN}} = 200$ GeV Au + Au minimum bias collisions as a function of p_T for $\pi^{\pm}$, $K^{\pm}$, $p^{\pm}$ and $\Lambda + \overline{\Lambda}$. Curves are hydrodynamics predictions.

The v_2 measured for K^0_s, $\Lambda + \overline{\Lambda}$, $\Xi + \overline{\Xi}$, and $\Omega + \overline{\Omega}$ at higher p_T are displayed in Fig. 5 for $\sqrt{s_{NN}} = 200$ GeV Au + Au minimum bias collisions.[17] The lighter mesons begin to deviate from the predictions of hydrodynamics at p_T greater than approximately 1 GeV/c, while heavier baryons deviate significantly at somewhat higher p_T. The baryons continue to have higher values of v_2 than the mesons at the largest p_T measured. Larger values of v_2 for baryons than mesons extending to larger transverse momenta may result from particles being created in soft processes and boosted to higher p_T by collective flow. Alternatively or in addition, coalescence of quarks to form composite particles occurs. The overall saturation of v_2 for larger momenta may reflect effects of the energy loss due to the large gluon densities created in these collisions,[18] which will be discussed later in this presentation.

Displayed in Fig. 6 is v_2 per quark measured by STAR[17,19] for K^0_s, Λ

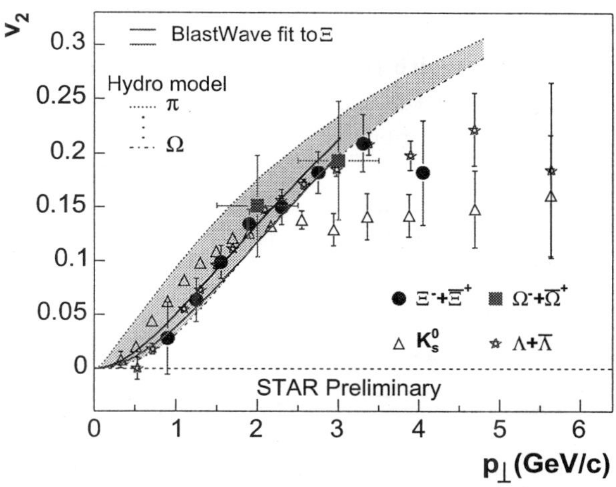

Figure 5. Elliptic flow v_2 for $K^0{}_s$, $\Lambda + \overline{\Lambda}$, $\Xi + \overline{\Xi}$, and $\Omega + \overline{\Omega}$.

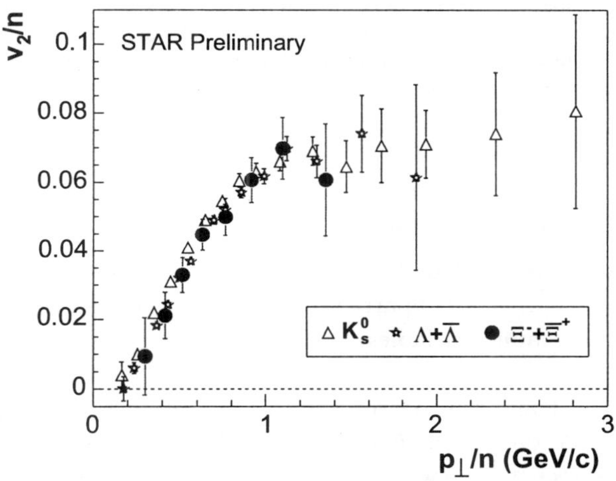

Figure 6. Elliptic flow per quark (v_2/n) for $K^0{}_s$, $\Lambda + \overline{\Lambda}$, $\Xi + \overline{\Xi}$, and $\Omega + \overline{\Omega}$ as a function of p_T per quark.

$+ \overline{\Lambda}$, and $\Xi + \overline{\Xi}$ as a function of p_T per quark in $\sqrt{s_{NN}} = 200$ GeV Au + Au minimum bias collisions. When v_2 is plotted per quark (v_2/n) for baryons and mesons, the values of v_2/n scale with p_T/n at large p_T. This is consistent with a quark coalescence picture for hadrons at quark $p_T >$

0.5 GeV/c [20] and is evidence for early collective flow at the quark level. This observation coupled with the extremely low to non-existent viscosity have led to descriptions of the system in terms of a nearly perfect liquid (non-viscous) of quarks and gluons.[21]

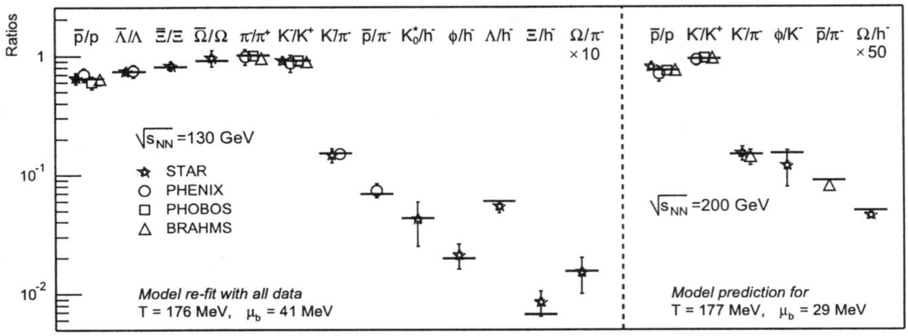

Figure 7. Particle ratios measured in RHIC experiments denoted by symbols in the legend for central collisions of Au + Au at $\sqrt{s_{NN}} = 130$ GeV (left panel) and 200 GeV (right panel). Results from a statistical-thermal model fit to the entire set of data are shown as a horizontal line for each ratio. Parameters (μ_B, T) for the best fit at each energy are shown at the bottom of each panel (see text).

4. Is the System Thermalized?

If the system can be described in terms of equilibrium thermodynamics, the ratios of the various types of particles must be reproduced with a consistent set of thermodynamic variables. Statistical and thermodynamic models reproduce the measured ratios using as variables the chemical freeze-out temperature (T) and the baryo-chemical potential (μ_B). These models employ hadronic degrees of freedom in a grand canonical ensemble. See Ref. [22] for a recent review. Displayed in Fig. 7 are the particle ratios measured at RHIC along with results of a statistical-thermal model fit.[23,24] The particle ratios for $\sqrt{s_{NN}} = 130$ GeV Au + Au can be fit with the parameters T = 176 MeV and $\mu_B = 41$ MeV. For $\sqrt{s_{NN}} = 200$ GeV Au + Au, T = 177 MeV and $\mu_B = 29$ MeV are required. The statistical-thermal model fits reproduce the data extremely well. When the same approach is used for the SPS Pb + Pb data at $\sqrt{s_{NN}} = 17.3$ GeV reasonable fits are found with T = 164 MeV and $\mu_B = 274$ MeV.[25,26] A similar approach[27] applied to particle production data from NA49 in Pb + Pb collisions, yields

T = 148 MeV and μ_B = 377 MeV at $\sqrt{s_{NN}}$ = 8.73 GeV, and T = 154 MeV and μ_B = 293 MeV at $\sqrt{s_{NN}}$ = 12.3 GeV, with an additional strangeness saturation parameter[c] of γ_s = 0.75 and 0.72, respectively. Thus, chemical freeze-out follows a curve in (μ_B, T) space as depicted in Fig. 8 where μ_B decreases from 293 MeV at the lowest SPS energy to 29 MeV at the highest RHIC energy, and the chemical freeze-out temperature increases gradually from 154 MeV to 176 MeV. When drawn on a (μ_B, T) plane, these values of (μ_B, T) from statistical model fits to the experimental data approach the deconfinement phase transition boundary predicted by lattice QCD.[28]

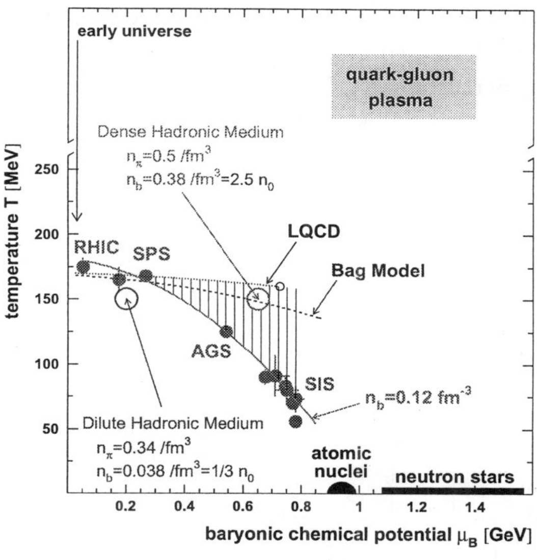

Figure 8. Nuclear phase diagram extracted from a thermal model analysis of experimental results at various energies and heavy ion facilities (SIS, AGS, SPS, and RHIC). The dotted curve labeled "LQCD" corresponds to the deconfinement phase boundary predicted by lattice QCD. The solid curve corresponds to freeze-out at a constant baryon density.

---

[c] A strangeness saturation parameter is necessary in a canonical approach or when additional dynamical effects are present affecting extraction of true chemical equilibrium values.

5. Suppression of Large Transverse Momentum Particles

Hard scattering can be used to probe the medium through which the hard-scattered partons propagate. The radiation energy loss of a parton traversing a dense medium is predicted to be significant and is sensitive to the gluon density of the medium.[29,30,31] In order to investigate parton energy loss in the medium, the RHIC experiments have measured hadron spectra and azimuthal correlations of hadrons with large transverse momentum.

To compare results from relativistic heavy ion collisions with those of elementary (p + p) interactions, a nuclear modification factor R_{AA} is defined as $R_{AA}(p_T) = \frac{dN^2/dp_T dy\ (AA)}{\langle N_{binary}\rangle dN^2/dp_T dy\ (NN)}$. $\langle N_{binary}\rangle$ is the number of binary collisions in a geometrical model in order to scale from elementary nucleon-nucleon (NN) collisions to nucleus-nucleus (AA) collisions. When $R_{AA} = 1$, AA collisions can be described as an incoherent superposition of NN collisions, as predicted by perturbative QCD (pQCD). This corresponds to scaling with the number of binary collisions (binary scaling).

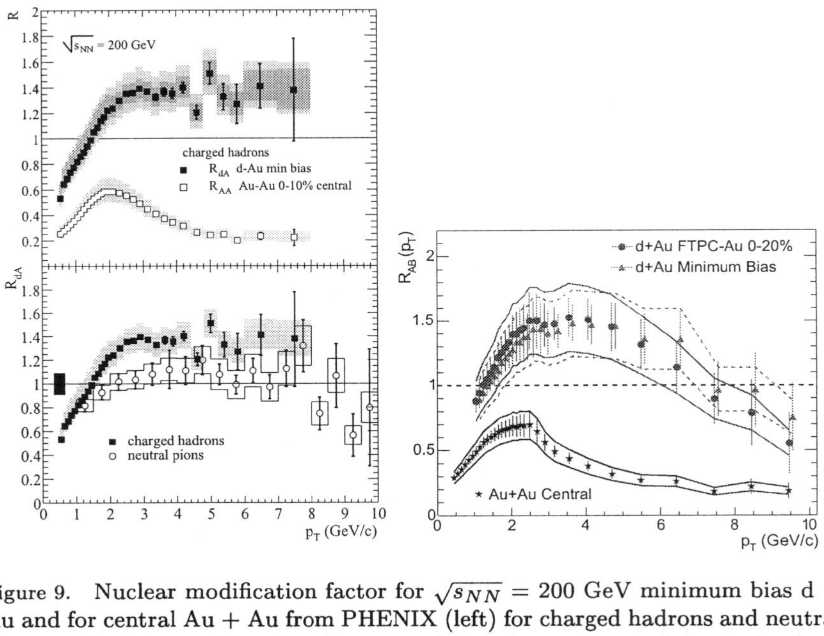

Figure 9. Nuclear modification factor for $\sqrt{s_{NN}} = 200$ GeV minimum bias d + Au and for central Au + Au from PHENIX (left) for charged hadrons and neutral pions; and minimum bias d + Au, central d + Au and central Au + Au from STAR (right) for charged hadrons.

5.1. *Light Hadrons*

Suppression of the hadron spectra at large transverse momenta $p_T > 2$ GeV/c has been observed in the nuclear modification factors measured in central Au + Au collisions at RHIC.[32,33,34] Displayed in Fig. 9 is the nuclear modification factor as a function of transverse momentum for central Au + Au collision data and for d + Au collisions from PHENIX (left panel)[35] and STAR (right panel).[36] The ratios are taken relative to measured p + p collision data scaled by the number of binary collisions. These data exhibit a clear suppression by a factor 4-5 in the central Au + Au case. The central Au + Au ratio remains rather flat up to the statistical limits of the data at 10 GeV/c transverse momentum, where no sign of reaching the pQCD limit of binary scaling is observed. The peripheral collision data (not shown) exhibit no nuclear modification, i.e. $R_{AA} = 1$, within errors.

The Au + Au relative to p + p yields can be reproduced by pQCD calculations incorporating parton energy loss in dense matter[37] or by a model incorporating initial-state gluon saturation.[38] To distinguish these two theoretical approaches, final state energy loss versus initial state gluon saturation, measurements were made in d + Au where no parton energy loss in a dense medium (final state suppression) is expected. The ratio $R_{dAu}(p_T) > 1$ in d + Au indicates a Cronin enhancement (initial state multiple scattering) for $1 < p_T < 7$ GeV/c with no suppression. The d + Au charged hadron data when compared to scaled p + p data[35,36,39,40] rule out initial-state gluon saturation leaving only final state effects as a cause of the suppression in the Au + Au data at mid-rapidity at RHIC. In order to describe the suppression of light hadrons at RHIC a gluon density of $dn_{gluon}/dy \sim 1000$ is required.[41] This is equivalent to an energy loss per unit length that is approximately 15 times that of normal nuclear matter.

Displayed in Figs. 10 and 11 are the ratios R_{CP} of identified particles (specified in the legends) produced in central collisions relative to those produced in peripheral collisions scaled by the number of binary collisions in each data set. The data in Fig. 10 are strange particle data and charged hadrons from STAR.[42] The R_{CP} ratios for mesons reach a maximum at p_T between 1 - 2 GeV/c with ratios R_{CP} considerably less than 1 and continue to be suppressed ($R_{CP} < 1$) up to $p_T \sim 7$ GeV/c. The R_{CP} ratios for baryons peak near unity and are suppressed relative to binary-scaling only above around 3 GeV/c. The PHENIX data on π^0 and $p + \overline{p}$ are shown in Fig. 11.[43] The π^0 are suppressed by a factor of $\sim$ 4 - 5, while the protons do not deviate from binary scaling up to $p_T \sim 4$ GeV/c.

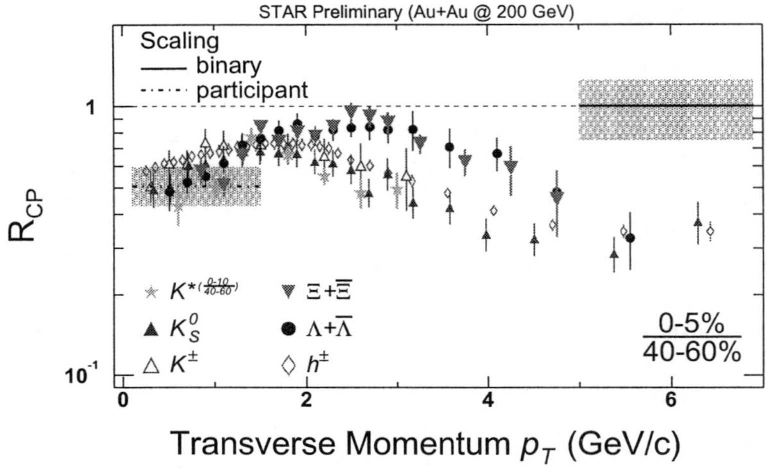

Figure 10. R_{CP} ratios for $\sqrt{s_{NN}}$ = 200 GeV Au + Au collision data measured in STAR.

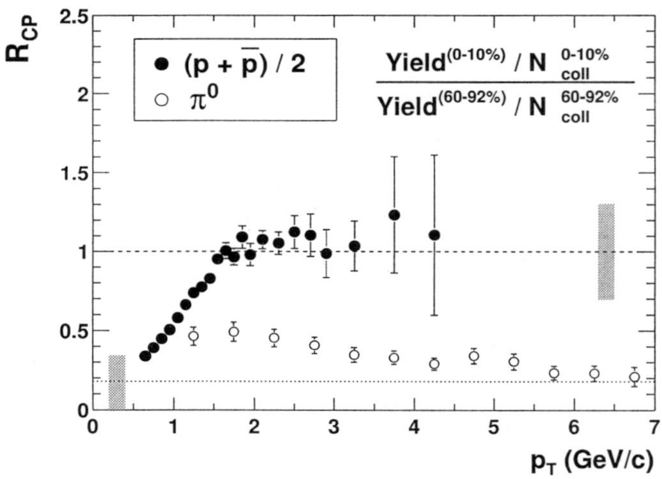

Figure 11. R_{CP} ratios for $\sqrt{s_{NN}}$ = 200 GeV Au + Au collision data measured in PHENIX.

5.2. *Charm and Beauty Hadrons*

Heavy (charm and beauty) quarks are expected to lose less energy while propagating through a dense colored medium. The large heavy quark mass reduces the available phase space for gluon radiation. This has been called the "dead cone effect".[44] PHENIX and STAR have measured the spectra of non-photonic electrons in $\sqrt{s_{NN}} = 200$ GeV p + p and Au + Au collisions over a range of impact parameters. After subtraction of electrons from photon conversions and light hadron decays, the resulting non-photonic electron spectra are predominantly from semi-leptonic decays of heavy quarks (D- and B-meson decays). The R_{AA} for non-photonic electrons measured in central collisions are displayed in Figs. 12 and 13 for PHENIX[45] and STAR,[46] respectively, as a function of p_T. The non-photonic electrons from decays of D- and B- mesons are clearly suppressed to about the same degree as the light hadron spectra.

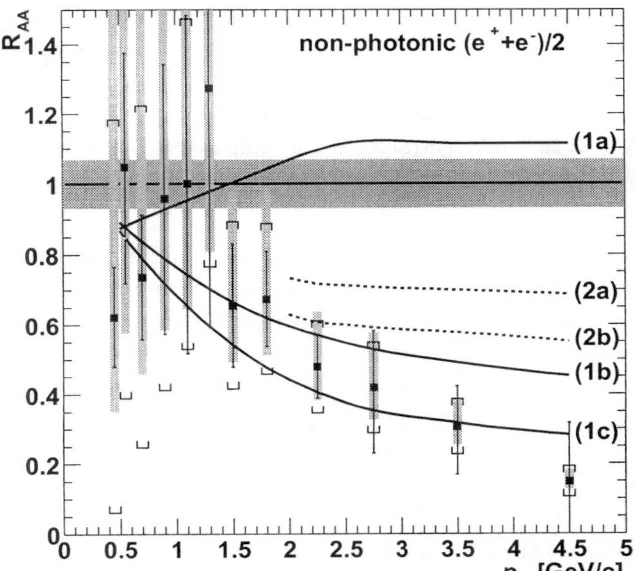

Figure 12. R_{AA} for non-photonic electrons as a function of transverse momentum (p_T) in $\sqrt{s_{NN}} = 200$ GeV central Au + Au compared to p + p in PHENIX for 0-10% centrality. See text for description of model curves.

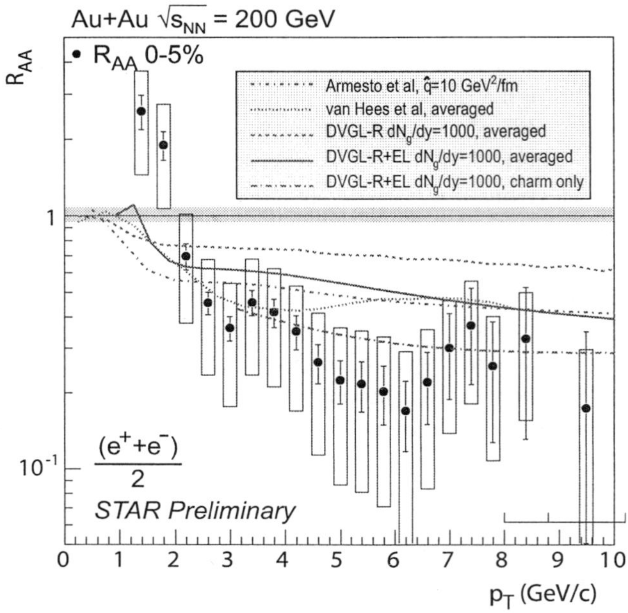

Figure 13. R_{AA} for non-photonic electrons as a function of transverse momentum (p_T) in $\sqrt{s_{NN}} = 200$ GeV central Au + Au compared to p + p in STAR for 0-5% centrality. See text for description of model curves.

The curves in Figs. 12 and 13 are predictions of various models. In Fig. 12 the curves 1 a, b , and c are from Ref. [47] and include decays of D-mesons only, whereas curves 2a and b include B-decays also. Curves 1 (a, b, and c) are for various values of the transverse momentum transfer squared per unit length (0 - no medium, 4, 14 GeV^2/fm) of the parton traversing the medium and is proportional to the density of scatterers. Curves 2 (a and b) are for various values of the gluon density of the medium (1000 and 3500). Clearly only curve 1c approximates what is seen in the data, but this curve does not include B-meson decays that become important above about 3 GeV/c. In Fig. 13 the parameters for transverse momentum squared labeled "Armesto" from Ref. [47] and gluon density labeled DVGL from Refs. [48,49] for the calculations are listed in the legend. The curve labeled "DVGL-R" corresponds to calculations including induced gluon radiation, while those labeled "DVGL-R+EL" correspond to gluon radiation and elastic scattering energy loss. The label "averaged" corresponds to charm and beauty decays while "charm only" is self-evident. The calculation labeled

"Armesto" includes only charm decays. Only the curve "DVGL-R+EL" for "charm only" with a value of gluon density consistent with the light hadron suppression fits the data. Also shown is a curve labeled "van Hees" from Ref. [50] that utilizes c and b quark resonances in a strongly coupled medium, but this result also does not fit the data. Clearly the lack of influence of B-meson decays on the spectra when comparing to theory is perplexing and this calls into question our understanding of the propagation of heavy quarks in the medium. Further experimental results separating charm and beauty are necessary to resolve this dilemma, future theoretical breakthroughs not-withstanding.

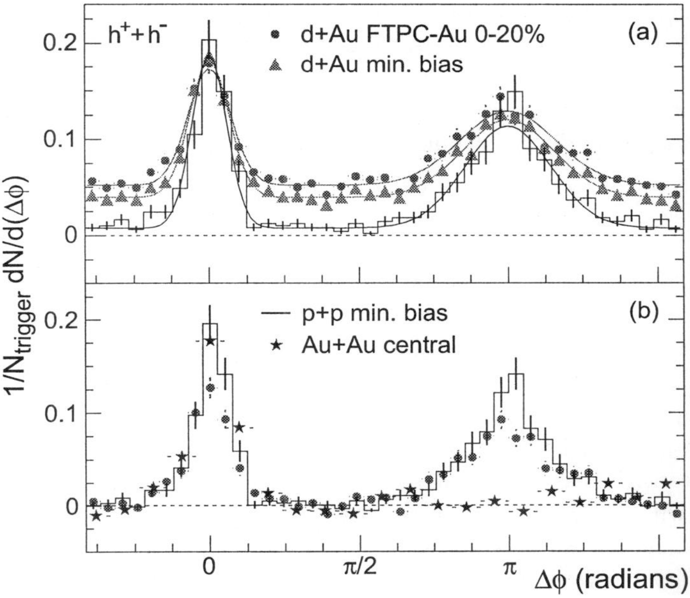

Figure 14. a) Charged hadron correlation functions (see text for details) for p + p interactions, and minimum bias and central collisions of d + Au at $\sqrt{s_{NN}} =$ 200 GeV. b) Comparison of correlations for p + p, central d + Au, and central Au + Au collisions with background subtracted.

6. Effects of Jet Quenching on Azimuthal Correlations

Two-particle correlations of high p_T hadrons are used to extract the hard scattering component from the soft background in relativistic heavy ion collisions. The fragmentation of partons produces jets, a correlation of particles in a small region of relative azimuthal angle ($\Delta\phi$) and pseudorapidity ($\Delta\eta$). As seen in Fig. 14 an enhancement is observed at $\Delta\phi = 0$ in azimuthal correlations near midrapidity upon correlating charged hadrons having $4 < p_T^{trig} < 6$ GeV/c with all other charged hadrons having 2 GeV/c $< p_T < p_T^{trig}$ in the same event. This has been seen in collisions of Au + Au at $\sqrt{s_{NN}}$ = 130 and 200 GeV ranging from minimum bias to central collisions, and in p + p and d + Au at $\sqrt{s}$ = 200 GeV.[12,51] The peak at small relative azimuthal angle is indicative of short range correlations, i.e. jets. Data from p + p, d + Au, and peripheral Au + Au (not shown) exhibit a distinct back-to-back correlation, a characteristic of hard-parton scattering, with peaks at $\Delta\phi = 0$ and π, indicating di-jets. This is shown in Fig. 14 for the p + p and d + Au reactions. However, for the most central Au + Au collisions the backward ($\Delta\phi = \pi$) correlation disappears. These results provide further evidence that in central Au + Au collisions a strong interaction results in quenching of high p_T partons.

7. J/ψ

An early prediction of a deconfinement signature was the suppression of J/ψ production in a deconfined medium.[52] J/ψ suppression has been observed in central nucleus-nucleus collisions at the SPS [53] and initial measurements at RHIC.[54]

A suppression of J/ψ production has been observed in $\sqrt{s_{NN}}$ = 17.3 GeV Pb + Pb collisions at the SPS [55] and attributed to a deconfinement phase transition as predicted by [52]. Alternative models based on interactions with hadronic co-movers have been proposed, but have problems reproducing the suppression as a function of centrality. Expectations at RHIC vary between the different models. If the suppression at the SPS is due to deconfinement, then the J/ψ should also be suppressed at RHIC energies. On the other hand, coalescence of $c\bar{c}$ pairs [56] and statistical-thermal models [57,58] predict significant J/ψ production at RHIC. First measurements of J/ψ have been reported by PHENIX at RHIC in $\sqrt{s_{NN}}$ = 200 GeV p + p and Au + Au collisions.[54] These results disfavor enhancement due to $c\bar{c}$ coalescence[56] and thermal $c\bar{c}$ production[58], while still not yet being able to distinguish between models for suppression relative to binary scaling. The high statistics measurements from PHENIX for J/$\psi \to e^+e^-$ at mid-rapidity and J/$\psi \to \mu^+\mu^-$ at forward rapidities will provide data to adequately address the J/ψ production and evolution mechanisms.

8. Summary and Conclusions

The completely unexpected observation of strong elliptic flow in Au + Au collisions on the first day at RHIC, the subsequent detailed measurements of the variation of the elliptic flow for different particles, and the scaling of the flow with the numbers of quarks in the various mesons and baryons have led to a description of the medium created in these collisions as a nearly perfect liquid of quarks and gluons. The observed ratios of the many different hadrons produced at RHIC represent equilibrium abundances predicted from thermal models. The final state hadrons are formed and thermally distributed at the universal hadronization temperature (175 MeV) with a collective flow derived from their quark predecessors. The suppression of hadrons at large transverse momentum in Au + Au collisions, absent in d + Au, provides evidence for a high density medium, consistent with the initial determination of large energy densities created at RHIC. The observed quenching of the away-side jet (and its absence in central collisions) in Au + Au collisions requires a strongly-coupled colored medium, rather than the perturbative one expected from the initial lattice calculations. This nearly perfect liquid of quarks and gluons has behavior similar to that of other strongly-coupled systems such as degenerate fermi gases of atoms at low temperatures.[59] Furthermore, the RHIC liquid of quarks and gluons has a viscosity (divided by the entropy density) that approaches a quantum lower bound for shear viscosity derived using strongly-coupled N = 4 supersymmetric Yang-Mills theory.[60]

However, much still needs to be understood. The new results on suppression of heavy quark hadrons defy predictions. Addressing the question of deconfinement by measuring the extent to which charmonium is suppressed is just getting underway. A complete quarkonium program (measuring J/ψ, ψ' and Upsilon states) at the LHC and upgraded RHIC luminosities will be instrumental in determining the initial temperatures and degree of deconfinement via comparisons to more sophisticated future lattice calculations. So far, experimental signatures of chiral symmetry restoration have been elusive at RHIC. This will be investigated further through measurements of low mass di-leptons to search for medium modification of hadronic resonances. New insight may also be gained from a better understanding of hadronization and fragmentation in vacuum and in medium.

The strongly-coupled quark-gluon liquid is an exciting and unexpected phenomenon that must be understood. The RHIC results require a more sophisticated, dynamical approach in order to extract the nuclear matter equation of state of the liquid, its transport properties (e.g. shear viscosity, heat conductivity, speed of sound), and determine its constituents. Similar theoretical techniques are being used in attempts to understand results (e.g. strongly-interacting atomic gases, the strongly-coupled quark-gluon liquid, and black holes) from vastly different fields. On the experimental side, new measurements at RHIC and the LHC will be undertaken.

References

1. F. Karsch, Nuc. Phys. A698 (2002) 199c.
2. K. Kanaya, Nuc. Phys. A715 (2003) 233c.
3. see for example M. Alford J. Phys. G: Nucl. Part. Phys. 30 (2004) S441.
4. for review of recent results at finite temperature and density see Z. Fodor, Nuc. Phys. A715 (2003) 319c.
5. "The Relativistic Heavy Ion Collider and its Detectors", eds. M. Harrison, T. Ludlam, and S. Ozaki, Nuc. Inst. Meth. A499 (2003) p. 235-824.
6. K. Adcox et al. (PHENIX) Phys. Rev. Lett. 87 (2001) 052301.
7. J. Adams et al. (STAR), nucl-ex/0407003.
8. C. Adler et al. (STAR), Phys. Rev. Lett. 87 (2001) 112303.
9. J.D. Bjorken, Phys. Rev. D27 (1983) 140.
10. B. Back et al (PHOBOS), Phys. Rev. C65 (2002) 061901R.
11. B. Back et al (PHOBOS), Phys. Rev. Lett. 87 (2001) 102303.
12. C. Adler et al. (STAR) Phys. Rev. Lett. 90 (2003) 032301.
13. C. Adler et al (STAR), Phys. Rev. Lett. 87 (2001) 182301 and Phys. Rev. Lett. 89 (2002) 132301.
14. P.F. Kolb et al., Phys. Lett. B500 (2001) 232; P.F. Kolb et al., Nucl. Phys. A696 (2001) 197.
15. P. Huovinen et al., Phys. Lett. B 503 (2001) 58.
16. D. A. Teaney, J. Phys. G30 (2004) S1247.
17. J. Castillo et al (STAR), nucl-exp/0403037.
18. M. Gyulassy, I. Vitev, and X.N. Wang, Phys. Rev. Lett. 86 (2001) 2537.
19. J. Adams et al (STAR), Phys. Rev. Lett. 92 (2004) 052302.
20. C. Nonaka, R. Fries, and S. Bass, nucl-th/0308051.
21. T. Hirano and M. Gyulassy, nucl-th/0506049v2.
22. P. Braun-Munzinger, K. Redlich, and J. Stachel, preprint nucl-th/0304013.
23. P. Braun-Munzinger, D. Magestro, K. Redlich, and J. Stachel, Phys. Lett. B518 (2001) 41 and D. Magestro (private communication).
24. for another statistical approach see F. Becattini, J. Phys. G28 (2002) 1553.
25. P. Braun-Munzinger, I. Heppe, and J. Stachel, Phys. Lett. B465 (1999) 15.
26. see also F. Becattini et al., Phys. Rev. C64 (2001) 024901.
27. F. Beccatini, M. Gazdzicki, and J. Sollfrank, Eur. J. Phys. C5 (1998) 143.
28. Z. Fodor and S.D. Katz, preprint hep-lat/0106002 (2001).
29. M. Gyulassy and M. Plümer, Phys. Lett. B243 (1990) 432.
30. X. N. Wang and M. Gyulassy, Phys. Rev. Lett. 68 (1992) 1480.
31. R. Baier, D. Schiff, and B. G. Zakharov, Annu. Rev. Part. Sci. 50 (2000) 37.

32. K. Adcox et al (PHENIX), Phys. Rev. Lett. 88 (2002) 022301.
33. C. Adler et al (STAR), Phys. Rev. Lett. 89 (2002) 202301.
34. B.B. Back et al. (PHOBOS), nucl-exp/0302015v1.
35. K. Adcox et al (PHENIX), Phys. Rev. Lett. 91 (2003) 072303.
36. J. Adams et al (STAR), Phys. Rev. Lett. 91 (2003) 072304.
37. I. Vitev and M. Gyulassy, Phys. Rev. Lett. 89 (2002) 252301.
38. D. Kharzeev, E. Levin and L. McLerran, Phys. Lett. 561 (2003) 93.
39. B.B. Back et al. (PHOBOS), Phys. Rev. Lett. 91 (2003) 072302.
40. I. Arsene et al. (BRAHMS), Phys. Rev. Lett. 91 (2003) 072305.
41. for a review of jet energy loss see M. Gyulassy et al., nucl-th/0302077v1.
42. M.A.C. Lamont et al (STAR), nucl-ex/0403059; J. Adams et al (STAR), nucl-ex/0306007.
43. S.S. Adler et al (PHENIX), nucl-ex/0307022.
44. Y.L. Dokshitzer and D.E. Kharzeev, Phys. Lett. B519 (2001) 199.
45. S.S. Adler et al. (PHENIX), Phys.Rev.Lett. 96 (2006) 032301).
46. J. Bielcik et al. (STAR), nucl-ex/0511005.
47. N. Armesto et al., Phys. Rev. D71 (2005)054027.
48. M. Djordjevic et al., Phys. Lett. B (in press), nucl-th/0507019.
49. S. Wicks et al., nucl-th/0512076.
50. R. Rapp et al., hep-ph/0510050v2.
51. C. Adler et al. (STAR) Phys. Rev. Lett. 89 (2002) 202301.
52. T. Matsui and H. Satz, Phys. Lett. B 178 (1986) 416.
53. M.C. Abreu et al (NA50), Phys. Lett. B 477 (2000) 28; ibid. Phys. Lett. B 521 (2001) 195.
54. S.S. Adler et al (PHENIX), Phys. Rev. C69 (2004) 014901.
55. L. Ramello et al (NA50), Nucl. Phys. A715 (2003) 243c.
56. R.L. Thews, M. Schroedter, and J. Rafelski, Phys. Rev. C 63 (2001) 054905.
57. L. Grandchamp and R. Rapp, Nucl. Phys. A709 (2002) 415.
58. A. Andronic et al, nucl-th/0303036.
59. R.M. OHara et al., Science 298 (2002) 2179.
60. P.K. Kovtun, D.T. Son and A.O. Starinets, Phys. Rev. Lett. 94 (2005) 111601.

FUNDAMENTAL EXPERIMENTS AT LOW ENERGIES

H.-JŰRGEN KLUGE
Gesellschaft für Schwerionenforschung (GSI)/Darmstadt and University of Heidelberg

Abstract

In my two lectures, delivered at the 2006 Lake Louise Winter Institute, two topics were addressed: firstly, atomic physics techniques which provide extremely high accuracy as well as sensitivity, and, secondly, tests of symmetries, studies of the forces acting on quantum mechanical systems, and the determination of the fundamental properties of fermions – all performed at low energies. In this contribution to the proceedings of the Winter Institute, mainly the second part is documented whereas only some references are given to the atomic-physics techniques applied.

Introduction

Over the past decades, the accuracy of measurements in many areas of physics has developed in an amazing way. A well known example is the atomic clock, which permits the measurement of frequency, time, and length to fourteen significant digits, with ever increasing accuracy. An accuracy of 10^{-18} seems to be in reach in the near future. For realizing highest accuracy, storing and cooling are very often applied. Storing enables an extended time of observation and results, therefore, as a consequence of the Heisenberg uncertainty principle, in an increased resolving power and often also in an improved accuracy. Cooling eliminates experimental imperfections to a large extent since the amplitudes of the motion can be limited to smaller dimensions. Furthermore, cooling reduces the Doppler broadening and, for particles confined in traps, also the Doppler shift. Finally, the sensitivity of the experiments is increased by cooling and storing so that very often only one single atom, ion, or particle is sufficient to perform the experiment. Working with a single particle has further advantages: A single particle is always polarized; its degrees of freedom are well under control and can easily be manipulated; a single particle is free of disturbing interactions caused by other particles. Generally, but not always, low-energy experiments are limited by systematic uncertainties rather than statistical ones.

Quite a number of Nobel prizes were awarded in recent years for the development of the general concept of cooling and storing: 1984 to J. van der Meer and C. Rubbia (storage and cooling of antiprotons); 1989 to H. Dehmelt and W. Paul (storage and cooling of ions); 1997 to S. Chu, C. Cohen-Tannoudji, and W. D. Phillips (storage and cooling of atoms); 2001 to E. Cornell, W. Ketterle, and C. Wieman (cooling of atoms to lowest temperatures and Bose-Einstein condensation).

The accuracy and sensitivity of measurements have reached a level where a number of very basic questions of physics and cosmology come into reach for low-energy experiments [1-3] Although the investigated effects are generally very tiny at low energies, this is often compensated by the extreme level of accuracy accessible today by use of cold or ultra-cold neutrons, cold or ultra-cold ions or atoms, protons, electrons, and their antiparticles, all either in-beam or confined in traps or storage rings.

Symmetry tests by low-energy experiments

CPT invariance and parity violation

Symmetry operations such as charge conjugation (C), parity inversion (P), and time reversal (T) are linked to conservation laws and provide severe tests of the concepts in physics. The most basic symmetries are CPT and Lorentz invariance. In the Standard Model, all interactions should be invariant under the combined CPT operation. Any deviation would signal new physics beyond the Standard Model.

In the low-energy sector, the usual way to test CPT invariance is to compare fundamental properties of particles with those of antiparticles such as mass, magnetic moment, charge, or lifetime. Sophisticated experiments have been performed in the past by atomic physics techniques, for example, for electrons, positrons, negative and positive muons, proton and antiprotons. The best limit on the level of 10^{-12} was obtained by a comparison of the g-factor of the electron with that of the positron [4]. A six order of magnitude more stringent test of CPT invariance is obtained in the K^0 system by comparing the relative masses of particle and antiparticle [5].

The ALPHA (former ATHENA), ASACUSA, and ATRAP collaborations started at the Antiproton Decelerator (AD) at CERN a program to test CPT invariance by comparing the hyperfine splitting in the ground state (ASACUSA) or the 1s-2s transition frequency (ALPHA and ATRAP) of hydrogen with that of antihydrogen [6-11]. These experiments will be continued at the Facility for Antiproton and Ion Research (FAIR) at Darmstadt [12] once this facility is operational. FAIR will provide the newly formed FLAIR collaboration [13] with antiproton intensities two orders of magnitude higher than presently available at the AD.

Parity non-conservation (PNC) experiments in atomic systems have reached a remarkable accuracy. Here, mixing of atomic states of opposite parity lead to tiny but still observable and measurable effects. The most precise result was obtained by C. Wieman et al. for ^{133}Cs [14] and is in perfect agreement with the Standard Model. Atomic PNC effects are caused by neutral currents mediated by the Z^0 boson. In order to isolate the weak electron-nucleon interaction the electronic wave functions have to be known very precisely. Aiming at accuracies of 1% or below this is presently only possible for the case of simple systems such as the alkaline elements. Different asymmetries observed for different hyperfine levels led also to the first observation of an anapole moment [14].

Presently there are efforts to extend atomic PNC experiments to heavier systems such as francium confined in a magneto-optical laser trap [15,16] or the radium ion confined in an ion trap [17-19]. The most advanced experiment was performed at Stony Brook and achieved to confine on the average 60 000 ^{210}Fr atoms in the magneto-optical trap with a trapping efficiency of 1% [15]. This experiment is being stopped and will be moved to TRIUMF where hopefully soon an actinide target will be available for the production of francium isotopes.

Since the PNC effect scales with Z^3 an increase in sensitivity of these experiments on Fr or Ra^+ by a factor of 18 is expected over that achieved with cesium. However, beside the drawback of the required production of radionuclides in the case of Fr, a disadvantage of the planned experiments is the more complex electronic structure of these heavy elements. This is circumvented by L. Labzowsky et al. [20] who proposed to use helium-like heavy ions where a strong enhancement of the PNC effect is caused by a near degeneracy of levels of opposite parity, i.e. $1s2s\ ^1S_0$ and $1s2p^3P_0$.

Electric Dipole Moment

CP violation was observed in the K^0 as well as in the B^0 particle-antiparticle system and can be accounted for by the Standard Model. If CPT is conserved, then time reversal must be violated, too. A powerful way to look for such an effect is the search for permanent electric dipole moments (EDM) of fundamental particles. The discovery of a finite EDM would suggest physics beyond the Standard Model. A permanent EDM is aligned along the spin axis and is detected by changing the polarity of the applied electric field. A dependence of the transition frequency from the polarity indicates a finite EDM, which violates the symmetries both for time reversal as well as for parity inversion. Its existence is deeply connected to the matter-antimatter asymmetry in the universe. The best limits for EDM´s have been obtained in atomic systems with $d < 2.1 \cdot 10^{-28}$ e·cm by Fortson et al. [21] by use of ^{199}Hg in cells and with $d < 1.6 \cdot 10^{-27}$ e·cm by Commins et al. [22] by use of a thallium atomic beam. Strong enhancement is expected in the octupole-deformed region of radon or radon nuclei due to close-lying parity doublets. Experiments are planned by Lu/Argonne [23] and Jungmann/KVI on ^{225}Ra in a magneto-optical trap [24] and by Chupp/Michigan on ^{223}Rn at TRIUMF in a cell [25]. Farley et al. propose a new highly sensitive method of looking for electric dipole moments of charged particles in a storage ring [26].

Using the molecule thallium fluoride [27], the group of E. A. Hinds aims at a limit for the proton of $d < 2 \cdot 10^{-28}$ e·cm with a prospect to decrease this limit further by another factor of five by using a beam of decelerated molecules [28].

Ultra-cold neutrons (UCN) are used to search for a neutron EDM. Presently, a limit of $d < 3.0 \cdot 10^{-26}$ e·cm is reported by [29]. The current neutron EDM experiments are limited by statistics as well as systematics [29, 30]. Required are more powerful sources of ultra-cold neutrons with much higher densities than presently available. Such sources are under development at PSI, KEK, ILL, Los Alamos, FRM II, TRIGA-Mainz, and other places and will provide UCN densities of up to 10^4 cm^{-3} which is more than two orders of magnitude higher than presently, for example, at the exit of the UCN turbine at ILL in Grenoble. The reduction of systematic uncertainty which amounts to $\approx 1.6 \cdot 10^{-26}$ e·cm [31], is the second challenge in order to obtain more tight limits for a neutron EDM. Essential is here to get a better control of the fluctuations of the magnetic field, especially those correlated with the switching of the polarity of the applied electric field, for which the sensitivity of magnetometers must be improved from presently 200 fT to better than 10 fT.

Absolute mass of the neutrino

Neutrino oscillations as observed in recent years can only be explained if masses are attributed to the neutrinos. However, neutrino oscillations are only sensitive to the differences of squared masses of different neutrino mass eigenstates: Absolute masses can only be determined by studying weak decays [32]. Most popular and giving the best upper limit for the mass of the electron antineutrinos is the study of the beta decay of tritium as performed at Mainz [32-34] and Troitsk [35]. Both experiments apply very similar techniques. In the Mainz experiment, the tritium source is placed in a magnetic field of 6 Tesla. Adiabatic magnetic guiding is employed to transport the β^- particles resulting from the tritium decay in the source to an electrostatic retardation spectrometer placed in a very weak magnetic field of $3\cdot10^{-4}$ T and operating at a potential near the corresponding end point energy of the β-decay spectrum (18.6 keV). The transmitted β-particles are detected and recorded as a function of the voltage applied to the retardation electrodes. Such a spectrometer provides excellent energy resolution, long-term stability, and low background rate.

The present upper limit for the mass of the electron antineutrino is $m(\nu_e) < 2.3$ eV at a confidence level of 95%. In order to decrease this limit by nearly an order of magnitude, KATRIN, a spectrometer scaled up by a factor of 7, is being set up at the Forschungszentrum Karlsruhe. With a diameter of 10 m and a length of 24 m, a lower limit of 0.3 eV (3σ) is expected after three years of data taking [36].

Scalar or Tensor Contributions to Weak Interaction

Soon after the discovery of parity violation in weak interaction it was found that vector and axial vector terms are predominantly required to describe the effects of weak interaction. The maximum contribution by scalar terms is now determined to be 7% [37,38], whereas tensor terms contribute maximum 13% [39]. Today, laser traps as well as ion traps are applied to study the correlations in weak decay and by this to fix more precisely possible scalar or tensor contributions. Experiments in which the recoil ion is detected in coincidence with the beta particle and the beta-neutrino correlation *a* is determined, have been pioneered at TRIUMF [40] by use of a magneto-optical trap (MOT). An overview on the experiments in progress or preparation is given in Table 1.

Unitarity of the CKM Matrix

As in the case of the neutrino sector, where oscillations are observed, weak interaction mixes the quark flavours. As a consequence, the observed eigenstates of the quarks are different from the weak interaction eigenstates. The mixing is described by the Cabbibo-Kobayashi-Maskawa (CKM) matrix. The Standard Model implies that the CKM matrix is unitary, and deviations from unitarity would therefore be a clear signal for physics beyond the Standard Model. Unitarity tests are most sensitively performed by checking the sum of the squared CKM matrix elements of the first row $V_{ud}^2 + V_{us}^2 + V_{ub}^2$ which should be equal to one. V_{ud} can be obtained from beta decays, from the decay of the neutron, or of the pion. V_{us} can be extracted from K-meson decays while the contribution of V_{ub} can be neglected.

Table 1: Overview on weak Interaction experiments in atom and ion traps.

Facility	Nuclear System, Research Topic	Ref.
	ATOM TRAPS	
TRINAT-ISAC/TRIUMF	^{38m}K, νβ-correlation	40
LBNL & UC Berkeley	^{21}Na, νβ-correlation	41
LANL - Los Alamos	^{82}Rb, β-asymmetry	42, 43
KVI - Groningen	Na, Ne, Mg, D-coefficient; EDM Ra	19
	ION TRAPS	
LPC - Caen	^{6}He, νβ-correlation	44
WITCH Leuven - ISOLDE	^{35}Ar, νβ-correlation	45
CPT-trap - Argonne	^{14}O, νβ-correlation	46
ISOLTRAP - CERN	^{74}Rb, ^{22}Mg, ^{34}Ar masses for $0^+ \rightarrow 0^+$ decays	47-50
CPT - Argonne	^{46}V mass for $0^+ \rightarrow 0^+$ decays	51
JVL-Trap - Jyväskylä	^{62}Ga mass for $0^+ \rightarrow 0^+$ decays	52
LEBIT - MSU	^{38}Ca mass for $0^+ \rightarrow 0^+$ decays	53
SHIPTRAP – GSI	masses for $0^+ \rightarrow 0^+$ decays	54
TITAN - TRIUMF	masses for $0^+ \rightarrow 0^+$ decays	55

The most precise value for the V_{ud} element can be extracted from the mean Ft value of super-allowed nuclear β-decay [56, 57], $F\,t \equiv f\,t\,(1 + \delta_R)(1 - \delta_C) = K\,/\,(2|V_{ud}|^2\,G_\mu^{\,2}\,(1 + \Delta_R))$, where t is the lifetime of the decay in question, f is the phase space factor, δ_R is the nucleus-dependent radiative correction, δ_C is the isospin-symmetry breaking correction, and Δ_R is the nucleus-independent radiative correction, G_μ is the Fermi coupling constant from muon decay, and K is a constant.

Experimentally, Ft is accessible via the following measured quantities [57]: the Q-value of the decay, which enters to the fifth power into the calculation of the statistical rate function f, the half-life $T_{1/2}$, and the branching ratio R. The uncertainty in the derivation of V_{ud} here is dominated by the uncertainties in the theoretical correction terms, and current nuclear experiments [49] are focused on testing and refining those correction terms that depend on nuclear structure.

Such correction terms are avoided entirely in neutron β-decay [58]. The unitarity check then is based solely on particle data, and nuclear structure effects are absent.

Only two measurements are necessary to extract V_{ud} with high precision: The neutron lifetime τ and a neutron decay correlation coefficient. The neutron lifetime τ is given by $f{\cdot}\tau\ (1 + \delta_R) = K/(|V_{ud}|^2\ G_\mu^2\ (1 + 3\ \lambda^2)\ (1 + \Delta_R))$. Note that the radiative correction δ_R is model-independent and well known, while $\Delta_R = 0.0240(8)$ is model-dependent and the same as in the nuclear Ft above. At present, in this context the ratio λ of the axial vector to vector coupling constant is derived from the neutron decay β-asymmetry A. Alternatively, it will also be derived from the electron-neutrino correlation a. The two coefficients A and a depend on λ as $A = -2\lambda(\lambda + 1)/(1 + 3\lambda^2)$ and $a = (1 - \lambda^2)/(1 + 3\lambda^2)$, where λ is assumed to be real. The uncertainty in the derivation of V_{ud} here is dominated by experimental errors, and current neutron decay experiments are focused on suppressing such experimental errors. The error due to A is by far the dominant error in the derivation of V_{ud}.

A critical survey of the results obtained by use of super-allowed $0^+ \rightarrow 0^+$ nuclear beta decays was published by Hardy and Towner in 2005 [57]. They obtained $V_{ud} = 0.9738(4)$ and the unitarity test on the top row of the matrix became $|V_{ud}|^2 + |V_{us}|^2 + |V_{ub}|^2 = 0.9966(14)$ using the Particle Data Group's currently recommended values for V_{us} and V_{ub}. This result corresponds to a deviation from the unitarity condition by 2.2 standard deviations.

A slightly larger deviation from the unitarity condition was found in studying the neutron decay [58]. Note, that a recent measurement of the neutron lifetime using a gravitational trap [59,60] shows a 6 sigma deviation from unitarity which is in contradiction with all previous experiments but would restore the CKM unitarity.

Test of QED and Fundamental Constants

Quantum electrodynamics (QED) is the currently most precisely investigated theory in physics. It describes the interaction of electrical charges by exchange of photons and serves also as an underlying concept for all other existing field theories. Experimental studies have been carried out with precision of up to 10^{-14}. For a few simple systems, the experimental accuracy is matched by nearly equally exact theoretical calculations, which show the power of the underlying mathematical framework of regularisation, renormalisation and covariant formulation. Where the theory cannot compete with the experiment, this is nowadays often caused by insufficient knowledge of other parameters such as nuclear size and structure. Even fundamental constants limit the predictive power of QED which in turn, together with high-precision experiments, has been capable to provide the most precise values for the fine structure constant and the mass of the electron [61].

For such precision studies only simple systems which do not consist of more than an atomic nucleus and a few electrons at most have to be investigated. Here, few-electron systems with a heavy nucleus is particularly important, because it allows to access the regime of the highest electromagnetic fields observable in the laboratory, more than 10^{18} V/m in a range of distances in the order of the electron's Compton wavelength where the perturbative treatment of QED corrections, which describes the hydrogen atom with uttermost precision, breaks down [62]. The energy contained in these fields is very close to that required for the spontaneous creation of an electron-positron pair out of the vacuum.

Highly-charged ions (HCI), combining very strong static fields and a simple electronic structure are ideal testing grounds for these investigations. HCI can be produced and stored in traps and rings, and sensitive techniques such as mass or laser spectroscopy in storage rings, ion traps and electron beam ion traps (EBIT) are being developed that hold great potential in this branch of physics. However, the experimental precision for highly-charged ions does not (yet) match that for hydrogen or simple light systems like the free electron or positronium. In the region of high charges, also the predictive power of QED is somewhat hampered: The relativistic electronic wave functions are located very close to the atomic nucleus and thus are considerably affected by nuclear size and structure where the latter are not sufficiently well known.

The 1s Lamb shift of the heaviest hydrogen and helium-like atoms has been addressed at the ESR by x-ray spectroscopy on stored, cooled, and decelerated ions. The deduced 1s Lamb shift of 458.8(4.2) eV for $^{238}U^{91+}$ provides the hitherto most precise test of QED for a single-electron system in the strong field regime [63]. The measured two-electron contribution to the ionization potential of helium-like $^{238}U^{90+}$ of 2248(9) eV represents the most accurate determination of two-electron effects in the domain of high-Z helium-like ions and already reaches the size of the specific two-electron radiative QED corrections [64]. In very heavy lithium-like ions the $2s_{1/2}$-$2p_{1/2}$ Lamb shift has been determined with an uncertainty of less than 100 meV by exploiting dielectronic recombination at the electron cooler of the ESR [65]. The most accurate Lamb shift measurements were performed very recently by use of the Livermore EBIT by Beiersdorfer et al. [66].

The hyperfine structure splitting (HFS) of heavy hydrogen-like ions in their 1s atomic ground state, which increases with Z^3 and lies in the optical regime for heavy elements such as $^{209}Bi^{82+}$ or $^{207}Pb^{81+}$, has been studied both at the experimental storage ring ESR at GSI by collinear laser spectroscopy [67,68] and at the SuperEBIT at Livermore [69-72].

Within a Mainz-GSI collaboration, the anomalous magnetic moment of the electron bound in the hydrogen-like ions $^{12}C^{6+}$ [73] and $^{16}O^{7+}$ [74] have been measured by storing a single ion in a Penning trap [75].The g-factor found for the 1s electron in these systems enabled a test of bound-state QED on a 0.25% level. Because of the good agreement of the *g*-factors of the electron bound in $^{12}C^{5+}$ and $^{16}O^{7+}$ with theory these data can be used for a more accurate determination of the electron mass [76].

A similar procedure can be applied to determine the fine-structure constant α (Fig. 1) from *g*-factor measurements on highly charged ions in the medium-heavy mass range. The relative uncertainty in α is related to the uncertainty in the *g*-factor by $\delta\alpha/\alpha \approx 1/(Z^2\alpha^2)\cdot\delta g/g$. Thus, it seems attractive to measure the *g*-factor at the highest possible nuclear charge *Z*. Nuclear structure effects and uncalculated higher-order QED contributions may, however, limit the determination of α. For the case of $^{40}Ca^{19+}$, e.g., we have $1/(Z^2\alpha^2) \approx 50$ and a measurement of the electronic *g*-factor similar to the previous ones [73, 74], will allow a determination of α with a relative uncertainty of $3\cdot 10^{-9}$, which is the accuracy of the present CODATA value. However, in order to reach this goal, the two-loop uncertainties of bound-state QED must be reduced and the atomic mass of ^{40}Ca has to be determined better than 10^{-10} as stated in Ref. [77]. Alternatively, g-factor measurements, both in boron-like and hydrogen-like uranium with an accuracy as presently achieved for $^{12}C^{5+}$, would result in an uncertainty of better than the currently accepted CODATA value of the fine structure constant [78].

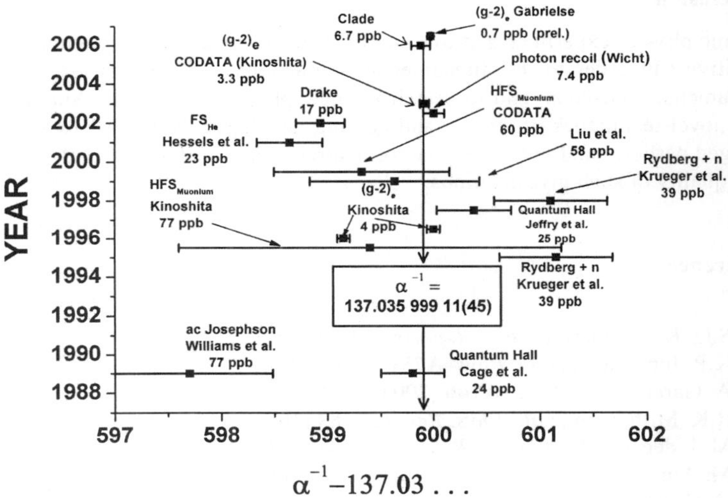

Fig. 1: Value of the fine structure constant as function of the year.

The dimensionless fine-structure constant $\alpha = e^2/(4\pi\varepsilon_0\hbar c)$ determines the strength of the electromagnetic interaction and, in a truly universal way, enters equations from many different subfields of physics. If the values of □ measured by different techniques are compared, as shown in Fig. 1, one finds that they significantly deviate within the given error bars. The reasons have to be clarified and it is mandatory to measure α more accurately with different techniques to see if one gets better agreement with the so far most accurate value, which was derived from g-2 in QED [79,80]. In July 2006, a six times more accurate value of α, determined from the electron g-value and QED calculations by the group of G. Gabrielse et al. was published [81,82]. This data point is included in Fig. 1.

The fine structure constant □can be expressed in terms of other quantities, for example [83,84] by

$$\alpha^2 = \left(\frac{2R_\infty}{c}\right)\left(\frac{h}{m_e}\right) = \left(\frac{2R_\infty}{c}\right)\left(\frac{h}{m_{Cs}}\right)\left(\frac{m_{Cs}}{m_p}\right)\left(\frac{m_p}{m_e}\right).$$

In the most recent listing of the fundamental physical constants [85], the Rydberg constant $R_\infty = m_e e^4/(8\,\varepsilon_0^2 h^3 c)$ is given with a relative uncertainty of $3.3\cdot 10^{-9}$. The velocity of light is defined by c = 299792458 m/s. Therefore, the accuracy of the ratio (h/m_e) to within $6.7 \cdot 10^{-9}$ presently determines the accuracy of α. However, it may be related to other ratios, for example to mass ratios. Recently, a measurement of the photon recoil on cesium (h/m_{Cs}) [86] and rubidium (h/m_{Rb}) [87] atoms was performed, but other systems resulting in other mass ratios are of interest as well.

Conclusion

Atomic physics experiments at low energy provide since long excellent accuracy and sensitivity in order to test fundamental interactions and symmetries, to determine fundamental constants, and to search for new physics beyond the Standard Model. The novel techniques of storing and cooling developed in the last two decades for charged particles and in the last one for neutral atoms, enable now and in the future a new quality of such investigations.

References

1 S.G. Karshenboim, *Phys. Reports* **422** (2005): 1.
2 K.P. Jungmann, *Nucl. Phys.* **A751** (2005): 87C.
3 A. Garcia, *Nucl. Phys.* **A746** (2004) 298C.
4 R.K. Mittleman *et al., Phys. Rev. Lett.* **83** (1999): 2116.
5 M. Fidecaro, H.J. Gerber, *Rep. Prog. Phys.* **69** (2006): 1713.
6 M. Amoretti *et al., Nature* (London) 419 (2002): 456.
7 G. Gabrielse *et al., Phys. Rev. Lett.* **89** (2002): 233401.
8 C. Storry *et al., Phys. Rev. Lett.* **93** (2004): 263401.
9 G. Gabrielse *et al., Phys. Rev. Lett.* **93** (2004): 073401.
10 N. Madsen *et al., Phys. Rev. Lett.* **94** (2005): 033403.
11 E. Widmann *et al., Nucl. Phys.* **A752** (2005): 87c.
12 http//www.gsi.de/fair/index_e.html
13 http//www.gsi.de/fair/experiments/flair/index_e.html
14 M.C. Noecker, B.P. Masterson, C.E. Wieman, *Phys. Rev. Lett,* **61** (1988): 310.
15 E. Gomez, L.A. Orozco, G.D. Sprouse, *Rep. Prog. Phys.* **69** (2006): 79.
16 G. Stancari *et al., Rev. Sci. Instr.* **77** (2006): 03A701.
17 J.A.Sherman *et al., Phys. Rev. Lett.* **94** (2005): 243001.
18 B.K. Sahoo *et al., Phys. Rev. Lett.* **96** (2006): 163003.
19 G.P. Berg *et al., Nucl. Instr. Meth.* **B204** (2003): 532.
20 L.N Labzowsky, G. Plunien, G. Soff, *Phys. Rev.* **A67** (2003): 012503.
21 M.V. Romalis *et al., Phys. Rev. Lett.* **86** (2001): 2505.
22 B.C. Regan *et al., Phys. Rev. Lett.* 88 (2002): 071805.
23 Z.-T. Lu, private communication 2006.
24 K. Jungmann, *Eur. Phys. J.* **A25** (2205): 677.
25 Proposal for an experiment at TRIUMF, Vancouver.
26 F.J.M. Farley *et al., Phys. Rev. Lett.* 93 (2004): 052001.
27 J.J. Hudson *et al., Phys. Rev. Lett.* **89** (2002) 023003.
28 E.A. Hinds, Talk at the Int. Conf on Atomic Phys 2006 (ICAP 2006), Innsbruck, Plenum Press, to be published.

29 C.A. Baker *et al.*, arXiv.org, hep-ex/0602020.
30 P.G. Harris *et al.*, *Phys. Rev. Lett.* **82** (1999): 904.
31 J.M. Pendlebury *et al.*, *Phys. Rev.* **A70** (2004): 032102.
32 C. Weinheimer, *Prog. Part. Nucl. Phys.* **57** (2006): 22.
33 C. Kraus et al., *Eur. Phys. J.* **C40** (2005): 447.
34 E.W. Otten, J. Bonn, C. Weonheimer, *Intern. J. Mass Spectr.* **251** (2006): 173.
35 V.M. Lobashev, *Nucl. Phys.* **A719** (2003): 153C.
36 B. Bornschein, *Prog. Part. Nucl. Phys.* **57** (2006): 38.
37 E.G. Adelberger *et al.*, *Phys. Rev. Lett.* **70** (1993): 2856; erratum *Phys. Rev. Lett.* **71** (1993): 469.
38 E.G. Adelberger *et al.*, *Phys. Rev. Lett.* **83** (1999): 1299.
39 C.H. Johnson *et al.*, *Phys. Rev.* **132** (1963): 1149.
40 A. Gorelov *et al.*, *Phys. Rev. Lett.* **94** (2005: 142501.
41 N.D. Scielzo *et al.*, *Phys. Rev. Lett.* **93** (2004): 102501.
42 M. Hausmann *et al.*, *Nucl. Phys.* **A746** (2004): 669C.
43 S.G. Crane *et al.*, *Phys. Rev. Lett.* **86** (2001): 2967.
44 D. Rodriguez *et al.*, *Eur. Phys. J.* **A25** (2005): 705.
45 V.Y. Kozlov *et al.*, *Int. J. Mass Spectr.* **251** (2006): 159.
46 G. Savard *et al.*, *Nucl. Phys.* **A654** (1999): 961c.
47 F. Herfurth *et al.*, *Phys. Rev. Lett.* **87** (2001): 142051.
48 M. Mukherjee *et al.*, *Phys. Rev. Lett.* **93** (2004): 150801.
49 A. Kellerbauer *et al.*, *Phys. Rev. Lett.* **93** (2004): 072502.
50 K. Blaum, *Phys. Rep.* **25** (2006): 1.
51 G. Savard *et al.*, *Phys. Rev. Lett.* **95** (2005): 102501.
52 T. Eronen *et al.*, *Phys. Lett.* **B636** (2006): 191.
53 G. Bollen *et al.*, *Phys. Rev. Lett.* **96** (2006): 152501.
54 G. Vorobjev *et al.*, Proc. Intern. Conf. Nuclei in the Cosmos (NIC-7), Geneva, Switzerland, 2006, Proc. of Science, http//pos.sissa.it.
55 J. Dilling *et al.*, *Nucl. Instr. Meth.* **B204** (2003): 492.
56 J.C. Hardy, I.S. Towner, G. Savard, *Intern. J. Mass Spectr.* **251** (2006): 95.
57 J.C. Hardy, I.S. Towner, *Phys. Rev.* **C71** (2005): 055501.
58 H. Abele *et al.*, *Phys. Rev. Lett.* **88** (2002): 211801.
59 A. Serebrov *et al.*, *Phys. Lett.* **B605** (2005): 72.
60 A. Serebrov *et al.*, *J. Research NIST* **110** (2005): 333.
61 J.L. Flowers, H.A. Klein, H.S. Margolis, *Contemporary Phys.* **45** (2004): 123.
62 P.J. Mohr, G. Plunien, G. Soff, *Phys. Rep.* **293** (1998): 227.
63 A. Gumberidze *et al.*, *Phys. Rev. Lett.* **94** (2005): 223001.
64 A. Gumberidze *et al.*, *Phys. Rev. Lett.* **92** (2004): 203004.
65 C. Brandau *et al.*, *Phys. Rev. Lett.* **91** (2003): 073202.
66 P. Beiersdorfer *et al.*, *Phys. Rev. Lett.* **95** (2005): 233003.
67 I. Klaft *et al.*, *Phys. Rev. Lett.* **73** (1994): 2425.
68 P. Seelig *et al.*, *Phys. Rev. Lett.* **81** (1998): 4824.
69 J.R. Crespo López-Urrutia *et al.*, *Phys. Rev. Lett.* **77** (1996): 826.
70 J.R. Crespo López-Urrutia *et al.*, *Phys. Rev.* **A57** (1998): 879.
71 P. Beiersdorfer *et al.*, *Phys. Rev.* **A64** (2001): 032506.
72 P. Beiersdorfer *et al.*, *Nucl. Inst. & Meth.* **B205** (2003): 62.
73 H. Häffner *et al.*, *Phys. Rev. Lett.* **85** (2000): 5308.
74 J. Verdú *et al.*, *Phys. Rev. Lett.* **92** (2004): 093002.

75 H. Häffner *et al.*, *Eur. Phys. J.* **D22** (2003): 163.
76 T. Beier *et al.*, *Phys. Rev. Lett.* **88** (2002): 011603.
77 U.D. Jentschura *et al.*, *Int. J. Mass Spectr.* **251** (2006): 102.
78 V.M Shabaev *et al.*, *Phys. Rev. Lett* **96** (2006): 253002.
79 R.S. VanDyck, P.B. Schwinberg, H.G. Dehmelt, *Phys. Rev. Lett.* **59** (1987): 26.
80 T. Kinoshita, *Phys. Rev. Lett.* **75** (1995): 4728.
81 B. Odom *et al.*, *Phys. Rev. Lett.* **97** (2006): 030801.
82 G. Gabrielse *et al.*, *Phys. Rev. Lett.* **97** (2006): 030802.
83 M.P. Bradley *et al.*, *Phys. Rev. Lett.* **83** (1999): 4510.
84 C. Carlberg, T. Fritioff, I. Bergström, *Phys. Rev. Lett.* **83** (1999): 203.
85 P.J. Mohr, B.N. Taylor, *Rev. Mod. Phys.* **77** (2005): 1.
86 V. Gerginov *et al.*, *Phys. Rev.* **A73** (2006): 032504.
87 P. Cladé *et al.*, *Phys. Rev. Lett.* **97** (2006): 033001.

NEUTRINO PHYSICS: A SELECTIVE OVERVIEW *

SCOTT M. OSER

University of British Columbia
Department of Physics & Astronomy
6224 Agricultural Road
Vancouver BC V6T 1Z1, Canada
E-mail: oser@phas.ubc.ca

Neutrinos in the Standard Model of particle physics are massless, neutral fermions that seemingly do little more than conserve 4-momentum, angular momentum, lepton number, and lepton flavour in weak interactions. In the last decade conclusive evidence has demonstrated that the Standard Model's description of neutrinos does not match reality. We now know that neutrinos undergo flavour oscillations, violating lepton flavour conservation and implying that neutrinos have non-zero mass. A rich oscillation phenomenology then becomes possible, including matter-enhanced oscillation and possibly CP violation in the neutrino sector. Extending the Standard Model to include neutrino masses requires the addition of new fields and mass terms, and possibly new methods of mass generation. In this review article I will discuss the evidence that has established the existence of neutrino oscillation, and then highlight unresolved issues in neutrino physics, such as the nature of three-generational mixing (including CP-violating effects), the origins of neutrino mass, the possible existence of light sterile neutrinos, and the difficult question of measuring the absolute mass scale of neutrinos.

1. Neutrinos In The Standard Model

A neutrino can be defined as a chargeless, colourless fermion. As such, neutrinos have only weak interactions, with tiny cross-sections, and are exceedingly difficult to detect. In the Standard Model of particle physics, there is one massless neutrino associated with each charged lepton (e, μ, or τ), and lepton flavour is rigorously conserved, so that for example the total number of "electron"-type leptons (charged or otherwise) is unchanged in

*Proceedings of the Lake Louise Winter Institute 2006. Slides available at http://www.phas.ubc.ca/~oser/ Due to length restrictions I have been forced to be selective and emphasize only recent results, and apologize to the many excellent researchers whose work has been neglected as a result.

all interactions. Indeed, an electron neutrino can be defined simply as the kind of neutrino produced when a W particle couples to an electron. Weak interactions are never observed to couple a charged lepton ℓ to the wrong type of neutrino. Nor do neutral current (Z-mediated) interactions couple together two neutrinos of different flavours. Interestingly, although no Standard Model process violates lepton flavour number, there is no associated symmetry of the Lagrangian that requires this to be so—that is, the absence of lepton-flavour-changing terms in the Lagrangian seems to be "accidental", and not the result of a deeper symmetry.

One of the most characteristic features of neutrinos in the Standard Model is that weak interactions couple only to left-handed neutrinos, or to right-handed antineutrinos. That is, in all cases the spin of a (massless) neutrino is observed to be antiparallel to its direction of motion. This characteristic is associated with the V-A nature of weak interactions. Whereas the electromagnetic current of an electron is given by

$$j^{\mu}_{EM} \propto \bar{e}\gamma^{\mu}e \tag{1}$$

the weak current that couples a ν_e to an electron has the form

$$j^{\mu}_{weak} \propto \bar{e}\gamma^{\mu}(1-\gamma^5)\nu_e. \tag{2}$$

The presence of the $1-\gamma^5$ factor (a V-A term) in the current projects out the left-handed chirality component of the ν_e. The result is that weak interactions only couple to left-handed neutrino states.

The failure to observe right-handed neutrinos suggests a plausibility argument as to why neutrinos could be expected to be massless. The apparent absence of right-handed neutrinos implies either that no $\nu_R \equiv (1+\gamma^5)\nu$ state exists, or if a ν_R does exist, then it happens to be a "sterile" state, having no couplings to any vector gauge bosons. Rather than postulate the existence of a ν_R state that has never been seen and lacks even weak interactions, an appeal to Ockham's razor suggests the more economical solution that the right-handed field ν_R not exist at all. However, in the Standard Model, a mass term is a term in the Lagrangian that couples left-handed and right-handed states:

$$\mathcal{L} = -m\bar{\psi}\psi = -m(\bar{\psi_L}\psi_R + \bar{\psi_R}\psi_L) \tag{3}$$

Accordingly, if no ν_R exists, then one cannot form such a mass term, and so the neutrino must be massless. The alternative is seemingly to postulate the existence of right-handed neutrino states which don't participate in even weak interactions but which provide the fields needed to produce neutrino masses. This unpalatable situation, as much as the fact that experimentally neutrino masses turned out to be immeasurably small, provided justification for assuming the neutrino mass to be zero in the Standard Model. That assumption turns out to be wrong, but is less of an *ad hoc* assumption than is sometimes claimed when one keeps in mind that the simplest alternative forces us to introduce sterile fermion fields even more ethereal that the neutrino itself!

2. Phenomenology Of Neutrino Oscillation

The Standard Model neutrinos strike me as rather dismal particles in the end. With no mass and very limited interactions, the major practical import that neutrinos seem to have is to provide a "junk" particle to balance a number of conservation laws such as 4-momentum, angular momentum, lepton number, and lepton flavour. Given this situation, and the difficulties associated with neutrino experiments to begin with, it is perhaps not surprising that neutrinos were for a long time a neglected area of particle phenomenology.

Some progress was made in 1962 when Maki, Nakagawa, and Sakata proposed (in true theorist fashion, on the basis of zero experimental evidence) a new phenomenon now known as neutrino oscillation.[1] The inspiration for this proposal was the observation that charged current interactions on quarks produce couplings between quark generations. For example, while naively we would expect the interactions of a $W^{\pm}$ to couple u to d, s to c, or t to b, weak decays such as $\Lambda^0 \to p\pi^-$ are also observed in which an s quark gets turned into a u quark, thus mixing between the second and first quark generations. We describe this by saying that there is a rotation between the mass eigenstates (e.g. u,d,s ...) produced in strong interactions, and the weak eigenstates that couple to a W boson. In this language, a W does not simply couple a u quark to a d quark, but rather it couples u to something we can call d', which is a linear superposition of the d, s, and b quarks. We describe this "rotation" between the strong and weak eigenstates by a 3×3 unitary matrix called the CKM matrix:

$$\begin{pmatrix} d' \\ s' \\ b' \end{pmatrix} = \begin{bmatrix} V_{ud} & V_{us} & V_{ub} \\ V_{cd} & V_{cs} & V_{cb} \\ V_{td} & V_{ts} & V_{tb} \end{bmatrix} \begin{pmatrix} d \\ s \\ b \end{pmatrix} \tag{4}$$

The off-diagonal elements of this matrix allow transitions between quark generations in charged current weak interactions, and through a complex phase in matrix V also produce CP violation in the quark sector. The measurement of the CKM matrix elements and exploration of its phenomenology has been one of the most active fields in particle physics for the past four decades.

Maki, Nakagawa, and Sakata (hereafter known as MNS) proposed that something similar could happen in the neutrino sector.[1] Once the muon neutrino was discovered in 1962[2], it became possible to suppose that neutrino flavour eigenstates such as ν_e or ν_μ might not correspond to the neutrino mass eigenstates. That is, the particle we call "ν_e", produced when an electron couples to a W, might actually be a linear superposition of two mass eigenstates ν_1 and ν_2. In the case of 2-flavour mixing, we can write:

$$\begin{pmatrix} \nu_e \\ \nu_\mu \end{pmatrix} = \begin{bmatrix} +\cos\theta & +\sin\theta \\ -\sin\theta & +\cos\theta \end{bmatrix} \begin{pmatrix} \nu_1 \\ \nu_2 \end{pmatrix} \tag{5}$$

While the formalism is exactly parallel to that used for quark mixing, with angle θ in Equation 5 playing the role of a Cabibbo angle for leptons, the resulting phenomenology is somewhat different. In the case of quarks, mixing between generations can be readily seen by producing hadrons through strong interactions, and then observing their decays by weak interactions. For example, we can produce a K^+ in a strong interaction, then immediately observe the decay $K^+ \to \pi^0 e^+ \nu_e$, in which an $\bar{s}$ turns into a $\bar{u}$. Neutrinos, however, have *only* weak interactions, and so we cannot do the trick of producing neutrinos by one kind of interaction and then detecting them with a different interaction. In other words, a rotation between neutrino flavour eigenstates and neutrino mass eigenstates such as in Equation 5 has no direct impact on weak interaction vertices themselves. W bosons will still always couple an e to a ν_e and a μ to ν_μ even if there is a rotation between the flavour and mass eigenstates.

To observe the effects of neutrino mixing we therefore must resort to some process that depends on the properties of the mass eigenstates. While the flavour basis is what matters for weak interactions, the mass eigenstate is actually what determines how neutrinos propagate as free particles in a vacuum. Imagine, for example, that we produce at time $t = 0$ a ν_e state with some momentum $\vec{p}$:

$$|\nu_e(t=0)\rangle = \cos\theta|\nu_1\rangle + \sin\theta|\nu_2\rangle \tag{6}$$

As this state propagates in vacuum, each term picks up the standard quantum mechanical phase factor for plane wave propagation:

$$|\nu(\vec{x},t)\rangle = \exp(i(\vec{p}\cdot\vec{x} - E_1 t))\cos\theta|\nu_1\rangle + \exp(i(\vec{p}\cdot\vec{x} - E_2 t))\sin\theta|\nu_2\rangle \tag{7}$$

Here the energy E_i of the ith mass eigenstate is given by the relativistic formula $E_i = \sqrt{\vec{p}^2 + m_i^2}$, and $\hbar \equiv c \equiv 1$. If the two mass eigenstates ν_1 and ν_2 have identical masses, then the two components will have identical momenta and energy, and so share a common phase factor of no physical significance. However, suppose that $m_1 \neq m_2$. If $m_i \ll p \equiv |\vec{p}|$, then we can expand the formula for E_i as follows:

$$E_i = \sqrt{p^2 + m_i^2} = p\sqrt{1 + m_i^2/p^2} \approx p + m_i^2/(2p) \tag{8}$$

At some time $t > 0$, the neutrino's state will be proportional to the following superposition:

$$|\nu(t)\rangle \propto \cos\theta|\nu_1\rangle + e^{i\phi}\sin\theta|\nu_2\rangle \tag{9}$$

with the phase difference ϕ being given by

$$\phi = \left(\frac{m_1^2}{2p} - \frac{m_2^2}{2p}\right) t \tag{10}$$

The net result is that at time t, the neutrino that originally was in a pure ν_e state is no longer in a pure ν_e state, but due to the phase difference ϕ will have acquired a non-zero component of ν_μ! We therefore can determine the probability that our original ν_e will interact as a ν_μ, which by Equation 10 depends on $\Delta m^2 \equiv m_1^2 - m_2^2$, $p \approx E$, and $t \approx L/c$ in the relativistic limit:

$$P(\nu_e \to \nu_\mu) = |\langle\nu_\mu|\nu(t)\rangle|^2 = \sin^2 2\theta \sin^2\left(\frac{1.27\Delta m^2 L}{E}\right) \tag{11}$$

In this formula Δm^2 is given in eV^2, L is the distance the neutrino has travelled in km, and E is the neutrino energy in GeV. The oscillation probability in Equation 11 has a characteristic dependence on both L and E that is a distinctive signature of neutrino oscillations. Figure 1 shows the oscillation probability vs. energy for representative parameters.

While Equation 11 suffices to describe oscillations involving two neutrino flavours in vacuum, the presence of matter alters the neutrino propagation, and hence the oscillation probability.[3] The reason for this is that ordinary matter is flavour-asymmetric. In particular, normal matter contains copious quantities of electrons, but essentially never any μ's or τ's. As a result, ν_e's travelling through matter can interact with leptons in matter by both W and Z boson exchange, while ν_μ or ν_τ can interact only by Z exchange. This difference affects the amplitude for forward scattering (scattering in which no momentum is transferred). Electron neutrinos pick up an extra interaction term, proportional to the density of electrons in matter, that acts as a matter-induced potential that is different for ν_e's than for other

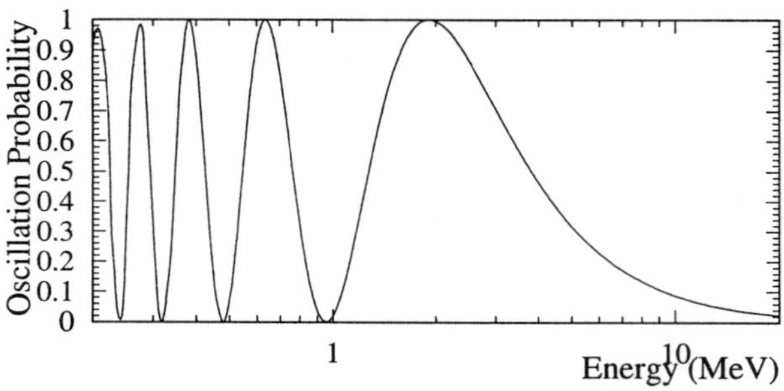

Figure 1. Oscillation probability as a function of neutrino energy for a fixed value of $\Delta m^2 L$, with $\sin^2 2\theta = 1$.

flavours. Effectively ν_e's travelling through matter have a different "index of refraction" than the other flavours. Equation 12 shows the time evolution of the neutrino flavour in the flavour basis including both mixing and the matter-induced potential:

$$i\frac{d}{dt}\begin{pmatrix}\nu_e\\ \nu_\mu\end{pmatrix} = \begin{pmatrix}-\frac{\Delta m^2}{4E}\cos 2\theta + \sqrt{2}G_F N_e & \frac{\Delta m^2}{4E}\sin 2\theta \\ \frac{\Delta m^2}{4E}\sin 2\theta & \frac{\Delta m^2}{4E}\cos 2\theta\end{pmatrix}\begin{pmatrix}\nu_e\\ \nu_\mu\end{pmatrix} \tag{12}$$

The additional term $\sqrt{2}G_F N_e$ appearing in Equation 12 is the matter-induced potential, which is proportional to the electron number density N_e and is linear in G_F. This effect, known as the MSW effect after Mikheev, Smirnov, and Wolfenstein[3], gives rise to a rich phenomenology in which oscillation probabilities in dense matter, such as the interior of the Sun, can be markedly different from those seen in vacuum. Of the experimental results to date, only in solar neutrino oscillations does the MSW effect play a significant role, although future long-baseline neutrino oscillation experiments also may have some sensitivity to matter effects.

The generalization of neutrino mixing and oscillation to three flavours is straightforward. Instead of a 2×2 mixing matrix, as in Equation 5, we relate the neutrino flavour eigenstates to the neutrino mass eigenstates by a 3×3 unitary matrix, completely analogous to the CKM matrix for quarks. The neutrino mixing matrix is known as the MNS matrix for Maki, Nakagawa, and Sakata, and occasionally as the PMNS matrix when acknowledging

Pontecorvo's early contributions to the theory of neutrino oscillations.[1]

$$\begin{pmatrix} \nu_e \\ \nu_\mu \\ \nu_\tau \end{pmatrix} = \begin{bmatrix} U_{e1} & U_{e2} & U_{e3} \\ U_{\mu 1} & U_{\mu 2} & U_{\mu 3} \\ U_{\tau 1} & U_{\tau 3} & U_{\tau 3} \end{bmatrix} \begin{pmatrix} \nu_1 \\ \nu_2 \\ \nu_3 \end{pmatrix}$$
$$\approx \begin{bmatrix} 0.9 & 0.5 & U_{e3} \\ -0.35 & 0.6 & 0.7 \\ 0.35 & -0.6 & 0.7 \end{bmatrix} \begin{pmatrix} \nu_1 \\ \nu_2 \\ \nu_3 \end{pmatrix} \quad (13)$$

Equation 13 gives the approximate values of the MNS matrix elements. The values of all of the elements except U_{e3} have been inferred at least approximately. The most striking feature of the MNS matrix is how utterly non-diagonal it is, in marked contrast to the CKM matrix. Neutrino mixings are in general large, and there is not even an approximate correspondence between any mass eigenstate and any flavour eigenstate. (Therefore it really does not make any sense to talk even approximately about the "mass" of a ν_e, except as a weighted average of its constituent mass eigenstates.) Only the unknown matrix element U_{e3} is observed to be small, with a current upper limit of $|U_{e3}|^2 < 0.03$ (90% confidence limit).[4] Section 3 will enumerate the many lines of evidence that demonstrate that neutrinos do in fact oscillate, and describe how the mixing parameters are derived.

3. Evidence For Neutrino Flavour Oscillation

Since 1998 conclusive evidence has been found demonstrating neutrino flavour oscillation of both atmospheric neutrinos and solar neutrinos.[5,6] In each case the oscillation effects have been confirmed by followup experiments using man-made sources of neutrinos.[8,9] Here I review the experimental situation, with a strong bias towards recent results.

3.1. *The Solar Neutrino Problem, With Solution*

The earliest indications of neutrino oscillations came from experiments designed to measure the flux of neutrinos produced by the nuclear fusion reactions that power the Sun. The Sun is a prolific source of ν_e's with energies in the ~0.1-20 MeV range, produced by the fusion reaction

$$4p + 2e^- \rightarrow {}^4\text{He} + 2\nu_e + 26.731 \text{ MeV}. \quad (14)$$

The reaction in Equation 14 actually proceeds through a chain of sub-reactions called the *pp* chain, consisting of several steps.[10] Each neutrino-producing reaction in the *pp* chain produces a characteristic neutrino energy spectrum that depends only on the underlying nuclear physics, while the rates of the reactions must be calculated through detailed astrophysical models of the Sun. Experimentally the *pp*, ^{8}B, and ^{7}Be reactions are the most important neutrino-producing steps of the *pp* chain.

The pioneering solar neutrino experiment was Ray Davis's chlorine experiment in the Homestake mine near Lead, South Dakota.[11] This experiment measured solar neutrinos by observing the rate of Ar atom production through the reaction $\nu_e+^{37}\text{Cl}\rightarrow^{37}\text{Ar}+e^-$. By placing 600 tons of tetrachloroethylene deep underground (to shield it from surface radiation), and using radiochemistry techniques to periodically extract and count the number of argon atoms in the tank, Davis inferred a solar neutrino flux that was just $\sim$1/3 of that predicted by solar model calculations.[11,12]

This striking discrepancy between theory and experiment at first had no obvious particle physics implications. Both the inherent difficulty of looking for a few dozen argon atoms inside 600 tons of cleaning fluid, and skepticism about the reliability of solar model predictions, cast doubt upon the significance of the disagreement. A further complication is that the reaction that Davis used to measure the ν_e flux was sensitive to multiple neutrino-producing reactions in the *pp* chain, making it impossible to determine which reactions in the Sun are not putting out enough neutrinos.

When scrutiny of both the Davis experiment and the solar model calculations failed to uncover any clear errors, other experiments were built to measure solar neutrinos in other ways. The Kamiokande and Super-Kamiokande water Cherenkov experiments have measured elastic scattering of electrons by ^{8}B solar neutrinos, using the directionality of the scattered electrons to confirm that the neutrinos in fact are coming from the Sun.[13] The measured elastic scattering rate is just $\sim$47% of the solar model prediction. The SAGE and GNO/GALLEX experiments have employed a different radiochemical technique to observe the $\nu_e+^{71}\text{Ge}\rightarrow^{71}\text{Ge}+e^-$ reaction, which is primarily sensitive to *pp* neutrinos, and have measured a rate that is $\sim$55% of the solar model prediction.[14]

Multiple experiments using different techniques have therefore confirmed a deficit of solar ν_e's relative to the model predictions. Although interpretation of the data is complicated by the fact that each kind of experiment is sensitive to neutrinos of different energies produced by different reactions in the *pp* fusion chain, in fact there is apparently no self-consistent way to modify the solar model predictions that will bring the astrophysical predictions into agreement with the experimental results. This situation suggested that the explanation of the solar neutrino problem may not lie in novel astrophysics, but rather might indicate a problem with our understanding of neutrinos.

While it was realized early on that neutrino oscillations that converted solar ν_e to other flavours (to which the various experiments wouldn't be sensitive) could explain the observed deficits, merely observing deficits in the overall rate was generally considered insufficient grounds upon which to establish neutrino oscillation as a real phenomenon. It was left for the Sudbury Neutrino Observatory (SNO) to provide the conclusive evidence that solar neutrinos change flavour by directly counting the rate of all active neutrino flavours, not just the ν_e rate to which the other experiments were primarily sensitive.

SNO is a water Cherenkov detector that uses 1000 tonnes of D_2O as the target material.[15] Solar neutrinos can interact with the heavy water by three different interactions:

$$\begin{array}{lll} (CC) & \nu_e + d & \rightarrow p + p + e^- \\ (NC) & \nu_x + d & \rightarrow p + n + \nu_x \\ (ES) & \nu_x + e^- & \rightarrow \nu_x + e^- \end{array} \tag{15}$$

Here ν_x is any active neutrino species. The reaction thresholds are such that SNO is only sensitive to ^{8}B solar neutrinos.[a] The charged current (CC) interaction measures the flux of ν_e's coming from the Sun, while the neutral current (NC) reaction measures the flux of all active flavours. The elastic scattering (ES) reaction is primarily sensitive to ν_e, but ν_μ or ν_τ also elastically scatter electrons with $\sim 1/6$th the cross section of ν_e.

SNO has measured the effective flux of ^{8}B neutrinos inferred from each reaction. In units of 10^6 neutrinos/cm^2/s the most recent measurements are[7]:

$$\begin{aligned} \phi_{CC} &= 1.68 \pm 0.06\ (\text{stat.})^{+0.08}_{-0.09}\ (\text{sys.}) \\ \phi_{NC} &= 4.94 \pm 0.21\ (\text{stat.})^{+0.38}_{-0.34}\ (\text{sys.}) \\ \phi_{ES} &= 2.34 \pm 0.22\ (\text{stat.})^{+0.15}_{-0.15}\ (\text{sys.}) \end{aligned} \tag{16}$$

In short, the NC flux is found to be in good agreement with the solar model predictions, while the CC and ES rates are each consistent with just $\sim 35\%$ of the ^{8}B flux being in the form of ν_e's.

This direct demonstration that $\phi_e < \phi_{total}$ provides dramatic proof that solar neutrinos change flavour, resolving the decades-old solar neutrino problem in favour of new neutrino physics. The neutrino oscillation model gives an excellent fit to the data from the various solar experiments, with

[a]The tiny flux of higher-energy neutrinos from the *hep* chain may be neglected here.

mixing parameters of $\Delta m^2 \approx 10^{-4} - 10^{-5}$ eV^2 and $\tan^2\theta \approx 0.4 - 0.5$. This region of parameter space is called the Large Mixing Angle solution to the solar neutrino problem. In this region of parameter space, the MSW effect plays a dominant role in the oscillation, and in fact ^{8}B neutrinos are emitted from the Sun in an almost pure ν_2 mass eigenstate.

3.2. *KamLAND*

Although neutrino oscillations with an MSW effect are the most straightforward explanation for the observed flavour change of solar neutrinos, the solar data by itself cannot exclude more exotic mechanisms of inducing flavour transformation. However, additional confirmation of solar neutrino oscillation has recently come from an unlikely terrestrial experiment called KamLAND.

KamLAND is an experiment in Japan that counts the rate of $\bar{\nu}_e$ produced in nuclear reactors throughout central Japan.[8] If neutrinos really do oscillate with parameters in the LMA region, then the standard oscillation theory predicts that reactor $\bar{\nu}_e$'s, with a peak energy of $\sim$ 3 MeV, should undergo vacuum oscillations over a distance of $\sim$ 200 km.[b] By integrating the flux from multiple reactors, KamLAND achieves sensitivity to this effect. Figure 2 shows the L/E dependence of the measured reactor $\bar{\nu}_e$ flux divided by the expected flux at KamLAND.[8] The observed flux is lower than the "no oscillation" expectation on average by $\sim$1/3, with an energy-dependent suppression of the $\bar{\nu}_e$ flux. The pattern of the flux suppression is in good agreement with the neutrino oscillation hypothesis with oscillation parameters in the LMA region.

That KamLAND observes an energy-dependent suppression of the reactor $\bar{\nu}_e$ flux, just as predicted by fits of the oscillation model to solar neutrino data, is dramatic confirmation of the solar neutrino results and demonstrates that neutrino oscillation is the correct explanation of the flavour change of solar neutrinos observed by the SNO experiment.

The solar experiments and KamLAND provide complementary constraints on the mixing parameters. Figure 3 demonstrates that solar neutrino experiments provide reasonably tight constraints on the mixing parameter $\tan^2\theta$, while the addition of KamLAND data sharply constrains the Δm^2 value.[7] This is because in the LMA region the solar neutrino

[b]At these low energies matter effects inside the Earth are negligible.

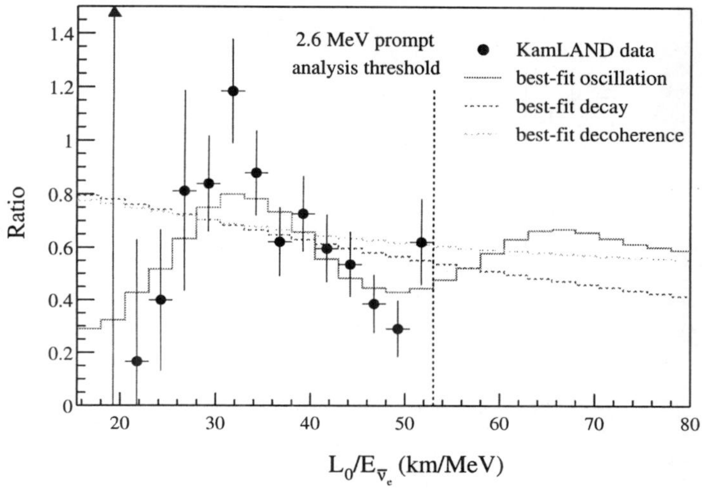

Figure 2. Ratio of the $\bar{\nu}_e$ reactor antineutrino flux measured by KamLAND to the expected flux without oscillations, as a function of L/E.

survival probability determines the mixing angle through

$$|U_{e2}|^2 \approx \sin^2\theta_{12} \approx \frac{\phi_{CC}}{\phi_{NC}} \tag{17}$$

while the observation of a distortion in the reactor antineutrino energy spectrum fixes Δm^2_{21}. Here the subscripts on θ_{12} and Δm^2_{21} reflect the fact that solar neutrino oscillations involve the first and second mass eigenstates.

3.3. *Atmospheric Neutrinos*

Although the solar neutrino problem provided early indications that the Standard Model's description of neutrinos is incomplete, resolution of the solar neutrino problem was a long time coming, and the first *conclusive* demonstration of neutrino oscillation actually came from studies of atmospheric neutrinos. Atmospheric neutrinos are produced when cosmic rays (primarily protons) collide in the upper atmosphere to make hadronic showers. These showers contain charged pions, which decay leptonically by $\pi^\pm \to \mu^\pm \nu_\mu$. The muons in turn generally decay in flight by $\mu^\pm \to e^\pm \nu_\mu \nu_e$, where I've ignored differences between ν and $\bar{\nu}$ states. A robust conclusion that follows from the decay sequence is that the ratio of ν_μ to ν_e in the atmospheric neutrino flux should be 2:1.

In 1998 the Super-Kamiokande collaboration reported results showing

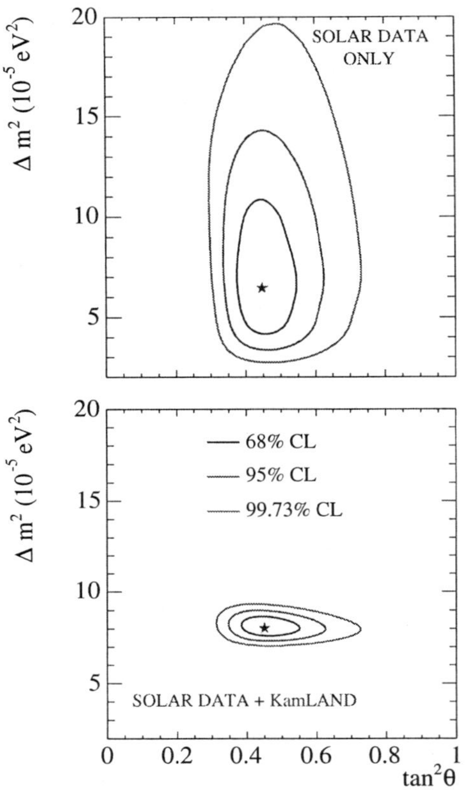

Figure 3. Oscillation parameter contours for solar neutrino data (top), and solar data + KamLAND data (bottom).

that that ratio of the flux of ν_μ to ν_e in fact is not 2:1, but is closer to 1:1.[5,16] Closer examination revealed that while the ν_e flux in fact is in good agreement with Monte Carlo predictions, the ν_μ flux shows a marked deficit. The size of this deficit varies with neutrino energy, and with the zenith angle of the event. This latter point is significant in that downgoing neutrinos are produced in the atmosphere just overhead, and have travelled < 10 km before reaching Super-Kamiokande, while upgoing neutrinos are produced in the atmosphere on the far side of the Earth, and have travelled $\sim$13,000 km before reaching the detector. As seen in Figure 4, the deficit between the expected and measured number of ν_μ is largest at low energy

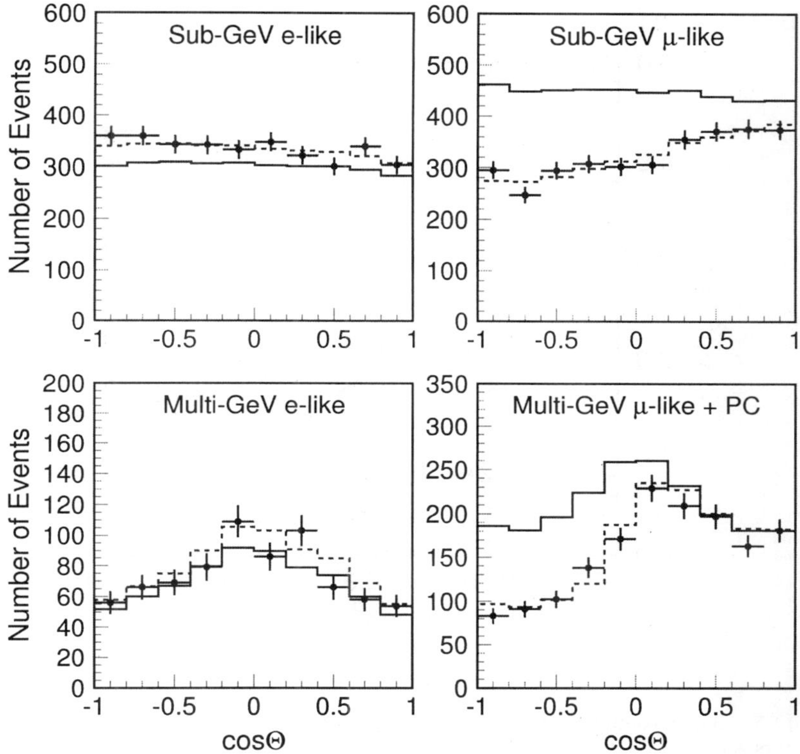

Figure 4. Fluxes of atmospheric ν_e and ν_μ as a function of zenith angle, as measured by Super-Kamiokande. The solid lines show the no oscillation prediction, while the dashed line passing through the data points is the best-fit oscillation prediction.

and at negative $\cos\theta$ (upward-going events).[17] This dependence on energy and on the distance travelled by the neutrino is characteristic of neutrino oscillations, and excludes a simple normalization error. These results were the first to establish conclusively that atmospheric neutrinos oscillate. The oscillation seems to be of the type $\nu_\mu \to \nu_\tau$. The atmospheric neutrino effect has been confirmed by a number of other experiments.[18]

Figure 5 shows the inferred mixing parameters from fitting a two-flavour oscillation model to the atmospheric neutrino data.[16] The data favour $\Delta m^2 \approx 2.5\times10^{-3}$ eV^2 and, surprisingly, a maximal mixing angle of $\theta \approx 45°$. (The term "maximal mixing" refers to the fact that each flavour eigenstate contains equal proportions of the two mass eigenstates if $\theta = 45°$.)

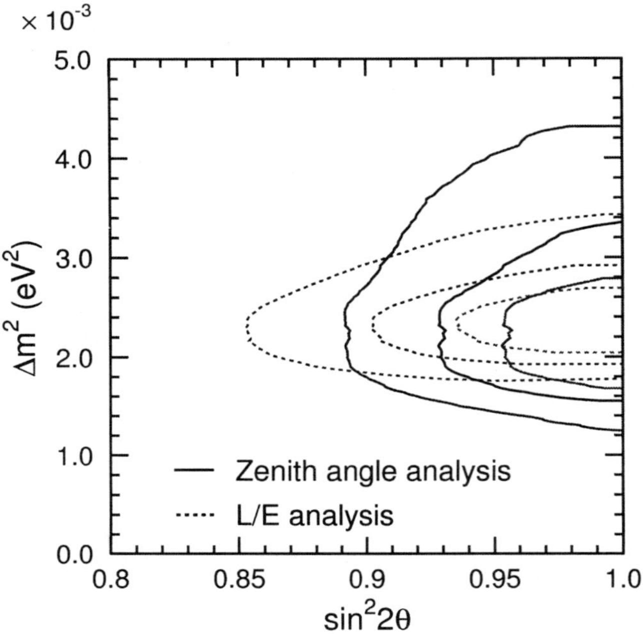

Figure 5. Super-Kamiokande atmospheric neutrino mixing contours. The two sets of contours are the 68%, 90%, and 99% contours from two different analysis techniques.

3.4. *Long-Baseline Neutrino Oscillation Experiments*

Just as solar neutrino oscillations have been confirmed with terrestrial (anti)-neutrinos by KamLAND, atmospheric neutrino oscillations have recently been confirmed by the K2K long-baseline neutrino oscillation experiment.[9] K2K produced a collimated beam of ν_μ by colliding a 12 GeV proton beam with an aluminum target, thereby producing π^+'s. These pions were then collected and focused with a set of magnetic horns, and the collimated pion beam then decayed in a long evacuated decay pipe by $\pi^+ \to \nu_\mu \mu^+$. The mean neutrino energy was 1.3 GeV, and the beam was aligned with the direction of the Super-Kamiokande detector, located 250 km away. A set of near neutrino detectors measured the neutrino beam's energy spectrum, interaction, and relative cross sections at a point 300 m from the pion production target. By comparing the neutrino energy spectrum and rate at the near detector to those measured at Super-

Kamiokande, the effects of neutrino oscillation over the 250 km baseline can be inferred. If the atmospheric neutrino effect is really explained by neutrino oscillations, then K2K should see an apparent "disappearance" of ν_μ, which oscillate into ν_τ that are too low in energy to be detected in Super-K through charged current interactions.

Data collected by K2K between 1999 and 2004 in fact show a deficit of muon-like events, and some indication of an energy dependence to the ν_μ disappearance effect as predicted for neutrino oscillations.[9] A combined maximum likelihood fit to the spectrum and rate excludes the null hypothesis of no oscillations at the 4.0σ level. The best-fit oscillation parameters are $\Delta m^2 = 2.8 \times 10^{-3}$ eV^2 and $\sin^2 2\theta = 1$, which are in excellent agreement with the values inferred from the atmospheric neutrino data.

The MINOS experiment is a conceptually similar long-baseline experiment in the United States. MINOS uses the NUMI neutrino beam produced by Fermilab's Main Injector, with a far detector located $\sim$ 730 km way in the Soudan mine in northern Minnesota, to study oscillations of ν_μ. MINOS should confirm K2K's results with somewhat higher statistics, and at the time of writing results are expected imminently[c].

3.5. *The Three-Flavour Picture*

In the previous sections, the solar and atmospheric neutrino oscillation effects were each analyzed separately in terms of oscillations between two neutrino mass eigenstates. In reality, we know there are (at least) three flavour eigenstates, and so three mass eigenstates. Properly speaking we need to consider the 3×3 MNS matrix, completely analogous to the CKM matrix for quarks, which can be parameterized as:

$$U = \begin{pmatrix} 1 & 0 & 0 \\ 0 & c_{23} & s_{23} \\ 0 & -s_{23} & c_{23} \end{pmatrix} \begin{pmatrix} c_{13} & 0 & e^{i\delta} s_{13} \\ 0 & 1 & 0 \\ -e^{-i\delta} s_{13} & 0 & c_{13} \end{pmatrix} \begin{pmatrix} c_{12} & s_{12} & 0 \\ -s_{12} & c_{12} & 0 \\ 0 & 0 & 1 \end{pmatrix} \tag{18}$$

Here $c_{ij} \equiv \cos\theta_{ij}$ and $s_{ij} \equiv \sin\theta_{ij}$.

The θ_{12} term in this parameterization of the MNS matrix is that which controls solar neutrino oscillations, which involve the first and second mass eigenstates. Experimentally $\theta_{12} \approx 32°$.[7] For comparison, the equivalent angle in the CKM matrix is the Cabibbo angle, which has the value $\theta_C \approx 13°$.

[c]As this paper went to press the MINOS collaboration released its first results, which confirmed ν_μ disappearance in the NUMI beamline with $\Delta m^2 \approx 3 \times 10^{-3}$ eV^2 (publication pending).

The mixing between the first and second generations of leptons is thus much larger than the mixing between the quark generations. Similarly, θ_{23}, which determines the amplitude of atmospheric neutrino oscillations, is consistent with maximal mixing ($\theta_{23} \approx 45°$), even though its quark counterpart equals just $\sim 2°$! It is unknown at present by how much θ_{23} actually deviates from maximal mixing angle, or whether this value is indicative of some kind of flavour symmetry between the second and third generations.

By comparison, the middle part of Equation 18 is poorly constrained. Limits on oscillations of reactor neutrinos at short baselines (~ 1 km) tell us that $\theta_{13} < 9°$.[4] In fact, current measurements of θ_{13} are consistent with zero. Presently nothing is known about the complex phase δ in the MNS matrix, which if non-zero would result in different oscillation patterns for neutrinos than for antineutrinos. This latter topic is of considerable interest. Recalling that all observed instances of CP violation in physics can be explained by a single complex phase in the CKM matrix, it is exciting to realize that the observation that neutrinos oscillate implies the possible existence of an entirely new source of CP violation—one involving leptons rather than quarks!

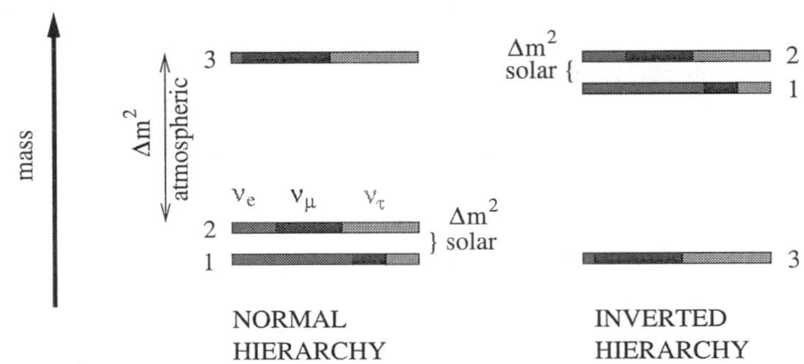

Figure 6. Normal and inverted neutrino mass hierarchies.

Measurements of atmospheric and solar neutrino oscillations also provide a partial determination of the pattern of the neutrino masses. Solar and reactor neutrino data have determined that $\Delta m^2_{21} \equiv m^2_2 - m^2_1 \approx 8.0 \times 10^{-5}$ eV^2 (see Figure 3)[7], while atmospheric and long baseline neutrino experiments[16,9] fix $|\Delta m^2_{32}| \approx 2.5 \times 10^{-3}$ eV^2. The solar neutrino experiments have successfully inferred the sign of Δm^2_{21} because the sign of the MSW effect in the Sun, which dominates in solar neutrino oscillations, depends on the sign of Δm^2. The atmospheric neutrino data however has no significant sensitivity at present to matter effects, and therefore it is not known whether $m_2 < m_3$ or rather $m_2 > m_3$. The result is that there are two possible mass hierarchies for the neutrino mass eigenstates. The so-called "normal" hierarchy has two light states and one heavier state, with $m_1 < m_2 < m_3$, while in the "inverted" hierarchy m_3 is the lightest state, with m_1 and m_2 being almost degenerate in mass (see Figure 6). Note that neutrino oscillation experiments are sensitive only to *differences* in m^2, and do not measure the absolute mass scale, although lower limits on the neutrino masses can be obtained by assuming the mass of the lightest mass eigenstate to be zero.

3.6. *The LSND Result and the MiniBooNE Experiment*

Until this point I have put off discussion of one other neutrino oscillation result that must be addressed. The LSND collaboration has reported 3.8σ evidence for $\bar{\nu}_\mu \to \bar{\nu}_e$ oscillations of $\bar{\nu}_\mu$ produced from the decay of stopped μ^+ in a beam dump, over a propagation distance of ~ 30 meters.[19] Other experiments, notably the KARMEN experiment, have failed to confirm this effect[20], but do not rule out the entire range of mixing parameters allowed by the LSND result. Mixing parameters with $\sin^2 2\theta \approx 10^{-3} - 10^{-2}$ and $\Delta m^2 \sim 0.1 - 1$ eV^2 are consistent with all data.[21]

The inferred value of Δm^2 from the LSND result is much larger than those seen in solar and atmospheric neutrino experiments. If there are only three light neutrinos, then one can only form two independent mass differences Δm^2. If the LSND effect is due to neutrino oscillation, then it implies a third independent value of Δm^2, and so requires a fourth neutrino mass eigenstate. However, the LEP measurements of the Z boson's invisible decay width confirm that there are only three active light neutrinos.[22] A fourth light neutrino, if it exists, must be sterile! Even worse, more detailed analyses of solar and atmospheric neutrinos show no indication of any sterile neutrino admixtures, and are difficult to reconcile with the existence of a single sterile neutrino.[23] By adding more than one sterile flavour, enough wiggle room can be introduced to explain all of the oscillation results.

The LSND result presents a particular problem for neutrino physics. Because this result has not yet been confirmed by an independent experiment, and because it has relatively drastic consequences such as implying the existence of one or more sterile neutrino flavours, there is widespread skepticism regarding its correctness. That being said, no fundamental flaw in the LSND experiment has been demonstrated, and it is very possible that the result is correct. Neutrinos may then be more bizarre than anyone would have guessed! At present the MiniBooNE experiment at Fermilab is attempting to definitively check the LSND result[24], and is expected to produce first results for $\nu_\mu \to \nu_e$ oscillations sometime in 2006. Because the LSND result has not yet been confirmed and cannot easily be accommodated within the standard 3-flavour oscillation model, it is most often ignored. Only more data can determine whether it can be ignored without great peril.

4. Future Directions In Neutrino Oscillation

In less than a decade we have evolved from a situation in which we had no direct evidence that neutrinos oscillate to the present day, in which both Δm^2 parameters are known to ~10-20%, and two of the three neutrino mixing angles are known at least approximately. One obvious way to proceed is to complete our picture of the MNS matrix by attempting to measure the unknown mixing parameters θ_{13} and δ_{CP}, along with the sign of Δm^2_{32} that determines whether neutrinos have a normal or inverted mass hierarchy.

4.1. *Measuring θ_{13}, The Mass Hierarchy, and CP Violation At Long-Baseline Experiments*

The Super-K and K2K oscillation results seem to be of the type $\nu_\mu \to \nu_\tau$, and are well described by a two-flavour mixing model.[16,9] However, in the full 3×3 mixing picture, there should be some probability that ν_μ's will instead oscillate into ν_e's in these experiments. For an L/E value tuned to Δm^2_{32}, this probability is given by[25]:

$$P(\nu_\mu \to \nu_e) \approx \sin^2 2\theta_{13} \sin^2 \theta_{23} \approx \frac{1}{2} \sin^2 2\theta_{13} \tag{19}$$

Current limits on θ_{13} bound this probability to $< 5\%$.

Because atmospheric neutrinos contain a significant fraction of ν_e, observing the small $\nu_\mu \to \nu_e$ transition probability is not feasible. Long baseline experiments however can produce almost 100% pure beams of ν_μ. By searching for the appearance of a small ν_e component in the beam at the oscillation maximum, the value of θ_{13} may be inferred.

Equation 19 is only approximate, and the true ν_e appearance probability is modified by other mixing parameters and by matter effects. In particular, it can be shown that at the first oscillation maximum, the ν_e appearance probability in vacuum is altered in the presence of matter according to[26]:

$$P_{matter}(\nu_\mu \to \nu_e) \approx \left(1 + 2\frac{E}{E_R}\right) P_{vacuum}(\nu_\mu \to \nu_e) \tag{20}$$

where E_R is a resonance energy given by $E_R = \Delta m^2_{32}/(2\sqrt{2}G_F N_e)$. This matter effect depends on the number density of electrons N_e, and also on the magnitude and the sign of Δm^2_{32}. This matter effect correction is more significant at large L or E values, and has the opposite sign for neutrinos and antineutrinos.

A second confounding effect comes from the CP-violating phase of the MNS matrix. CP symmetry requires that neutrinos and antineutrinos oscillate identically, so that $P(\nu_\mu \to \nu_e) = P(\bar{\nu}_\mu \to \bar{\nu}_e)$ in vacuum. However, a non-zero value of δ_{CP} can make these probabilities unequal. One can then define a CP asymmetry for ν_e appearance which, ignoring matter effects, is given by[25]:

$$\mathcal{A}_{CP} = \frac{P(\nu_\mu \to \nu_e) - P(\bar{\nu}_\mu \to \bar{\nu}_e)}{P(\nu_\mu \to \nu_e) + P(\bar{\nu}_\mu \to \bar{\nu}_e)} \simeq \frac{\Delta m^2_{21} L}{4E_\nu} \cdot \frac{\sin 2\theta_{12}}{\sin \theta_{13}} \cdot \sin \delta_{CP} \tag{21}$$

The CP effect both changes $P(\nu_\mu \to \nu_e)$ and creates a non-zero $\mathcal{A}_{CP}$. Notice that the size of $\mathcal{A}_{CP}$ as measured at the oscillation peak for the atmospheric neutrino Δm^2_{32} depends on the solar neutrino parameters Δm^2_{21} and θ_{12} as well. The reason for this is that, just as in the quark sector, CP violation in the neutrino sector is an interference effect: in this case, an interference between oscillations at the solar and atmospheric frequencies. To observe this effect, oscillations at both Δm^2 values must be of roughly comparable size, and θ_{13}, which has the effect of coupling the atmospheric and solar oscillations in Equation 18, must be non-zero. Fortunately for those of us interested in actually observing CP violation by neutrinos, recent solar neutrino results establishing the LMA solution imply that both solar mixing parameters are reasonably large relative to the atmospheric neutrino mixing parameters. If θ_{13} is not too small, then observation of non-zero $\mathcal{A}_{CP}$ may be possible.

Because the $\nu_\mu \to \nu_e$ oscillation probability depends on θ_{13}, sign(Δm^2_{32}), and δ_{CP}, multiple measurements at different energies and/or baselines will be needed to disentangle the different effects. Figure 7 illustrates the dependence of $P(\nu_\mu \to \nu_e)$ and $P(\bar{\nu}_\mu \to \bar{\nu}_e)$ on the different oscillation parameters, for monoenergetic (anti)neutrino beams with $E = 1.5$ GeV and $L = 732$ km. The sign of Δm^2_{32} defines the normal and inverted mass hierarchies, dividing the predicted probabilities into two separate "cones".

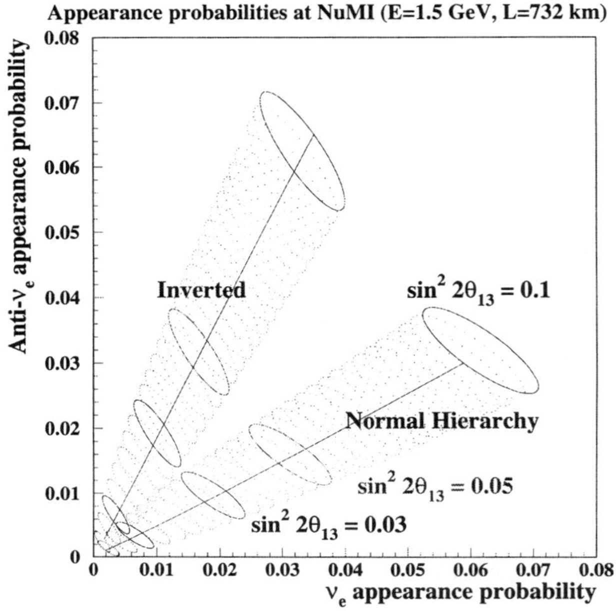

Figure 7. Oscillation probabilities for $\nu_\mu \to \nu_e$ vs. $\bar{\nu}_\mu \to \bar{\nu}_e$ for an off-axis experiment in the NUMI beamline. The solid diagonal lines correspond to $\delta_{CP} = 0$.

Increasing θ_{13} moves one out along either cone to larger oscillation probabilities. With θ_{13} and sign(Δm^2_{32}) fixed, varying δ_{CP} traces out an ellipse in the plane, as shown in the figure. A suitably precise measurement of the neutrino and antineutrino appearance probabilities could determine the mass hierarchy for largish θ_{13}, as well as defining an allowed region in the $\theta_{13} - \delta_{CP}$ plane. Measurements at different choices of L and E will have different sensitivity to matter effects and to δ_{CP} (see Equations 20 and 21), and can be used to break any remaining parameter degeneracies.

At present just one experiment to study ν_e appearance at the atmospheric Δm^2 has been funded. This is the T2K experiment in Japan.[25] T2K will use a megawatt-scale proton beam at the Japan Proton Accelerator Research Complex (JPARC) in Tokai to produce a ν_μ beam that will be directed towards Super-Kamiokande, 295 km away. By pointing the neutrino beam about 2° away from Super-K, T2K will take advantage of a trick called "off-axis focusing", which results in a nearly monoenergetic neutrino beam with a peak energy of $\sim$700 MeV. At these energies the dominant interactions are charged current quasi-elastic ($\nu_\ell + n \to p + \ell$).

A set of sophisticated near detectors will measure the beam properties before oscillation. T2K will have approximately 50 times greater statistics than K2K. With its relatively low beam energy and small baseline, T2K is relatively insensitive to matter effects.

The most important backgrounds to ν_e appearance at T2K are a small component of ν_e in the beam itself, and neutral current π^0 production at Super-K. The latter is only a background to ν_e appearance if Super-K fails to detect one of the two γ-rays. This could happen in very asymmetric decays in which one photon takes the bulk of the π^0's energy, or if optical scattering of Cherenkov light sufficiently obscures one of the two Cherenkov rings. For five years of running at nominal luminosity (5×10^{21} protons on target), T2K expects to achieve sensitivity to θ_{13} down to $\sin^2 2\theta_{13} \approx 10^{-2}$ (the exact limit depends on the value of δ_{CP}).[27] The measured value of θ_{13} is partially degenerate with δ_{CP}, and separating the two parameters will require additional measurements with antineutrinos and/or at other baselines. Assuming that T2K successfully detects ν_e appearance in the ν_μ beam, the natural followup is to switch the polarity of the beam and look for $\bar{\nu}_\mu \to \bar{\nu}_e$. With a beam power upgrade and possibly the construction of a larger far detector, this "phase 2" program could then begin to explore CP violation in the neutrino sector.

In addition to measuring the ν_e appearance probability, future long-baseline neutrino experiments such as T2K will measure the ν_μ disappearance probability with much higher statistics, allowing precision measurements of Δm^2_{32} and θ_{23}. Such measurements can test how close θ_{23} is to maximal mixing (45°), explore whether any fraction of the ν_μ flux is oscillating to a non-interacting (sterile) neutrino flavour, and test the energy dependence of the neutrino oscillation prediction with high precision.

Although T2K is currently the only funded new long-baseline experiment to search for ν_e appearance, the NOνA collaboration in the US has proposed building a new off-axis detector, optimized for detecting electron appearance, in Fermilab's NUMI beamline.[26] At a baseline of $\sim$ 730 km and a beam energy of $\sim$ 2 GeV, the NOνA experiment could have some sensitivity to matter effects and the sign of the mass hierarchy if θ_{13} is not too small, and would otherwise have similar sensitivity to ν_e appearance as T2K. The proposed far detector is a massive finely segmented liquid scintillator detector. The NOνA proposal is currently in the early stages of the approval process.

4.2. *Reactor Neutrino Experiments*

An alternate approach to measuring θ_{13} is to do precision reactor neutrino experiments at short baselines. The full 3-flavour formula for reactor $\bar{\nu}_e$ oscillation is[28]:

$$P(\bar{\nu}_e \to \bar{\nu}_e) = 1 - \sin^2 2\theta_{13} \sin^2 \left(\frac{1.27\Delta m^2_{13} L}{E}\right) - \cos^4 \theta_{13} \sin^2 2\theta_{12} \sin^2 \left(\frac{1.27\Delta m^2_{21} L}{E}\right) \tag{22}$$

The first term, which is proportional to $\sin^2 2\theta_{13}$ and depends on the larger Δm^2 value, dominates over the second at short baselines. The second term only becomes significant at reactor neutrino energies for $L \approx 200$ km. KamLAND was successfully able to use the second term to confirm the solar neutrino effect[8], but experiments at shorter baselines instead yield limits on θ_{13}. Currently the best limits on θ_{13} come from the CHOOZ reactor neutrino experiment, which limits $\sin^2 2\theta_{13} < 0.15$ at the 90% C.L.[4]

A new reactor neutrino experiment with high statistics and improved systematics may be able to achieve significantly improved θ_{13} sensitivity.[28] The keys to better sensitivity are to use a very intense reactor, with power in the gigawatt range, and to use both a near detector right next to the reactor and a far detector 1 or 2 km away in order to cancel systematics between the two detectors. A significant advantage of reactor θ_{13} experiments is that they are not sensitive to CP-violating effects (which can only be measured in an appearance measurement, not in a disappearance measurement), nor to matter effects, which are negligible at the relevant L and E values. A good reactor neutrino experiment therefore would provide a clean measurement of just θ_{13}. This provides significant complementarity to long-baseline ν_e appearance experiments, which are sensitive to a combination of θ_{13}, the mass hierarchy, and δ_{CP}.

An added advantage of reactor θ_{13} experiments is that they are relatively inexpensive, with a typical estimated price tag of ~$50M. For this reason, it seems that experimenters have proposed new experiments at virtually every reactor complex in the world with significant power output. Prominent sites for proposed experiments include Daya Bay in China, Braidwood in Illinois, and the Double CHOOZ proposal in France, although this list is far from exhaustive.[28] It seems likely that one or more of these proposals will be funded, but at present it is not clear which ones. The physics case for a sensitive reactor θ_{13} experiment seems compelling, however.

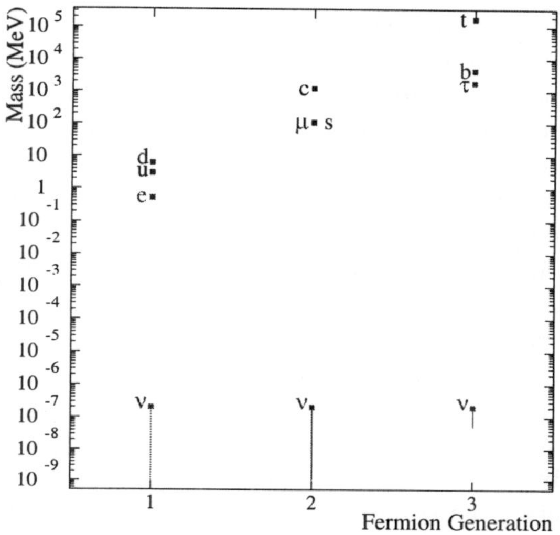

Figure 8. Masses of the Standard Model fermions. The purple lines indicate the range of allowed neutrino masses for ν_1, ν_2, and ν_3, assuming a normal mass hierarchy.

5. Altering The Standard Model To Accommodate Neutrino Mass

In the Standard Model, neutrinos have zero mass. This is not simply an *ad hoc* assumption, but a consequence of the fact that the Standard Model does not contain right-handed neutrino fields. Fermion mass terms in the Standard Model Lagrangian have the form $-m\bar{\psi}\psi = -m(\bar{\psi}_L\psi_R + \bar{\psi}_R\psi_L)$. Without a right-handed field, no such term can exist. In this section I shall examine possible ways in which the Standard Model may be extended to include non-zero neutrino mass.

The most obvious solution to this problem is to simply add right-handed neutrino states ν_R to the Standard Model and to give them Yukawa couplings to ν_L through the Higgs field, just like other fermions. This is called a Dirac mass term. While superficially this places neutrinos on the same footing as the other fermions, one striking difference is that ν_R, having neither charge, colour, nor couplings to $W^\pm$ or Z, are sterile fields (i.e. they don't couple to the vector gauge bosons)—making them in an important sense unlike all other Standard Model particles. An additional puzzle is that in order to explain the smallness of neutrino masses, their Yukawa couplings must be made anomalously small. As illustrated in Figure 8, within each generation the charged fermions are separated in mass by no more than 1 or 2 orders of magnitude, but the neutrino mass eigenstates are many orders of magnitude lighter than their charged counterparts. While it may rightfully be objected that we have no good explanations for the numerical values of the masses of *any* fermions, the disparity between neutrino and charged fermion masses suggests that neutrinos might not simply acquire mass in the same manner as other fermions.

Another possible way to add neutrino mass terms is to recognize that there already exists a right-handed neutral fermion in the Standard Model—namely, the antineutrino. Is it possible to identify ν_R with the antineutrino, and so generate mass terms of the form $\bar{\nu_L}\nu_R$ by combining a neutrino with its antineutrino? For charged fermions, the answer would clearly be no: since particles and antiparticles have opposite charges, a term that directly couples a fermion to its antifermion violates charge conservation! However, the situation is different with neutrinos, which are chargeless particles. The Majorana neutrino hypothesis takes advantage of this chargelessness by positing that an antineutrino is just a neutrino with its spin flipped by 180°! One might then form a "Majorana mass" term that couples a left-handed neutrino with the right-handed antiparticle.

Nonetheless, within the minimal Standard Model, Majorana mass terms are in fact forbidden. The reason is that although the Standard Model does not conserve either baryon number B or lepton number L non-perturbatively, it does conserve the quantity $B - L$ exactly. A Majorana mass term on the other hand results in $|\Delta(B - L)| = 2$. It turns out that without extending the Standard Model particle content in some manner, a $B - L$ violating term cannot be generated at any order, even as an effective operator.[29]

However, the addition of an *additional* right-handed Majorana field to the Standard Model can resolve the problem. Let ν_L be a 2-component field describing left-handed neutrino/right-handed antineutrino that couples to weak interactions. Let ν_R now denote an *additional* right-handed Majorana field, independent of ν_L, which does not couple to weak interactions. Because ν_R is an electroweak singlet, it can possess a bare Majorana mass term that couples ν_R to its antiparticle. We may also have a Dirac mass term (Yukawa coupling) between ν_R and the active light neutrino ν_L. The mass terms in the Lagrangian are then[30,29]:

$$-\Delta\mathcal{L} = m_D\bar{\nu_L}\nu_R + \frac{1}{2}m_R^*\nu_R^T C\nu_R + h.c. \tag{23}$$

The first term here is a Yukawa coupling between ν_L and ν_R, and is referred to as a Dirac mass term. For charged fermions, this is the only allowed mass term. The second term is a Majorana mass term that couples ν_R to its antiparticle. This term is allowed, and violates no gauge symmetries, provided that ν_R is chargeless—that is, that ν_R is its own antiparticle. It's evident that ν_L and ν_R should be thought of here as separate fields, with independent mass terms and in fact different masses.

Having written down Equation 23, some magic now results. We can rewrite the mass term in the Lagrangian as

$$-\Delta\mathcal{L} = \frac{1}{2}(\nu_L, \nu_R)\begin{pmatrix} 0 & m_D \\ m_D & m_R \end{pmatrix}\begin{pmatrix} \nu_L \\ \nu_R \end{pmatrix} \tag{24}$$

Equation 24, which is not obviously diagonal, can be diagonalized to yield the physical mass eigenstates. There are two eigenvalues:

$$M_{heavy} \approx m_R, \qquad M_{light} \approx \frac{m_D^2}{m_R} \tag{25}$$

Because ν_R is an electroweak singlet, its mass is not protected by any electroweak symmetry, and the theoretical expectation is that it should be quite massive—possibly at the GUT scale.[29,30] On the other hand, we would naively expect m_D to be similar in size to the Dirac masses of other fermions. If we take $m_R = 10^{15}$ GeV as a typical GUT-scale energy and $m_D = 200$ GeV as representative of the Yukawa coupling of the heaviest charged fermion, we would estimate the largest light neutrino mass to be $M_{light} = (200\text{ GeV})^2/(10^{15}\text{ GeV}) = 0.04$ eV. This value is exactly the right order of magnitude for the neutrino mass inferred by $\sqrt{\Delta m_{32}^2} \approx 0.05$ eV!

Something semi-miraculous has occurred. By introducing a right-handed neutrino with a mass near the GUT scale, as is motivated by GUT models, with a "normal" Dirac coupling m_D to ν_L, we naturally produce very light neutrino masses for ν_L, without having to fine-tune the Dirac mass coupling. The heavier that M_{heavy} is, the lighter that M_{light} becomes, which gives rise to the name "seesaw mechanism" for this method of generating light neutrino masses. Obviously the close numerical correspondence between $\sqrt{\Delta m_{32}^2}$ and M_{light} in the previous paragraph should not be taken too seriously, since we do not know the exact values of m_R or m_D to use in the calculation. The exact mass calculation in fact depends on the details of the physics at higher energy scales. Nonetheless, the seesaw mechanism provides as least a proof of principle as to how very light neutrino masses can be generated without fine-tuning the Dirac mass coupling, while providing a fascinating example of a novel method of generating masses for fundamental particles.

6. Determining The Absolute Mass Scale Of Neutrinos

Although neutrino oscillation measurements demonstrate the existence of neutrino masses, they cannot determine the absolute values of the masses, since oscillations are only sensitive to *differences* in Δm^2. One may make an educated guess of the absolute masses if one assumes that each mass eigenstate is much heavier than the previous one ($m_1 \ll m_2 \ll m_3$), reproducing the pattern of charged fermion masses. In this limit then $m_1 \approx 0$ eV, $m_2 \approx \sqrt{\Delta m^2_{21}} = 0.009$ eV, and $m_3 \approx \sqrt{\Delta m^2_{32}} = 0.05$ eV. For an inverted hierarchy we get $m_3 \approx 0$ eV and $m_1 \approx m_2 \approx 0.05$ eV.

Of course it is not clear whether a strict mass hierarchy should hold. As the mass of the lightest mass eigenstate increases, the three mass eigenstates approach a limit of degenerate masses. The best upper limits on neutrino mass come from cosmology. Massive neutrinos act as a form of hot dark matter that tends to wash out clustering at small angular scales during structure formation, since relativistic dark tends to "stream out" of small density perturbations, but not larger ones. This effect can leave signatures in the cosmic microwave background radiation, in weak lensing surveys, and in large scale structure surveys. While the exact limits obtained depend on which data sets are included in the fits and with what priors, the published limits[31] for the sum of the three mass eigenstates range from $\sim 0.4-0.7$ eV. It would not be far wrong to say that cosmology limits the mass of any individual mass eigenstate to be < 0.2 eV.

Less stringent but more model-independent limits come from measurements of the energies of the products of weak decays. Notable among these are studies of the endpoint of tritium beta decay. If neutrinos have non-zero mass, this mass has the effect of reducing the maximum energy available for the β particle in the decay. Careful measurements of the shape of the β energy spectrum at the endpoint limit the effective mass of a ν_e (the weighted average of its mass eigenstates) to < 2.5 eV at the 95% C.L. The KATRIN collaboration has proposed a next generation tritium endpoint measurement with sensitivity down to 0.2 eV, which might be able to measure or rule out the case of three quasi-degenerate masses.[32]

6.1. *Neutrinoless Double Beta Decay*

In a class by themselves are experiments to measure neutrinoless double beta decay. Normal double beta decay is a doubly weak process in which a nucleus decays by simultaneously emitting two electrons and two $\bar{\nu}_e$. Double beta decay can occur when single beta decay is energetically forbidden, but the $|\Delta Z| = 2$ process is energetically allowed. If neutrinos are Majorana particles (so that a neutrino is its own antiparticle), then instead of emitting two neutrinos, a Feynman diagram exists in which a virtual neutrino is emitted then reabsorbed as an antineutrino. The result is a beta decay in which two electrons but no neutrinos are emitted. Neutrinoless double beta decay violates lepton number by $|\Delta L| = 2$, and differs kinematically from ordinary double beta decay in that the two emitted electrons now contain all of the emitted energy of the transition. The experimental signature of neutrinoless double beta decay is therefore a peak right at the endpoint of the distribution of the sum of the two electrons' energies.

The rate of neutrinoless double beta decay depends on the available phase space and on nuclear matrix elements of the decaying nucleus, but can also be shown to depend an effective neutrino mass by[30]:

$$R \propto \langle m_\nu \rangle^2 = \left| \sum_i^N U_{ei}^2 m_i \right|^2 \tag{26}$$

The effective mass depends on the elements in the first row of the MNS matrix. The mass values enter because they control how much of the "wrong" chirality is mixed into each neutrino, determining the transition of a Majorana neutrino into an antineutrino. (Recall that by the Majorana neutrino hypothesis an antineutrino is just a neutrino of the opposite chirality.)

Positive detection of neutrinoless double beta decay would arguably be the most exciting possible result in neutrino physics, since it would simultaneously establish that neutrinos are Majorana particles, show that lepton number is violated, and settle what the absolute values of the neutrino masses are. This phenomenon has been searched for in many candidate nuclei, but no confirmed detections have been found. The best upper limits comes from the ^{76}Ge system, which limits $\langle m_\nu \rangle < 0.35$ eV at the 90% C.L.[33]

While neutrinoless double beta decay experiments are tremendously difficult due to the rarity of such decays and the existence of various potential backgrounds, many proposals for next generation experiments exist. These proposals rely on much larger exposures (kilograms of material $\times$ years of data-taking), and on sophisticated active or passive means to reject backgrounds. One such proposed experiment is the MAJORANA experiment, whose goal is to collect 2500 kg-years exposure of ^{76}Ge to achieve sensitivity down to $\langle m_\nu \rangle < 0.05$ eV.[34] The EXO experiment will look for neutrinoless double beta decay in 10 tonnes of ^{136}Xe in a liquid or gas TPC, and will attempt to tag the resulting barium ion using spectroscopic techniques to eliminate backgrounds, with a sensitivity goal of ~ 0.01 eV.[35] These and other next-generation experiments, if successful, have some hope of covering the expected range for $\langle m_\nu \rangle$ for degenerate neutrino masses and for the inverted hierarchy. A null result would tell us that neutrinos, if Majorana particles, must have a normal mass hierarchy, but by itself could not distinguish between the possibilities that neutrinos either have a normal mass hierarchy or are simply not Majorana particles. (In principle, though, a determination from long baseline neutrino oscillation experiments that neutrinos have an inverted mass hierarchy, combined with a null result from a sufficiently sensitive neutrinoless double beta decay experiment, could demonstrate that neutrinos are not Majorana particles!)

7. Conclusions

The past decade of neutrino physics has been revolutionary. We have gone from having no confirmed evidence for neutrino physics beyond the Standard Model to the current situation, in which oscillation has been observed in four separate systems, with reasonably precise measurements of two Δm^2 values and two of the four independent mixing parameters in the MNS matrix (assuming that this matrix really is unitary!) Neutrino oscillation is new physics beyond the Standard Model, and requires the addition of new fields and new parameters to the Standard Model. It may even point to the existence of new mechanisms of mass generation.

With the discovery of neutrino mixing, we are now entering an era of precision lepton flavour physics. Just as the study of the CKM matrix has been one of the most important areas in particle physics for decades, studies of lepton flavour may lead to new insights into the origins of flavour, CP violation, and the relationship between quarks and leptons.

In the near future, the experimental emphasis is likely to be on determining θ_{13} through long baseline or reactor neutrino experiments, as well as precisely testing the predictions of the neutrino oscillation model. Longer term we can aspire to looking for CP violation by neutrinos in long baseline oscillation experiments, searching for neutrinoless double beta decay in an attempt to answer the Majorana vs. Dirac neutrino question, and improving limits on neutrino mass from direct kinematic experiments or from cosmology. All the while anomalies like the controversial LSND result remind us that neutrinos may present other surprises that we have not even anticipated yet.

Clearly I'm an optimist about the future of neutrino research. Given that neutrino oscillations are the first new particle physics beyond the Standard Model, the (over?)abundance of new proposals for experiments, and the fact that even today neutrino experiments are probing new physics at a tiny fraction of the cost of large collider experiments, how can I not be an optimist for the future of our field? I hope in the end that the reader agrees with me in this regard.

Acknowledgments

I wish to thank the organizers of the Lake Louise Winter Institute for inviting me to speak at the Institute. John Ng and Maxim Pospelov provided valuable discussions about the theory of Majorana neutrino masses, but should be held blameless for all of my mistakes in presenting it.

References

1. Z. Maki, N. Nakagawa, and S. Sakata, Prog. Theor. Phys. 28, 870 (1962); V. Gribov and B. Pontecorvo, Phys. Lett. **B28** 493 (1969).
2. Danby et al., PRL 9, 36 (1962).
3. S. P. Mikheyev and A. Yu. Smirnov, Sov. J. Nucl. Phys. **42** 913 (1985); L. Wolfenstein, Phys. Rev. **D17** 2369 (1978).
4. M. Apollonio et al., Eur. Phys. J. C27, pp. 331-374 (2003).
5. Y. Fukuda et al., PRL 81 (1999) 1562-1567; Y. Fukuda et al., PRL 82 (1999) 2644-2648.
6. Q. R. Ahmad et al., PRL 87 (2001) 071301; Q. R. Ahmad et al., PRL 89 (2002) 011301; Q. R. Ahmad et al., PRL 87 (2001) 011302; S. N. Ahmed et al., PRL 92 (2004), 181301.
7. B. Aharmin et al., Phys. Rev. C72 (2005), 055502.
8. T. Araki et al., PRL 94 (2005) 081801; K. Eguchi et al., PRL 90 (2003) 021802.
9. E. Aliu et al., PRL 94 (2005), 081802; M. H. Ahn et al., PRL 90 (2003), 041801.
10. John Bahcall, *Neutrino Astrophysics*, Cambridge University Press, 1989.
11. B. T. Cleveland et al., Astrophys. J. 496, 505 (1998).
12. See for example Bahcall et al., Astrophys. J. 621 (2005) L85-L88.
13. J. Hosaka et al., hep-ex/0508053, submitted to PRD; M. B. Smy et al., Phys. Rev D69 (2004) 011104; S. Fukuda et al., Phys. Lett. B539, 179 (2002).
14. V. Gavrin, *Results from the Russian American Gallium Experiment (SAGE)*, VIIIth International Conference on Topics in Astroparticle and Underground Physics (TAUP 2003), Seattle, September 5–9, 2003; J.N. Abdurashitov *et al.*, J. Exp. Theor. Phys. **95**, 181 (2002); C. Cattadori, *Results from Radiochemical Solar Neutrino Experiments*, XXIst International Conference on Neutrino Physics and Astrophysics (Neutrino 2004), Paris, June 14–19, 2004.; E. Bellotti, *The Gallium Neutrino Observatory (GNO)*, VIIIth International Conference on Topics in Astroparticle and Underground Physics (TAUP 2003), Seat-

tle, September 5–9, 2003; M. Altmann *et al.*, Phys. Lett. B **490**, 16 (2000); W. Hampel *et al.*, Phys. Lett. B **447**, 127 (1999).
15. SNO Collaboration, Nucl. Instr. and Meth. A449 (2000), 1972.
16. Y. Ashie et al., Phys.Rev. D71 (2005) 112005; Y. Ashie et al., PRL 93 (2004) 101801.
17. M. Ishitsuka, Proceedings for the XXXIXth Recontres de Moriond on Electroweak Interactions (2004), hep-ex/0406076.
18. For example, see M. Ambrosio et al., Eur. Phys. J. C35 (2004), 323; M. Sanchez et al., Phys. Rev. D68 (2003), 113004; K.S. Hirata et al., Phys. Lett. B280, 146 (1992).
19. A. Aguilar et al., Phys. Rev. D64 (2001) 112007.
20. B. Armbruster et al., Phys. Rev. D65 (2002) 112001.
21. Church et al., Phys. Rev. D66 (2002) 013001.
22. ALEPH, DELPHI, L3, OPAL, and SLD Collaborations, accepted for publication in Physics Reports, CERN-PH-EP/2005-051, SLAC-R-774, hep-ex/0509008.
23. M. Maltoni et al., Nucl. Phys. B 643 (2002) 321.
24. H. Ray, Int. J. Mod. Phys. A20 (2005) 3062; M. H. Shaevitz, prepared for the Fujihara Seminar: Neutrino Mass and the Seesaw Mechanism, KEK, Japan, February, 2004, hep-ex/0407027.
25. Y. Itow et al., "The JHF-Kamioka neutrino project", hep-ex/0106019
26. D. Ayres et al., "Letter of Intent to build an Off-axis Detector to study $\nu_\mu \to \nu_e$ oscillations with the NUMI Neutrino Beam, hep-ex/0210005.
27. T2K Collaboration, "T2K ND280 Conceptual Design Report", T2K Internal Document
28. K. Anderson et al., "White Paper Report on Using Nuclear Reactors to Serach for a value of θ_{13}", January 2004, hep-ex/0402041.
29. For a good general discussion of neutrino masses, see E. Kh. Akhmedov, "Neutrino Physics", Lectures given at the Trieste Summer School in Particle Physics, June 7-July 9, 1999, hep-ph/0001264.
30. S. R. Elliott & P. Vogl, Ann. Rev. Nucl. Part. Sci 52 (2002), 115-151.
31. See for example O. Elgaroy & O. Lahav, New. J. Phys. 7 (2005), 61.
32. KATRIN collaboration, "KATRIN: A next generation tritium beta decay experiment with sub-eV sensitivity for the electron neutrino mass" (2001), hep-ex/0109033.
33. H.V. Klapdor-Kleingrothaus et al., Eur. Phys. J. A12 (2001), 147; C. E. Alseth et al,. Phys. Rev. C 59 (1999), 2108; C. .E. Alseth, Phys. Rev. D 75 (2002), 092007.
34. Majorana collaboration, "White Paper on the Majorana Zero-Neutrino Double-Beta Decay Experiment" (2003), http://majorana.pnl.gov/documents/WhitePaper.pdf
35. M. Danilov et al., Phys. Lett. B 480 (2000), 12-18.

LOW-ENERGY TESTS OF THE STANDARD MODEL

M. POSPELOV

Department of Physics and Astronomy,
University of Victoria, Victoria, BC, V8P 1A1, Canada
E-mail: pospelov@uvic.ca

This is a mini-series of two lectures where I introduce general concepts of low-energy tests of new physics beyond Standard Model, discuss the effective field theory approach, and review in some detail the current status of searches for permanent electric dipole moments of elementary particles and heavy atoms.

1. Lecture I

1.1. *Introduction*

Suppose you are supplied with infinite amount of money and resources, and asked to perform any experiments possible in your power to learn about the physics at extremely short-distance scales, shorter than the inverse of 100 GeV or so. Chances are that you would settle for one of the following of three strategies or for all of them: 1) build a powerful collider and sophisticated detectors; 2) perform some of the super-precise measurements at low energies; 3) learn about the underlying physics using cosmological and astrophysical observations. Modern science follows all these three paths, and each of those have their own limitations and advantages.

Collider physics learns about the short-distance scales by producing new particles/resonances in "real" time and studying the products of their decays. Typically, to probe or discover new physics at scale Λ_{NP} (NP stands for New Physics here), a typical energy of $E_{cm} \sim \Lambda_{NP}$ in the center of mass of colliding particles is required. This is the most direct way of learning about new scales, and arguably the most expensive.

Precision measurements at low energies, on the other hand, access the short-distance scales via a completely different strategy. They provide high precision measurements of energy shifts that receive corrections from *virtual* heavy particles, so that typically

$$\Delta\text{Energy} \sim \frac{m^{n+1}}{\Lambda_{NP}^n},$$

where n is larger than 0, and m is the characteristic energy/momentum scale of particles these experiments works with. Obviously, to access high-energy scale Λ_{NP} in the situation when $m << \Lambda_{NP}$ the high-resolution measurements of ΔE are required.

Finally, the "third way" consists in cosmological/astrophysical observations, that often provide strong probes of the short-distance physics, if it leaves some signatures in the sky. It may manifest itself in the alteration of the primordial spectrum of perturbations, seen through the CMB and lensing experiments, or modify the predictions of the Big Bang Nucleosynthesis, or contribute to the star cooling rate and so on.

These three ways are totally complementary! As time scales and money needed to run collider programs increase, precision measurements become more popular (and more expensive as well). In these lectures, I will focus on the "second way", and cover some generic issues pertinent to precision searches, before turning into more in-depth coverage of CP violation, and its search with the electric dipole moments (EDMs) of neutrons and heavy atoms.

Obviously, not every precise measurement of ΔE matters, but what kind of low energy experiments are sensitive to short distance physics? Let us take the example of two nucleons, proton and neutron and ask the question of what we can learn about weak interactions with them. Quoting the proton mass from the Particle Data Group book [1], $m_p = 938.27200 \pm 0.00004\text{MeV}$, we certainly can see that this is a precise measurement indeed (with even better accuracy for $m_n - m_p$). The last digits ...7200 "know" about the contribution of weak interactions to the proton mass, and could be used for the precision extraction of information about weak interaction, $\delta m_p^{\text{weak}} \sim G_F m_p^3 \sim 10^{-5} m_p$, if not the following obstacle: Strong and electromagnetic interactions inside the proton are not theoretically tractable. On account of this, the masses or hadrons cannot be calculated from the first principles. Perhaps the best hope is Lattice QCD, which has a long way to go, as at the moment it can hardly get beyond even the first digit. Therefore, there is nothing about the fundamental short-distance interactions that can be learned from high-precision measurement of hadron masses. However, it is obvious that in order to learn about weak interactions in $n-p$ system, one should simply look at processes or observables that cannot be induced by strong or electromagnetic forces, such as $n \to p + e + \bar{\nu}$.

This example is an obvious, if not too trivial, illustration to the following general rule: *Necessary condition for the success of any low-energy measurement in providing information about short-distance scales is to minimize/eliminate contributions of larger distance scales due to known physics and/or make sure that they are calculable.*

Before I list the particle physics topics one can study with the low-energy tools, I would like to show that the accuracy of modern measurements at low energy has a similar physics reach with the collider experiments. For example, a typical energy resolution in modern EDM experiments $\Delta\text{Energy} \sim 10^{-6}\text{Hz} \sim 10^{-21}\text{eV}$ translates into limits on EDMs, $|d| < \Delta\text{Energy}/\text{Electric field} \sim 10^{-25}\text{e} \times \text{cm}$. Assuming the 100% violation of CP at the energy scale Λ_{CP} related to some heavy sector that interacts with the SM fields, I get a simple theoretical dimensional estimate, $d \sim 10^{-2} \times (1\ \text{MeV}/\Lambda_{CP}^2)$, that translates into the sensitivity to $\Lambda_{CP} \sim 1$ TeV, which is quite competitive with future LHC measurements. However, again, this estimate assumes maximal breaking of CP, and a cautious person would immediately converge to a conclusion that one needs a good deal of luck in order to see a non-zero effect at low energies.

Below, I list some types of low-energy physics sensitive to $\Lambda_{NP} > 100$ GeV, and I call it "A" list because the physics searched here has very clear particle physics motivations:

(1) Proton decay
(2) neutrinoless double beta decay, neutrino oscillations
(3) CP violation via measuring electric dipole moments
(4) New flavor physics beyond CKM via rare decays
(5) Search for Z' through parity violation in atoms or Moller scattering. Search for $V + A$ currents.
(6) Non-SM contributions to $g - 2$ of muon and to $b \to s\gamma$
(7) Possible WIMPs via direct scattering
(8) ...

The experiments with $g - 2$ of the muon is unique on this list, as it is nearly the only one where electromagnetic and strong contributions can be calculated to a certain degree of accuracy. Even though this list is not meant to be exhaustive, it is important to note that there are not that many things that have been omitted. Indeed, the number of possibilities for probing ultra-short distance physics with low energy tools is limited. Under-scoring this, any new physics occurring at short distances that conserves flavor, CP, baryon and lepton number, etc. can be hardly extracted from the low energy data. A classic example for this would be the Higgs boson of the SM. Only collider physics would be able to probe the SM Higgs sector. In my opinion, we are entering the time when the efficiency of muon $g-2$, $b \to s\gamma$, tests for non-standard CP with K and B mesons as probes of New Physics cannot be further improved. Indeed, all these observables are in the reasonable agreement with the SM, and increase in accuracy on the experimental side would have to be accompanied by the better quality theoretical calculations. The latter is next-to-impossible to achieve because of the limitations set by nonperturbative QCD physics. Precision searches are becoming more and more concentrated on measuring channels where SM predicts strictly or nearly 0, which narrows the "A" list even further.

However, there is a rapid expansion of the number of subjects on what I would call a "B" list, which test less conventional things that find their motivations outside of particle physics.

(1) Change of couplings in time
(2) Lorentz/CPT breaking effects
(3) Tests of gravity at all distances, including sub-mm scales
(4) ...

For some time the interest to these things was purely accuracy-driven: I have such and such precise instrument, what else can I test with it? However, in the past decade or so this field got new motivations, as the attempts to unify particle physics with gravity intensified. It turned out, for example, that the modification of Newton's law might be a direct consequence of gravity becoming as strong as the rest of the interactions at the scale of 1 TeV. But the most solid piece of motivations comes from cosmology, that now has a firm evidence for the so-called dark energy. It is either a cosmological constant, and all observations to date agree with this hypothesis, or an ultra-light scalar field called quintessence. If quintessence interacts with matter, it may mimic such bizarre effects as the change of couplings in time, and create a preferred frame that would appear as the breaking of spatial isotropy.

1.2. *Effective field theory approach*

Effective field theory is the modern working language of particle physics. Its basic ideology follows the Born-Oppenheimer idea or separating a quantum mechanical system into the fast and slow degrees of freedom and "integrating out" fast degrees of freedom. Effective field theory approach has been instrumental in discerning the nature of weak interactions: from Pauli who correctly identified relevant degrees of freedom, to Fermi, who wrote down the first effective four-fermion interaction, to the discovery of parity non-conservation and elucidating the $V-A$ nature of the weak currents. A combination of the low-energy and collider experiments eventually led to the modern $SU(2) \times U(1)$ theory of electroweak interaction. It is important to note that effective theory was used even before the consistent renormalizable theory of electroweak interactions was written down, and the coefficients in front of different effective terms were determined experimentally. At energies much smaller than masses of electroweak bosons W and Z can be integrated out (I drop the quotation marks from now on) and the Standard Model Lagrangian reduces to QED and QCD parts corrected by the effective four-fermion operators

$$L_{SM} \longrightarrow L_{QED} + L_{QCD} + \frac{G_F}{\sqrt{2}} \sum J_\mu^{weak} J_\mu^{weak}. \tag{1}$$

In this formula $J_\mu^{weak} J_\mu^{weak}$ represent the collection of dimension six gauge invariant operators composed from the QED and QCD degrees of freedom. The dimension of an operator is an important concept, that allows to determine its scaling with energy. To remind the obvious, the dimension of the bosonic fields (four space-time dimensions) is 1, fermionic field is 3/2, Lagrange density is 4, and action is dimensionless.

Before I proceed with analyzing possible operators that can come from physics beyond SM, it is useful to recall what the building blocks of the SM are. Besides the gauge bosons that belong to the $SU(3) \times SU(2) \times U(1)$ gauge group, there is a large number of left- and right-handed fermions, and one complex doublet of scalars (the Higgs doublet).

Table 1. Standard Model field charge assignments

Name	Field	$SU(3)$	$SU(2)$	$U(1)$
Left-handed quarks doublets	$Q = (u_L, d_L)$	**3**	**2**	1/6
Right-handed up quarks	$U = u_R$	**3**	**1**	2/3
Right-handed down quarks	$D = d_R$	**3**	**1**	-1/3
Left-handed lepton doublets	$L = (\nu_L, e_L)$	**1**	**2**	-1/2
Right-handed leptons	$E = e_R$	**1**	**1**	-1
Higgs doublet	$H = (h_+, h_0)$	**1**	**2**	1/2

The interaction between gauge and matter fields are completely determined by gauge invariance and parametrized by the three gauge couplings.

The Yukawa interaction between fermions and Higgs bosons is in that sense "arbitrary", and chosen by hand to reproduce the observed fermion mass spectrum and matrix elements of quark mixing angles V_{CKM}. An important feature of the SM is the arbitrariness in the choice of the right-handed mixing matrix, which does not enter any observables. It is important to note that with the exception of QCD theta term, the presence of all interactions have been seen in nature, either directly as in the interactions of gauge bosons and fermions, or indirectly as the Higgs interactions inferred from mass terms.

1.3. *Standard Model as an effective field theory*

Most likely, the Standard Model itself is a low-energy limit of another theory(ies) with some New Physics at a scale $\Lambda > v_{SM}$:

$$L_{UT} \longrightarrow L_{SM} + \sum \frac{1}{\Lambda} O_{SM}^{dim=5} + \sum \frac{1}{\Lambda^2} O_{SM}^{dim=6} + ... \qquad (2)$$

Here "UT" stands for ultimate theory, and $O_{SM}^{dim=5,6..}$ are higher-dimensional operators composed from the SM fields and satisfying $SU(3) \times SU(2) \times U(1)$ gauge invariance. Not that all dimension 4 operators resulting from new physics can be subsumed into the existing terms in L_{SM}. Low-energy precision experiments are working with SM particles and try to "hunt" for these corrections. So far, the best evidence for physics beyond SM is the neutrino oscillations and cosmology. Neutrino oscillations prove that the SM Lagrangian is not an adequate description of the neutrino sector: either new non-renormalizable interactions (effective operators), or new light degrees of freedom (right-handed neutrinos) are required to be introduced. Cosmology also poses a formidable problem: besides already mentioned dark energy, there is definitely a presence of unknown type particles of dark matter. It is absolutely not clear what the dark matter is "made of", but it is all but certain that it is not composed from the SM fields. Unfortunately, this evidence for new physics does not allow to pinpoint exact scale Λ that should enter into our expansion (2). A clear indication on new energy scale contained in Newton's constant is rather discouraging: the Planck mass $M_{Pl} \sim G_N^{-1/2} \sim 10^{19}$ GeV is clearly unattainable for almost all high-precision measurements. Therefore, for now I would like to keep the scale Λ to be completely arbitrary, and analyze the sensitivity of low-energy experiments to it.

What are the general properties of effective operators $O_{SM}^{dim>4}$ that I include in (2)? These operators are

(1) satisfying gauge invariance of the SM
(2) not reducible to simpler operators on equations of motion
(3) reflecting symmetries of New Physics at scale Λ_{UV}
(4) organized in series of increasing dimension
(5) easily classified at a given dimension (5,6...)
(6) implying a cutoff $< \Lambda_{UV}$ when put inside loops
(7) not a good description if Energy$\sim \Lambda_{UV}$.

It is quite remarkable that in the Standard Model there is only one type of dimension 5 operators. Indeed, the combination of the left-handed lepton doublet and the Higgs field allows to form a composite fermionic operator of dimension 5/2 that is the total gauge singlet. Taking the products of two such operators results in the (famous) dimension five interaction

$$O^{dim=5} = \frac{Y_{ij}}{\Lambda_{(5)}}(L_i H)(L_j H)$$

that induces Majorana neutrino masses upon the Higgs condensation,

$$M_{ij} = \frac{Y_{ij} v^2}{\Lambda_{(5)}},$$

and eventually leads to neutrino oscillations. Needless to say that it is the most "economic" approach to neutrino mass problem beyond the SM.

At dimension 6 level we have a plethora of effective operators, and instead of listing them all I just quote some of the notable representatives:

- Baryon-lepton number violating: $(ud)(ue)$,...
- Flavor-violating: $(\bar{d}\gamma_\mu s)(\bar{d}\gamma_\mu s)$,...
- CP-violating: $(\bar{e}i\gamma_5 e)(\bar{q}q)$,...
- Chirality violating: $\overline{(ud)}(\bar{e}\nu)$,...
- Parity violating: $(\bar{e}\gamma_\mu\gamma_5 e)(\bar{q}\gamma_\mu q)$,...
- Violating nothing: $(H^\dagger H)^3$, $f^{abc}F^a_{\mu\nu}F^b_{\nu\lambda}F^c_{\lambda\mu}$, ...

Adding these operators to the Standard Model allows to calculate all relevant observables in terms of *Wilson coefficients* that parametrize these operators. Such type of calculations has been a chief occupation for many theorists working on physics beyond the SM over the years. Skipping the technical steps of calculation of matrix elements, and equating Wilson coefficients with the inverse Λ in appropriate power, I derive the limit of sensitivity for current experiments to different types of physics. The results are compiled in the table below, and all of them are rather order-of-magnitude estimates than exact answers. Given a specific model, these results can be, of course, refined.

Table 2. Sensitivity to Λ_{NP} in the Standard Model

Phenomenon	Limit/Reach in GeV	Source
p decay	$\Lambda_{\not{B}} \gtrsim few \times 10^{15}$	p lifetime
ν oscillations	$\Lambda_R \sim 10^{15} - 10^{16}$	Δm_ν^2
$\Delta F = 2$ meson mixing	$\Lambda_{QF} \gtrsim 10^7 - 10^8$	$\Delta m_{K(B)}$; ϵ_K
EDMs	$\Lambda_{CP} \gtrsim 10^6$	EDMs of n, Tl, Hg
lepton flavor	$\Lambda_{LF} \gtrsim 10^6$	$\mu \to e$ conversion
Parity non-conservation	$\Lambda_{Z'} \gtrsim 10^2 - 10^3$	PNC in Cs; Moller sc.

It is somewhat disappointing that none of these experiments is directly sensitive to the Planck scale physics. It means that *any* Planck scale suppressed interactions are allowed, and low-energy precision observables are not sensitive to the physics of quantum gravity, unless quantum gravity does more esoteric things such as *e.g.* breaking of Lorentz invariance.

1.4. *MSSM as an effective theory*

Supersymmetry (SUSY) has been entertained as a technical solution to the so-called gauge hierarchy problem, a problem of great disparity between electroweak and Planck scales. Among its theoretical successes, minimal supersymmetric standard model (MSSM) can exhibit the following: There are no large corrections to the electroweak precision observables, in agreement with LEP and other tests. (Compare that to a rival theory of technicolor, where corrections are large!) The lightest SUSY particle is stable and could explain the dark matter, provided, of course, that all superpartners can be created/annihilated in pairs (R-parity). Field content of the MSSM allows for better unification of gauge couplings, a theoretically very appealing perspective. Furthermore, SUSY allows for a connection to a consistent theory of quantum gravity, string theory. Admittedly, two last "advantages" are purely theoretical.

It would be unfair not to exhibit also the troubling spots of the MSSM: LEP data squeeze it to a corner where fine-tuning of the Higgs sector is necessary; it has 108 parameters and thus poor predictivity even if the number of parameters is cut drastically as in the so-called constrained MSSM. An arbitrary pattern of superpartner masses has several generic consequences: huge flavor changing neutral currents, huge EDMs that are not seen. Initially designed to solve the problem of hierarchy among dimensionful parameters, supersymmetric models still require tuning of the cosmological constant Λ_{cosm}. Finally, in a gauge-unified SUSY models, proton decay might be a huge problem, unless the model is carefully chosen.

Keeping all this in mind, I proceed with stating that MSSM, if it has anything to do with reality, is obviously not a "theory of everything", but rather again an effective SUSY field theory coming from a more fundamental theoretical structure. Then, it is instructive to disregard for now the subtleties of the soft-breaking, and analyze MSSM *with* higher dimensional corrections. In other words, I represent the superpotential of MSSM as a collection of "standard" terms plus the effective operators:

$$W_{UT} = W_{MSSM} + \text{linear combination of } \; O^{dim>4}_{MSSM}, \tag{3}$$

where the "normal" renormalizable terms are given by

$$W_{MSSM} = Y_u QUH_u - Y_d DQH_d - Y_e ELH_d - \mu H_d H_u \tag{4}$$

and there are additional dim=5 operators:

$$H_uLH_uL;\ UUDE;\ QQQL;\ H_dH_uH_dH_u;\ UQEL;\ UQDQ. \tag{5}$$

A profound difference with the SM is that because of the presence of the numerous scalar fields (squarks, sleptons), there are many more dimension 5 operators than in the SM. Below the scale of the SUSY breaking that theorists believe to be related to the electroweak vev, these operators must reduce to the SM operators. Nothing particularly new happens with the first operator: it would still give neutrino masses, but the second and third operators can give a tremendous enhanced compared to the SM case. These are "dangerous" operators that induce proton decay. In the SM the Wilson coefficients in front of these operators would scale as $1/\Lambda^2_{\not B}$, while in the MSSM this scaling is $\sim (\alpha_s/\pi) \times (\Lambda_{\not B} m_{soft})^{-1}$. Proton decay has a lot greater sensitivity to New Physics in the SUSY world! In particular, all $1/M_{Pl}$-sized operators that mediate proton decay *are experimentally excluded* if the weak-scale SUSY is correct. Similar type of enhancement in sensitivity can be demonstrated for lepton flavor violation and EDMs [2].

Table 3. Sensitivity to Λ_{NP} in the MSSM

Phenomenon	Limit/Reach in GeV	Source
p decay	$\Lambda_{\not B} \gtrsim 10^{24}$	SuperK
$\Delta F = 2$ meson mixing	$\Lambda_{QF} \gtrsim 10^7 - 10^8$	$\Delta m_{K(B)}$; ϵ_K
EDMs	$\Lambda_{CP} \gtrsim 10^8 - 10^9$	EDMs of n, Tl, Hg
lepton flavor	$\Lambda_{LF} \gtrsim 10^8$	$\mu \to e$ conversion
Parity non-conservation	$\Lambda_{Z'} \gtrsim 10^2 - 10^3$	PNC in Cs; Moller sc.

2. Lecture II

2.1. *Different types of EDMs*

Searches for flavor-diagonal CP-violation, while insensitive to the CKM phase, are essentially "background" free, probes of new physics. Indeed, SM predicts tiny EDMs, at least five orders of magnitude below modern experimental capabilities. EDMs, through continuous experimental development since the work of Purcell and Ramsey [3], remain our most sensitive probes of this sector. All existing searches have failed to detect any intrinsic EDM, and indeed the precision to which EDMs are now known to vanish is remarkable, and sufficient to render them some of the most important precision tests of the Standard Model. In this more general context, the strong CP problem, associated with the tuning of θ_{QCD}, becomes just the most highly tuned example among many possible CP-odd operators which could contribute to the observable EDMs of nucleons, leptons, atoms and molecules. Anticipating the presence of such CP-odd sources is not without motivation. Indeed, one of the strongest motivations comes from cosmology, where the success of the inflationary scenario, together with the observed cosmological dominance of baryons over antibaryons, suggests that a non-zero baryon number was generated dynamically in the early Universe. According to the Sakharov criteria [4], this requires a source of CP-violation, and within the Standard Model, neither the Kobayashi-Maskawa phase nor the θ–term can create conditions that would lead to the generation of an appreciable net baryon number. This strongly suggests the presence of another, yet to be discovered, source of CP violation in nature. Although exceptions exist, e.g. the leptogenesis scenario, EDMs generally provide a highly sensitive diagnostic for these new CP-odd sources.

The level of experimental precision achieved in EDM searches has improved dramatically since the early work of Purcell and Ramsey, and has been broadened to many atomic and nuclear quantities. Indeed, following significant progress throughout the past decade, the EDMs of the neutron [5], and of several heavy atoms and molecules [6,7,8,9,10,11] have been measured to vanish to remarkably high precision. From the present standpoint, it is convenient to classify the EDM searches into three main categories, distinguished by the dominant physics which would induce the EDM, at least within a generic class of models. These categories are: the EDMs of paramagnetic atoms and molecules; the EDMs of diamagnetic atoms; and the EDMs of hadrons, and nucleons in particular. For these three categories, the experiments that currently champion the best bounds on CP-violating parameters are the atomic EDMs of thallium and mercury and that of the neutron, as listed in Table 4.

Table 4. Current constraints within three representative classes of EDMs

Class	EDM	Current Bound
Paramagnetic	^{205}Tl	$\vert d_{\mathrm{Tl}}\vert < 9 \times 10^{-25} e\,\mathrm{cm}$ (90% C.L.) [6]
Diamagnetic	^{199}Hg	$\vert d_{\mathrm{Hg}}\vert < 2 \times 10^{-28} e\,\mathrm{cm}$ (95% C.L.) [7]
Nucleon	n	$\vert d_n\vert < 6 \times 10^{-26} e\,\mathrm{cm}$ (90% C.L.) [5]

The upper limits on EDMs obtained in these experiments can be translated into tight constraints on the CP-violating physics at and above the electroweak scale, with each category of EDM primarily sensitive to different CP-odd sources. For example, the neutron EDM can be induced by CP violation in the quark sector, while paramagnetic EDMs generally result from CP violating sources that induce the electron EDM. Despite the apparent difference in the actual numbers in Table 4, all three limits on d_n, d_{Tl}, and d_{Hg} actually have comparable sensitivity to fundamental CP violation, e.g. superpartner masses and CP-violating phases, and thus play complementary roles in constraining fundamental CP-odd sources. This fact can be explained by the way the so-called Schiff screening theorem [12] is violated in paramagnetic and diamagnetic atoms. The Schiff theorem essentially amounts to the statement that, in the nonrelativistic limit and treating the nucleus as point-like, the atomic EDMs will vanish due to screening of the applied electric field within a neutral atom. The paramagnetic and diamagnetic EDMs result from violations of this theorem due respectively to relativistic and finite-size effects, and in heavy atoms such violation is maximized. For heavy paramagnetic atoms, *i.e.* atoms with non-zero electron angular momentum, relativistic effects actually result in a net enhancement of the atomic EDM over the electron EDM. For diamagnetic species, the Schiff screening is violated due to the finite size of the nucleus, but this is a weaker effect and the induced EDM of the atom is suppressed relative to the EDM of the nucleus itself. These factors equilibrate the sensitivities of the various experimental constraints in Table 4 to more fundamental sources of CP violation.

2.2. *Anatomy of EDMs*

The effective CP-odd flavor-diagonal Lagrangian normalized at 1 GeV, which is taken to be the lowest perturbative quark/gluon scale, plays a special role in EDM calculations. At this scale, all particles other than the u, d and s quark fields, gluons, photons, muons and electrons can be considered heavy, and thus integrated out. As a result, one can construct an effective Lagrangian by listing all possible CP-odd operators in order of increasing dimension,

$$L_{\text{eff}} = L_{\text{dim}=4} + L_{\text{dim}=5} + L_{\text{dim}=6} + \cdots . \tag{6}$$

There is only one operator at dimension 4, the QCD theta term,

$$L_{\text{dim}=4} = \frac{g_s^2}{32\pi^2}\,\bar{\theta} G^a_{\mu\nu}\widetilde{G}^{\mu\nu,a}, \tag{7}$$

where on account of the axial U(1) anomaly, the physical value of θ – denoted $\bar{\theta}$ – also includes the overall phase of the quark mass matrix,

$$\bar{\theta} = \theta + \text{Arg}\,\text{Det}M_q. \tag{8}$$

The anomaly can be used to shuffle contributions between the θ-term and imaginary quark masses, but only the combination $\bar{\theta}$ is physical and I choose to place it in front of $G\tilde{G}$ taking $\text{Det}M_q$ to be real. In further discussion, I assume that the Peccei-Quinn symmetry removes the theta term at the price of introducing an axion.

At the dimension five level, there are (naively) several operators: EDMs of light quarks and leptons and color electric dipole moments of the light quarks,

$$L_{\text{dim}=5} = -\frac{i}{2}\sum_{i=u,d,s,e,\mu} d_i\,\overline{\psi}_i(F\sigma)\gamma_5\psi_i - \frac{i}{2}\sum_{i=u,d,s}\widetilde{d}_i\,\overline{\psi}_i g_s(G\sigma)\gamma_5\psi_i, \tag{9}$$

where $(F\sigma)$ and $(G\sigma)$ are a shorthand notation for $F_{\mu\nu}\sigma^{\mu\nu}$ and $G^a_{\mu\nu}t^a\sigma^{\mu\nu}$.

In fact, in most models these operators are really dimension-six operators in disguise. The reason is that, if we proceed in energy above the electroweak scale and assume the system restores SU(2)$\times U(1)$ as in the Standard Model, gauge invariance ensures that these operators must include a Higgs field insertion. In practice, this feature can also be understood in most models by going to a chiral basis, where we see that these operators connect left- and right-handed fermions, and thus require a chirality flip. This is usually supplied by an insertion of the fermion mass, i.e. $d_f \sim m_f/M^2$, again implying that the operators are effectively of dimension six.

Consequently, for consistency I should also proceed at least to dimension six where we encounter the CP-odd three-gluon Weinberg operator and a host of possible four-fermion interactions, $(\bar{\psi}_i\Gamma\psi_i)(\bar{\psi}_j i\Gamma\gamma_5\psi_j)$:

$$L_{\rm dim=6} = \frac{1}{3}w\ f^{abc}G^a_{\mu\nu}\widetilde{G}^{\nu\beta,b}G_\beta{}^{\mu,c} + \sum_{i,j} C_{ij}\ (\bar{\psi}_i\psi_i)(\bar{\psi}_j i\gamma_5\psi_j) + \cdots \tag{10}$$

In this formula, the operators with C_{ij} are summed over all light fermions. Going once again to a chiral basis, I can argue as above that the four-fermion operators, which require two chirality flips, are in most models effectively of dimension eight. Nonetheless, in certain models they may be non-negligible.

A lot of hard work has been invested over the years into the calculation of the EDM observables in terms of the Wilson coefficients. Here I quote the result of the QCD sum rule calculation for neutron EDM:

$$d_n(d_q, \tilde{d}_q) = (1.4 \pm 0.6)(d_d - 0.25 d_u) + (1.1 \pm 0.5)e(\tilde{d}_d + 0.5\tilde{d}_u). \tag{11}$$

Diamagnetic atoms are the most complex as they involve three types of calculations: atomic, nuclear and QCD. The current compilation of the mercury EDM as a function of dimension 5 operators is as follows [13]:

$$\begin{aligned} d_{\rm Hg} &= 7\times 10^{-3}\, e\,(\tilde{d}_u - \tilde{d}_d) + 10^{-2}\, d_e \\ &\quad -1.4\times 10^{-5}\, e\,{\rm GeV}^2 \end{aligned} \tag{12}$$

Finally, the easiest system to handle is the EDM of a paramagnetic atom. Its relation with the electron EDM and the electron-nucleon CP-violating interaction constant is given by

$$d_{\rm Tl} = -585 d_e - e\ 43\ {\rm GeV} C_S. \tag{13}$$

2.3. *EDMs and SUSY*

Here we would just like to demonstrate the main point implied by the SUSY EDM calculations in a simplistic model in which *all* soft-breaking parameters are taken equal to a unique scale $M_{\rm SUSY}$ at the electroweak scale, i.e. $M_i = m_Q = m_D = \cdots = |\mu| = |A_i| = M_{\rm SUSY}$. Even with this drastic reduction of SUSY parameter space, there is room for two CP-violating phases, $A = |A|\exp(i\theta_A)$ and $\mu = |\mu|\exp(i\theta_\mu)$, complex phases of the so-called trilinear soft-breaking parameter A, and the superpotential mass parameter μ. In a more humane language of actual particles, these phases would enter into the vertices of SM fermions with gauginos and sfermions. Working at leading order in $v^2/M^2_{\rm SUSY}$, we can then present the following compact results for all dimension 5 operators (with $q = d, u$),

$$\frac{d_e}{e\kappa_e} = \frac{g_1^2}{12}\sin\theta_A + \left(\frac{5g_2^2}{24} + \frac{g_1^2}{24}\right)\sin\theta_\mu \tan\beta,$$

$$\frac{d_q}{e_q\kappa_q} = \frac{2g_3^2}{9}\Big(\sin\theta_\mu[\tan\beta]^{\pm 1} - \sin\theta_A\Big) + O(g_2^2, g_1^2), \qquad (14)$$

$$\frac{\tilde{d}_q}{\kappa_q} = \frac{5g_3^2}{18}\Big(\sin\theta_\mu[\tan\beta]^{\pm 1} - \sin\theta_A\Big) + O(g_2^2, g_1^2).$$

The notation $[\tan\beta]^{\pm 1}$ implies that one uses the plus(minus) sign for $d(u)$ quarks, g_i are the gauge couplings, and $e_u = 2e/3$, $e_d = -e/3$. As usual, $\tan\beta$ stands for the ratio of the two Higgs vevs. For the quarks we quoted the explicit result only for the gluino-squark diagram that dominates in this limit. All these contributions to d_i are proportional to κ_i, a universal combination corresponding to the generic dipole size,

$$\kappa_i = \frac{m_i}{16\pi^2 M_{\rm SUSY}^2} = 1.3\times 10^{-25}{\rm cm}\times\frac{m_i}{1{\rm MeV}}\left(\frac{1{\rm TeV}}{M_{\rm SUSY}}\right)^2, \qquad (15)$$

which varies by a factor of a few for $i = e, d, u$ depending on the value of the fermion mass. The perturbative nature of the MSSM provides a loop suppression factor in (15).

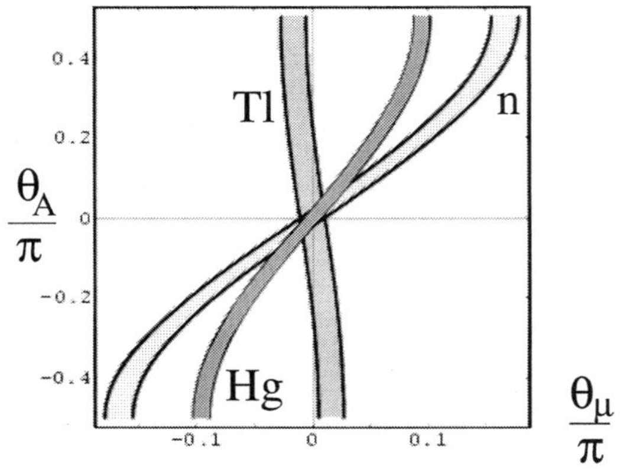

Figure 1. The combination of the three most sensitive EDM constraints, d_n, $d_{\rm Tl}$ and $d_{\rm Hg}$, for $M_{\rm SUSY} = 500$ GeV, and $\tan\beta = 3$. The region allowed by EDM constraints is at the intersection of all three bands around $\theta_A = \theta_\mu = 0$.

Plugging these results into the expressions for d_n, $d_{\rm Tl}$ and $d_{\rm Hg}$ and comparing them to the current experimental bounds, we arrive at a set of constraints on θ_A and θ_μ depending on $M_{\rm SUSY}$ and $\tan\beta$. In Figure 1, I plot these constraints in the (θ_μ,θ_A)–plane for $M_{\rm SUSY} = 500$ GeV and $\tan\beta = 3$. The region allowed by the EDM constraints is at the intersection of all three bands around $\theta_A = \theta_\mu = 0$. One can observe that the combination of all three constraints strengthens the bounds on the phases, and protects against the accidental cancellation of large phases that can occur within one particular observable. The uncertainty in the QCD and the nuclear calculations may affect the width of the $d_{\rm Hg}$ constraint band, but do not change its slope on the (θ_μ, θ_A) plane.

Figure 1 exemplifies the so-called SUSY *CP* problem: either the *CP*-violating phases are small, or the scale of the soft-breaking masses is significantly larger than 1TeV, or schematically,

$$\delta_{CP} \times \left(\frac{1\text{TeV}}{M_{\rm SUSY}}\right)^2 < 1. \tag{16}$$

The need to provide a plausible explanation to the SUSY *CP* problem has drawn a lot of theoretical attention.

- *Heavy superpartners.* If the masses of the supersymmetric partners exhibit certain hierarchy patterns the SUSY *CP* problem can be alleviated. One of the more actively discussed possibilities is an inverted hierarchy among the slepton and squark masses, i.e. with the squarks of the first two generations being much heavier than the stops, sbottoms and staus. This does not mean, however, that the EDMs in such models become comparable to $d_i(\delta_{\rm KM})$. Indeed the two-loop contributions to d_i and w involving the third generation sfermions are not small in this framework, and indeed are at (or sometimes above) the level of current experimental sensitivity.
- *Small phases.* A rather obvious possibility for suppressing EDMs is the assumption of an exact (or approximate) *CP* symmetry of the soft-breaking sector. This is essentially a "model-building" option, and various ways of avoiding the SUSY flavour and *CP* problem in this way have been suggested in the past fifteen years [14,15,16]. The idea of using low-energy supersymmetry breaking looks especially appealing, as it can also help in constructing an axion-less solution to the strong *CP* problem [17].
- *Accidental cancellations.* Another possibility entertained in recent years [18,19] is the partial or complete cancellation between the contributions of several *CP*-odd sources to physical observables, thus allowing for $\delta_{\rm CP} \sim O(1)$ with $M_{\rm SUSY} < 1$ TeV. Since the num-

ber of potential CP-odd phases is large, and the superpartner mass spectrum is clearly unknown, one cannot exclude this possibility in principle. However, as we illustrated in Figure 8, d_n, $d_{\rm Tl}$ and $d_{\rm Hg}$ depend on different combinations of phases, and the possibility of such a cancellation looks improbable. A more thorough exploration of the MSSM parameter space in search of acceptable solutions that pass the EDM constraints was performed in [20,21], and in the absence of additional parameter tuning did not identify any significant regions of cancellation.

- *No electroweak scale supersymmetry.* Of course, there is always the possibility that other mechanisms (or no easily identifiable mechanism at all) lie behind the gauge hierarchy problem and the SM is a good effective theory valid up to energy scales *much larger* than 1 TeV. In this case there is no SUSY CP problem by definition.

3. Conclusions

I have given an overview of principles that form the foundation of the low-energy probes of ultra-short distance physics. Introducing the notion of effective field theory, I described how the neutrino mass matrix can be linked with the effective dimension five operator in the Standard Model, and showed that the Planck-scale suppressed physics cannot be probed in the SM. This situation changes drastically in supersymmetric models, where many more of dimension five operators can exist. Arbitrary baryon and lepton number violating operators suppressed by the Planck scale is no longer allowed in the theory.

Next, I described in very general terms how the measurements of different types of EDMs constrain New Physics. After giving a synopsis of main EDM formulae, I specialized them to the case of the SUSY models with the breaking of CP in the soft-breaking sector. An absence of the non-zero signal at high level of precision creates a so-called SUSY CP problem, which is a serious challenge for the MSSM phenomenology.

To illustrate the reach of the future EDM experiments, I allow myself to show a slice of the constrained MSSM parameter space, with m_0 and $m_{1/2}$ parameters scanned up to 1 TeV [22]. In practice, due to the renormalization group evolution of SUSY masses, this scan covers squarks as heavy as 3 TeV. The lines of constant ratio of predicted EDMs to the expected experimental sensitivity are plotted. With the phase on order $\theta_\mu \sim 0.15$, the EDMs can reach even those corners of the SUSY parameter space that are not directly accessible at colliders. It also means, that should the SUSY particles be found at the LHC, the EDMs would provide a sensitive probes of the CP properties of their couplings.

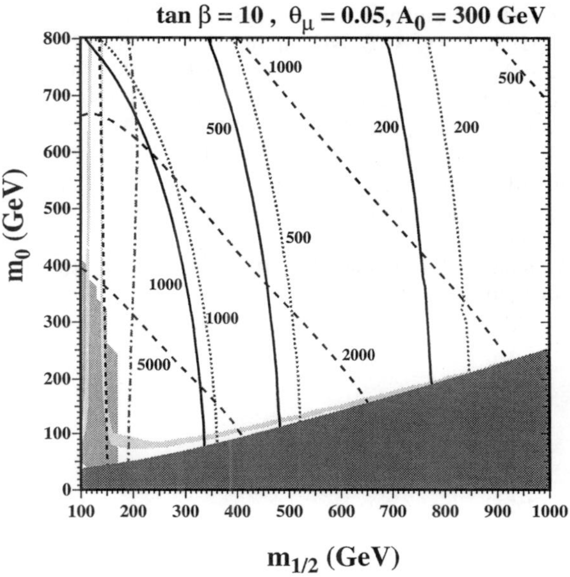

Figure 2. Scan of the $m_0 - m_{1/2}$ parameter space, with future EDM sensitivity. Deuteron (solid), neutron (dotted), and electron (dashed) EDM contours are plotted, weighted with the expected accuracy of 2×10^{-27}, 1×10^{-27}, and 3×10^{-29} $e$cm.

References

1. S. Eidelman *et al.* [Particle Data Group], Phys. Lett. B **592**, 1 (2004).
2. M. Pospelov, A. Ritz and Y. Santoso, Phys. Rev. Lett. **96**, 091801 (2006).
3. E. M. Purcell and N. F. Ramsey, Phys. Rev. **78**, 807 (1950).
4. A. D. Sakharov, Pisma Zh. Eksp. Teor. Fiz. **5**, 32 (1967) [JETP Lett. **5** 24 (1967 SOPUA,34,392-393.1991 UFNAA,161,61-64.1991)].
5. P. G. Harris *et al.*, Phys. Rev. Lett. **82**, 904 (1999).
6. B. C. Regan *et al.*, Phys. Rev. Lett. **88**, 071805 (2002).
7. M. V. Romalis, W. C. Griffith and E. N. Fortson, Phys. Rev. Lett. **86**, 2505 (2001).
8. D. Cho, K. Sangster, E.A. Hinds, Phys. Rev. Lett. **63**, 2559 (1989).
9. M. A. Rosenberry and T. E. Chupp, Phys. Rev. Lett. **86**, 22 (2001).
10. S.A. Murthy *at al.*, Phys. Rev. Lett. **63**, 965 (1989).

11. J. Hudson, B. E. Sauer, M. R. Tarbutt, and E. A. Hinds, Phys. Rev. Lett. **89**, 023003 (2002).
12. L.I. Schiff, Phys. Rev. **132**, 2194 (1963).
13. M. Pospelov and A. Ritz, Annals Phys. **318**, 119 (2005)
14. M. Dine, A. E. Nelson and Y. Shirman, Phys. Rev. D **51** (1995) 1362.
15. L. Randall and R. Sundrum, Nucl. Phys. B **557**, 79 (1999).
16. Z. Chacko, M. A. Luty, A. E. Nelson and E. Ponton, JHEP **0001** (2000) 003.
17. G. Hiller and M. Schmaltz, Phys. Lett. B **514**, 263 (2001).
18. T. Ibrahim and P. Nath, Phys. Lett. B **418**, 98 (1998).
19. T. Falk and K. A. Olive, Phys. Lett. B **375** (1996) 196; M. Brhlik, G. J. Good and G. L. Kane, Phys. Rev. D **59** (1999) 115004; M. Brhlik, L. L. Everett, G. L. Kane and J. Lykken, Phys. Rev. Lett. **83** (1999) 2124.
20. V. D. Barger, T. Falk, T. Han, J. Jiang, T. Li and T. Plehn, Phys. Rev. D **64** (2001) 056007.
21. S. Abel, S. Khalil and O. Lebedev, Phys. Rev. Lett. **86** (2001) 5850; Nucl. Phys. B **606** (2001) 151.
22. K. A. Olive, M. Pospelov, A. Ritz and Y. Santoso, Phys. Rev. D **72**, 075001 (2005).

CHARGED PARTICLE MULTIPLICITIES IN ULTRA-RELATIVISTIC AU+AU AND CU+CU COLLISIONS

B.B.BACK* for the PHOBOS Collaboration

B.ALVER[4], B.B.BACK[1], M.D.BAKER[2], M.BALLINTIJN[4], D.S.BARTON[2], R.R.BETTS[6], A.A.BICKLEY[7],R.BINDEL[7], W.BUSZA[4], A.CARROLL[2], Z.CHAI[2], V.CHETLURU[6], M.P.DECOWSKI[4], E.GARCIA[6], T.GBUREK[3], N.GEORGE[2], K.GULBRANDSEN[4], C.HALLIWELL[6], J.HAMBLEN[8], I.HARNARINE[6], M.HAUER[2], C.HENDERSON[4], D.J.HOFMAN[6], R.S.HOLLIS[6], R.HOLYNSKI[3], B.HOLZMAN[2], A.IORDANOVA[6], E.JOHNSON[8], J.L.KANE[4], N.KHAN[8], P.KULINICH[4], C.M.KUO[5], W.LI[4], W.T.LIN[5], C.LOIZIDES[4], S.MANLY[8], A.C.MIGNEREY[7], R.NOUICER[2], A.OLSZEWSKI[3], R.PAK[2], C.REED[4], E.RICHARDSON[7], C.ROLAND[4], G.ROLAND[4], J.SAGERER[6], H.SEALS[2], I.SEDYKH[2], C.E.SMITH[6], M.A.STANKIEWICZ[2], P.STEINBERG[2], G.S.F.STEPHANS[4], A.SUKHANOV[2], A.SZOSTAK[2], M.B.TONJES[7], A.TRZUPEK[3], C.VALE[4], G.J.VANNIEUWENHUIZEN[4], S.S.VAURYNOVICH[4], R.VERDIER[4], G.I.VERES[4], P.WALTERS[8], E.WENGER[4], D.WILLHELM[7], F.L.H.WOLFS[8], B.WOSIEK[3], K.WOZNIAK[3], S.WYNGAARDT[2], B.WYSLOUCH[4]

[1]*Argonne National Laboratory, Argonne, IL 60439, USA*
[2]*Brookhaven National Laboratory, Upton, NY 11973, USA*
[3]*Institute of Nuclear Physics PAN, Krakow, Poland*
[4]*Massachusetts Institute of Technology, Cambridge, MA 02139, USA*
[5]*National Central University, Chung-Li, Taiwan*
[6]*University of Illinois at Chicago, Chicago, IL 60607, USA*
[7]*University of Maryland, College Park, MD 20742, USA*
[8]*University of Rochester, Rochester, NY 14627, USA*

The PHOBOS collaboration has carried out a systematic study of charged particle multiplicities in Cu+Cu and Au+Au collisions at the Relativistic Heavy-Ion Collider (RHIC) at Brookhaven National Laboratory. A unique feature of the PHOBOS detector is its ability to measure charged particles over a very wide angular range from 0.5° to 179.5°corresponding to $|\eta| <5.4$. The general features of the charged particle multiplicity distributions as a function of pseudo-rapidity, collision energy and centrality, as well as system size, are discussed.

*E-mail: back@anl.gov

1. Introduction

The multiplicity of charged particles is a central observable in relativistic heavy-ion collisions, which provides information about the properties of the hot and dense fireball formed in such collisions. More detailed information about the system is embedded in identified particle spectra, but in most cases such spectra are not available over the full (pseudo)-rapidity range. The information that can be obtained from the non-identified charged particle measurements therefore provide unique opportunity to study the bulk properties of the system[1].

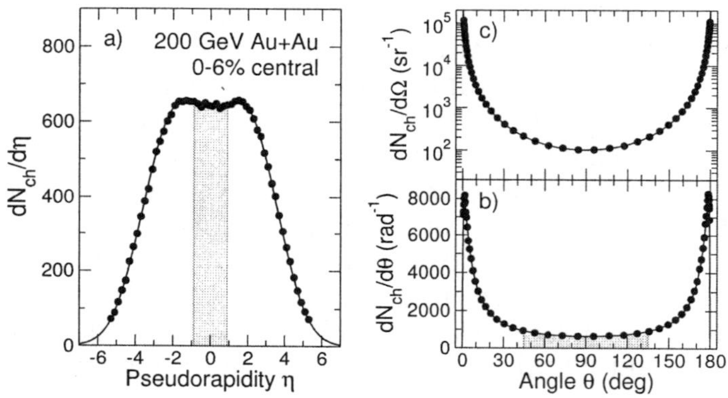

Figure 1. Illustration of the charged particle distribution for 0-6% central 200 GeV $Au + Au$ collisions. Panel a: The pseudorapidity density, $dN/d\eta$ is shown as a function of η. Panel b: The corresponding angular distribution $dN/d\theta$ is shown as a function of the angle θ relative to the beam axis. Panel c: same as for panel (b) but here $dN/d\Omega$ is shown. The shaded regions in panels (a) and (b) indicate the angular region where the transverse momentum p_t exceeds the longitudinal momentum $p_{||}$.

Figure 1a shows the pseudo-rapidity distribution of charged particles emitted in $Au + Au$ collisions at $\sqrt{s_{NN}}$=200 GeV measured by the PHOBOS collaboration[2]. The typical mid-rapidity plateau extending over about four units of pseudorapidity, leads to a steep fall-off on either side. The solid curve represents a fit to the data using three Gaussians, one centered at η=0 and two located symmetrically on either side of mid-rapidity. This fit serves to estimate the total charged particle multiplicity by extrapolation into the unmeasured region. In this case we find that over 99% of the charged particle yield falls into the PHOBOS acceptance region. The grey band encompasses the region, for which the transverse momentum exceeds the longitudinal momentum, $p_t > p_{||}$, which is most likely to reveal the signatures of the hot and dense fireball formed in the collision. Figs. 1b,c show, however, that the charged particle distribution is actually very forward-backward peaked in real space coordinates and that the mid-rapidity plateau corresponds to a minimum in the $dN/d\Omega$ distribution.

2. Multiplicity at mid-rapidity

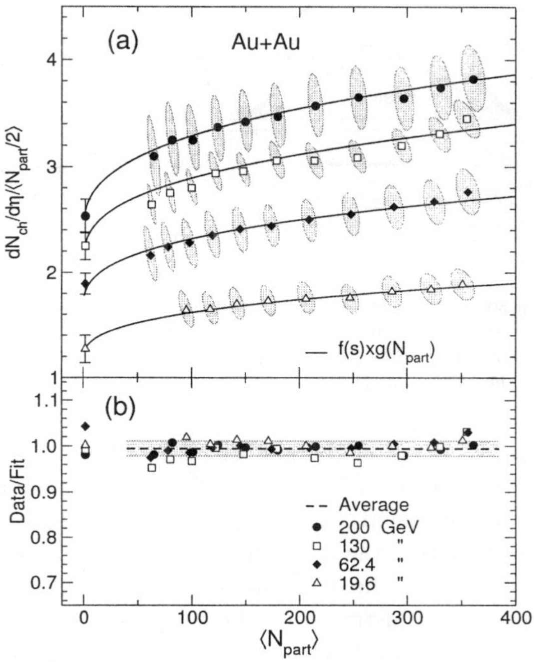

Figure 2. Illustration of charged particle distribution factorization for Au+Au collisions. Panel a: The pseudorapidity density, $dN/d\eta$ at $|\eta| < 1$ is shown as a function of N_{part} for $Au + Au$ and $pp/p\overline{p}$ collisions. Panel b: The ratio of the data to the fit function $f(s) \times g(N_{part})$ is seen to lie in a narrow band around unity.

The centrality dependence of the mid-rapidity multiplicity may be expressed in terms of the number of nucleons, N_{part}, of the initial Au nuclei that participate in the collision. This variable is determined from the energy deposited in two scintillator paddle counters located at $3 < |\eta| < 4.5$ using Glauber model simulations[3]. In Fig. 2a the measured charged-particle multiplicity at mid-rapidity[4,5,6], normalized by $N_{part}/2$ in order to facilitate a comparison to elemental nucleon-nucleon collisions, is shown as a

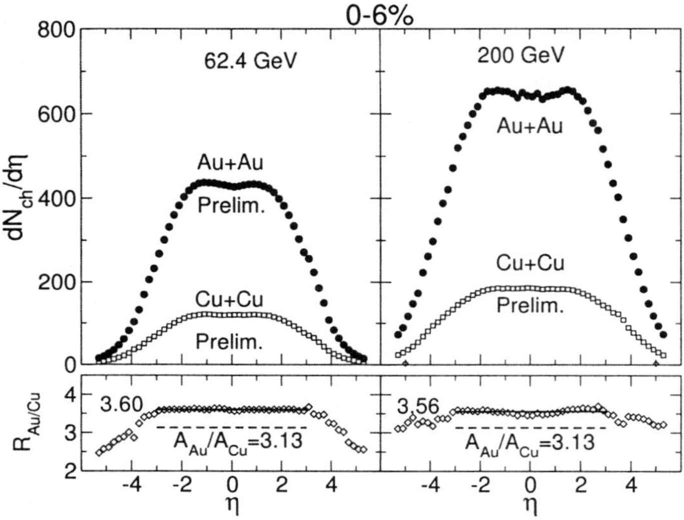

Figure 3. Comparison of charged particle multiplicity for $Au+Au$ and $Cu+Cu$ collisions at 62.4 GeV (left panels) and 200 GeV (right panels). The ratios of charged particle production density between $Au + Au$ and $Cu + Cu$ collisions are given as a function of pseudo-rapidity in the lower panels.

function of N_{part} for collision energies of $\sqrt{s_{NN}}$=19.6, 62.4, 130, and 200 GeV. One observes that $dN_{ch}/d\eta/\langle N_{part}/2\rangle$ increases smoothly with both the centrality of the collision, expressed in terms of $\langle N_{part}\rangle$, and collision energy. For all centralities, the multiplicities exceed those measured in pp[7] and $p\overline{p}$[8] collisions. In fact, the dependence on centrality and energy factorize to high accuracy as illustrated by the solid curves, which are given as the product two functions, *i.e.* $dN_{ch}/d\eta/\langle N_{part}/2\rangle = f(s) \times g(N_{part})$. The degree to which the factorization is valid is illustrated in Fig. 2b, where the ratio of the data to the fit function is seen to deviate from unity by less than about 1% (grey band). This factorization is surprising considering the significant changes in particle production mechanisms that are expected over the energy range of these data.

3. System size dependence

An alternative method of varying the size of the fireball is to collide ions of different size. Such a comparison has been carried out for central $Au + Au$ and $Cu+Cu$ collisions as illustrated in Fig. 3 at $\sqrt{s_{NN}}$=62.4 and 200 GeV. It is evident that similar shapes of the pseudorapidity distributions are found for both collision systems. Indeed, the lower panels of Fig. 3 show that the ratio of $dN_{ch}/d\eta$ between $Au + Au$ and $Cu + Cu$ collisions is essentially constant over the range $|\eta| < 3$ at a value of 3.60 for 62.4 GeV and 3.56 for 200 GeV. These values are slightly larger than the ratio of the total number of nucleons available in the initial state, namely A_{Au}/A_{Cu}=3.13. Outside of $|\eta| < 3$ the ratio falls off (steeply at 62.4 GeV and less so at 200 GeV) indicating that the high-$|\eta|$ tails fall off more steeply for $Au + Au$ than for $Cu + Cu$.

4. Summary

The PHOBOS experiment has a unique capability to measure nearly the full charged particle distribution as a function of pseudorapidity. At mid-rapidity the charged particle multiplicities have been measured for $Au + Au$ collisions as a function of both energy and centrality of the collisions. A surprising result is that the dependence on these two variables can be factorized to high accuracy ($\sim$ 1%). A comparison of the pseudorapidity distributions for central $Au + Au$ and $Cu + Cu$ collisions shows that these exhibit the same shape over six units of pseudorapidity, but they differ somewhat in the tails of the distributions at high values of $|\eta|$.

Acknowledgments

This work was partially supported by U.S. DOE grants DE-AC02-98CH10886, DE-FG02-93ER40802, DE-FC02-94ER40818, DE-FG02-94ER40865, DE-FG02-99ER41099, and W-31-109-ENG-38, by U.S. NSF grants 9603486, 0072204, and 0245011, by Polish KBN grant 1-P03B-062-27(2004-2007), by NSC of Taiwan Contract NSC 89-2112-M-008-024, and by Hungarian OTKA grant (F 049823).

References

1. B.B.Back, *Phys. Rev.* **C72**, 064906 (2005)
2. B.B.Back *et al.*, *Phys. Rev. Lett.* **91**, 52303 (2003)
3. R.S.Hollis *et al.*, *J.Phys Conf. Ser.* **5**, 46 (2005)
4. B.B.Back *et al.*, *Phys. Rev.* **C70**, 021902 (2004)
5. B.B.Back *et al.*, *Phys. Rev.***C65**, 061901 (2003)
6. B.B.Back *et al.*, *Submitted to Phys. Rev.* **C**, nucl-ex/0509034
7. W.Thome *et al.*, *Nucl. Phys.* **B129**, 365 (1977)
8. G.J.Alner *et al.*, *Z. Phys.***C33**, 1 (1989); R.E.Ansorge *et al.*, *ibid.*, **43** 357 (1989)

STANDARD MODEL PHYSICS AT CMS

S. BEAUCERON*
ON BEHALF OF CMS COLLABORATION.
DAPNIA/SPP,
CEA Saclay,
91191 Gif-sur-Yvette, FRANCE
E-mail: stephanie.beauceron@cea.fr

The LHC will provide proton-proton collisions at energies of about 14 TeV and at luminosity up to 10^{34} cm^{-2}s^{-1}. It will be possible to observe new physics but also to have precise measurements of Standard Model phenomena. The perspectives for top quark physics, electro-weak physics, gauge boson couplings and B physics with CMS are briefly discussed here. With the high rate and energy, new processes from the Standard Model, as for example single top, will become visible.

1. Introduction

The LHC is expected to provide proton-proton collisions at a centre-of-mass of 14 TeV, at an instantaneous luminosity ranging from 2×10^{33} cm^{-2}s^{-1} (low luminosity phase) to 10^{34} cm^{-2}s^{-1} (high luminosity phase). This frontier machine will open a new window on physics beyond the Standard Model. At the same time many improved measurements of Standard Model physics can be performed. In this short report some of these measurements are briefly reviewed and the sensitivity expected with the CMS detector [1] are discussed. Most of the results presented here are currently being updated in preparation of the Physics TDR volume II [2].

2. Top Quark Physics

The top quark mass is a fundamental Standard Model parameter. It allows to constraint the Standard Model Higgs mass through electroweak radiative corrections. The $t\bar{t}$ production is a major background for a lot of processes beyond Standard Model. At LHC, top quarks are mainly produced in pairs

*DAPNIA/SPP CEA–SACLAY.

via gluons fusion (90%) and quark annihilation (10%). A top quark will decay immediately in almost 100% of cases in $W^{\pm}b$. According to the decay of the Ws, selection of $t\bar{t}$ pairs can be classified as fully hadronic (both W decay hadronically, 46%), semileptonic (one W decays leptonically, the second decays hadronically, 29%) and dileptonic (both W decay leptonically).

2.1. *Top Quark Mass*

Two methods to compute the top mass have been developed so far at CMS. The first one presented here uses the semileptonic decay events. The lepton is used for the selection of events, while the mass is computed using the second top which is decaying hadronically. This analysis selects 3000 signal events for a integrated luminosity of 10fb^{-1} with a background level of 5% [3] shown on figure 1. The typical systematic error, mainly coming from the uncertainty on the jet energy scale, is less than 0.9 GeV. The second method uses also semileptonic decay events but requires also that the b-hadron from the same side of the $W \to l\nu$, decays via a J/Ψ [4]. The top mass is then computed using the linear relation between the maximum of the distribution of invariant mass of the lepton and J/Ψ as shown on figure 1. This method needs a lot of statistics but the systematic error on the top mass is reduced as it is independant of the jet energy scale. With 100fb^{-1} of integrated luminosity, 1000 signal events are expected and systematic errors on the top mass could be less than 1 GeV.

2.2. *Spin correlation and W polarization in top pairs*

Others properties of the top quark can be studied with the CMS detector. The top quark is decaying before hadronization so leptons coming from the W decay retain the information on the spin of the top quark. Looking at dileptonic decay of $t\bar{t}$ events, ones can use the differential angular distribution and measure the asymmetry of same sign spin and opposite sign spin. With 30 fb^{-1} of integrated luminosity, a statistical error on the asymmetry of 0.035 is expected as well as a systematic error of 0.028 [3].

The structure of the weak current allows the W boson to be longitudinally polarized in semileptonic decay. The differential angular distribution of the lepton is described as a function of left-handed component, longitudinal polarization and right-handed one. In the Standard Model the right-handed polarization must be zero. A statistical uncertainty of 0.023 and a systematic uncertainty of 0.022 on the longitudinal polarization can be obtained [3].

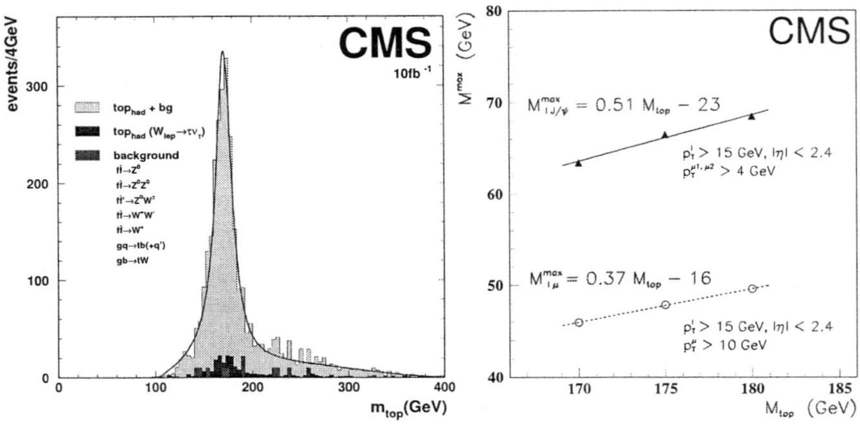

Figure 1. Left plot: Top Mass reconstructed in semileptonic events for 10fb^{-1} of integrated luminosity. Right Plot: Linear relation between Top Mass and maximum of invariant mass distribution of lepton J/Ψ system.

2.3. *Single Top*

The top quark can be produced individually with a lower cross section than the pair production. This process is predicted in Standard Model but not yet observed. It can be used for a direct mesurement of $|V_{tb}|$. This channel is also very sentitive to physics beyond Standard Model. A preliminary selection has been proposed to allow the observation of this process [5].

3. Gauge Boson Couplings

Measurements of vector boson couplings via the cross-section measurements is a test the non-Abelian nature of the Standard Model gauge theory. New physics can appear in anomalous gauge boson couplings. Studies were made in the $W\gamma$ channel [6] and in the $Z\gamma$ channel [7] and limits on the couplings can be set assuming 100 fb^{-1}. The transverse momentum spectrum of the photon is sensitive to anomalous couplings, as shown on figure 2.

4. Drell-Yan production of lepton pairs

The Drell-Yan process $pp \to l^+l^- + X$ and its forward-backward asymmetry is shown in figure 3. In this measurement, new physics can appear [8]. Also, the higher center of mass energy allows the production of dilepton masses greater than 1 TeV: new particles beyond the Standard Model can show up in this distribution (fig 3).

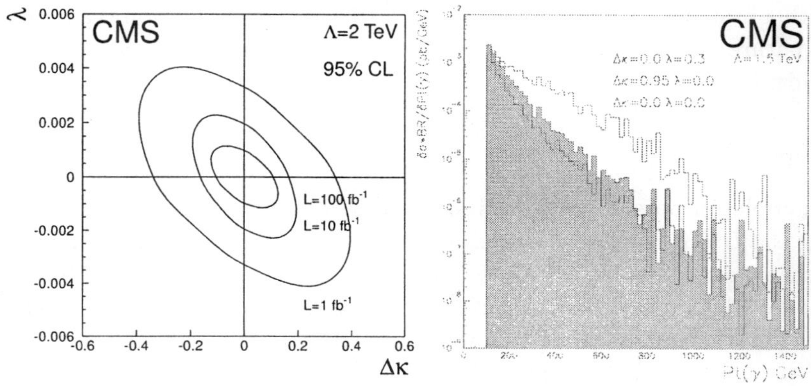

Figure 2. Left plot: 95% confidence level on the couplings $\lambda, \Delta\kappa$ for differents luminosity with a scale $\Lambda = 2$ TeV. Right Plot: Transverse momentum of photon produced in association with a Z-boson. Anomalous coupling will increase the number of events seen at high transverse momentum.

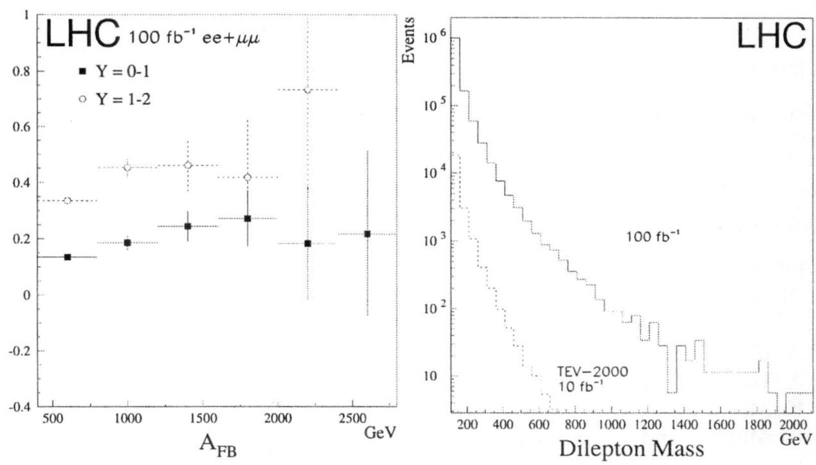

Figure 3. Left plot: Forward-backward asymmetry measurement. Right Plot: Invariant mass of dileptons.

5. B-Physics: $B_s \to \mu^+\mu^-$

The CMS detector has a rich potential for B-Physics because of its precise tracking and vertexing. It is possible to study CP violation, to look for rare decays, study B-lifetime, B_c mesons etc.

The process $B_s \to \mu^+\mu^-$ is highly suppressed in the Standard Model with a branching ratio of $(3.42\pm0.54)\times10^{-9}$ and any deviation is a potential signature for new physics. After 3 years of low luminosity(30 fb^{-1}) a 4σ evidence [9] is possible.

6. Conclusion

Examples of possible Standard Model physics studies have been given. Using the CMS detector the mass of the top could be measured with a systematic error less than 1 GeV, improving the constraint on the Standard Model Higgs mass. Spin correlation in $t\bar{t}$ will be studied and a systematic error on the spin asymmetry lower than 0.03 will be obtained with 30 fb^{-1}. W polarisation in $t\bar{t}$ events will be measured with a precision better than 0.023. Triple Gauge Boson Coupling will be constrainted to the percent level. The Drell-Yan process will allow to measure its forward-backward asymmetry and to study high dilepton masses. The B-physics program is rich and even a low rate process as $B_s \to \mu^+\mu^-$ will reach a significance of 4σ after 3 years of running at low luminosity.

Acknowledgments

Many thanks to the organizers for the wonderful conference and to all CMS colleagues who provided inputs to this presentation.

References

1. http://cmsinfo.cern.ch/ and references to Technical Design Reports therein.
2. Physics Technical Design Report Volume II in preparation.
3. L. Sonnenschein, *CMS Note* 2001/001.
4. A. Kharchilava, *CMS Note* 1999/065.
5. A.Giammanco, *CMS COnference Report* 2006/011.
6. K. Mackay and P. R. Hobson, *CMS Note* 2001/052.
7. K. Mackay and P. R. Hobson, *CMS Note* 2002/028.
8. D. Bourilkov, *CMS Note* 2000/035.
9. A. Nikitenko, A. Starodumov and N. Stepanov, *CMS Note* 1999/039, hep-ph/9907256

RECENT RESULTS IN DIFFRACTIVE ep SCATTERING AT HERA

M. BECKINGHAM*
DESY Hamburg,
Notkestrasse 85,
22607 Hamburg, Germany
E-mail: matthew.beckingham@desy.de

Recent diffractive results from the H1 and ZEUS experiments and their comparisons to pQCD models are presented. In particular results on inclusive diffraction as well as exclusive final states such as the diffractive production of D^* mesons, dijets and vector mesons are discussed.

1. Introduction

Deep inelastic diffractive scattering at HERA, the process $ep \to eXY$ where the final state systems X and Y are well separated in rapidity, provides a unique probe of the diffractive interaction and its interpretation within QCD. For the inclusive process, a hard scale is present when the virtuality of the exchanged photon, Q^2, is large - this is also the case for Deeply Virtual Compton Scattering (DVCS), $ep \to ep\gamma$. Diffractive vector meson production, $ep \to eVY$, may also provide a hard scale from a large vector meson mass or large four-momentum transfer at the the proton vertex, $|t|$. When such hard scales are present the data may be interpreted within perturbative QCD (pQCD).

2. Inclusive Diffraction

The inclusive diffractive DIS cross section may be defined as

$$\frac{d^3\sigma^{ep\to eXY}}{dx_{I\!P}d\beta dQ^2} = \frac{4\pi\alpha^2}{\beta Q^4}\left(1 - y + \frac{y^2}{2}\right)\sigma_r^{D(3)}(x_{I\!P}, \beta, Q^2), \qquad (1)$$

*On behalf of the H1 and ZEUS collaborations

where t is integrated over, as the proton system Y is usually not detected. The reduced diffractive cross section, $\sigma_r^{D(3)}$, is expressed in terms of the diffractive and longitudinal diffractive structure functions, $F_2^{D(3)}(x_{I\!P}, \beta, Q^2)$ and $F_L^{D(3)}(x_{I\!P}, \beta, Q^2)$ respectively, as

$$\sigma_r^{D(3)} = F_2^{D(3)} - \frac{y^2}{1+(1-y)^2} F_L^{D(3)}, \tag{2}$$

such that $\sigma_r^{D(3)} \sim F_2^{D(3)}$ at all but large values of y. QCD hard scattering factorisation states that the reduced cross section can be expressed in terms of a product of diffractive parton density functions (DPDFs), f_i^D, and partonic hard scattering matrix elements, $\sigma^{\gamma^* i}$, as $\sigma_r^{D(3)} \sim \sum \sigma^{\gamma^* i} \otimes f_i^D$.

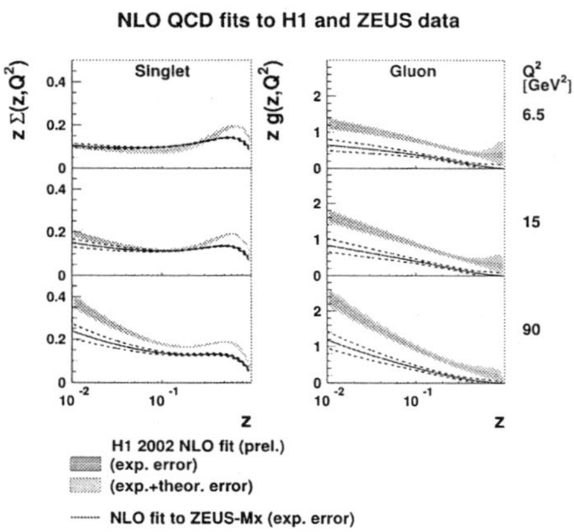

Figure 1. Diffractive quark singlet and gluon PDFs, as obtained from NLO DGLAP fits to the H1-LRG and ZEUS-M_X data.

Results have been released recently on inclusive diffraction by both the H1 and ZEUS collaborations. These each use one of three methods to identify diffractive events - by demanding a large rapidity gap (LRG) between the two hadronic systems in the final state, by using the distribution of events in the mass of the X system, M_X, or by tagging the proton system in Roman Pot detectors within the HERA beampipe.

A recent analysis[1] for the HERA-LHC workshop has extracted DPDFs by applying the same approach as used for the H1-2002-prelim NLO QCD fit (made to the H1 LRG data) to the inclusive diffractive data from ZEUS (using the M_X method). Figure 1 shows that while the singlet distributions agree reasonably at low Q^2, differences occur at larger Q^2. The gluon distribution from the ZEUS data is seen to be significantly smaller than that from H1. Both of these observations follow from the fits reflecting the different Q^2 dependences of the H1 LRG and ZEUS M_X data.

3. Diffractive Dijet and D^* Production

The universality of the DPDFs extracted from inclusive diffractive data, and hence factorisation in diffraction, may be tested by comparison to exclusive diffractive processes. Of particular interest are the diffractive production of dijets and D^* mesons, as these processes are directly sensitive to the gluon component of the diffractive exchange - the dominant contribution to the DPDFs.

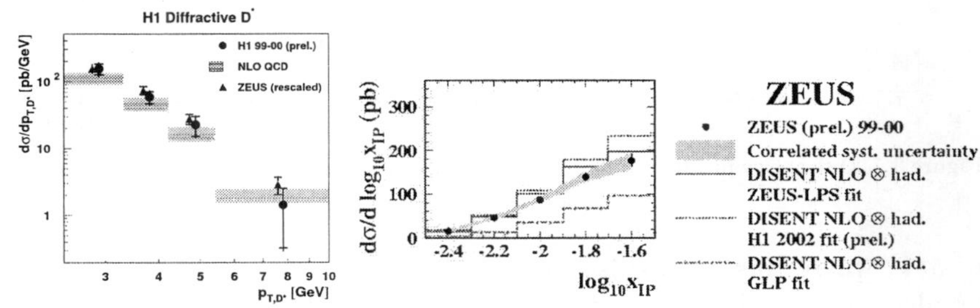

Figure 2. Diffractive production of D^* mesons (a), and dijets (b) in photoproduction

Figure 2a shows the differential cross section for diffractive D^* production in DIS[2] as a function of transverse momentum p_{T,D^*}. The ZEUS data are extrapolated to the H1 kinematic range for comparison. Both measurements agree and are described by the NLO QCD prediction using the H1 DPDFs. The differential cross section for diffractive dijets in DIS[3] as a function of $x_{I\!P}$ (Fig. 2b) is described reasonably by the predictions from both the H1 2002 fit and the ZEUS-LPS (Leading Proton Spectrometer) fit. However the prediction from the ZEUS GLP fit (using the M_X method) lies below the data. The description of the diffractive dijet and D^* cross sections by some of the DPDFs may show that factorisation holds in DIS. However, the differences between the DPDF predictions show that further understanding is needed. In particular, including diffractive dijet data in the DPDF fits will help to further constrain the diffractive gluon density.

Figure 3 shows the differential cross section for diffractive photoproduction of dijets[4] as a function of x_γ^{jets}, which, in a leading order approach, estimates the fractional momentum of the partons from the photon that enter the hard subprocess. The new prediction using the H1 2002 NLO DPDFs lies above the data. Figure 3a shows that when only the resolved photon is suppressed ($x_\gamma^{jets} < 0.9$) the H1 2002 NLO fit is unable to describe the data. However, Fig.3b shows the data are described by the H1 2002 NLO fit when scaled by a factor of 0.5. Figure 3c shows the differential cross section for diffractive photoproduction of D^* mesons[5] as a function

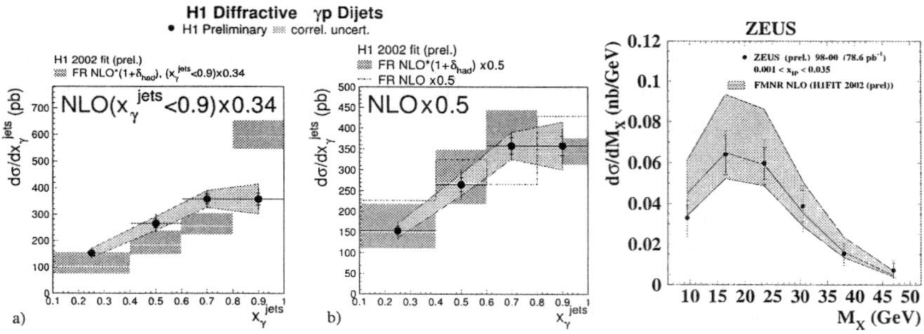

Figure 3. Diffractive production of dijets (a), (b) and D^* mesons (c) in photoproduction

of M_X, where this process is dominated by direct interactions. The data are described by the DPDFs from the H1 fit 2002, but within a large error band. That diffractive photoproduction of D^* mesons is described may suggest that factorisation also holds in direct photoproduction. However, the diffractive photoproduction of dijets is described by a global suppression, rather than by only a suppression due to the hadronic interaction of the photon, suggesting that factorisation is broken. Again, a better understanding of the DPDFs is need before firm conclusions are made.

4. Diffractive Vector Meson Production and DVCS

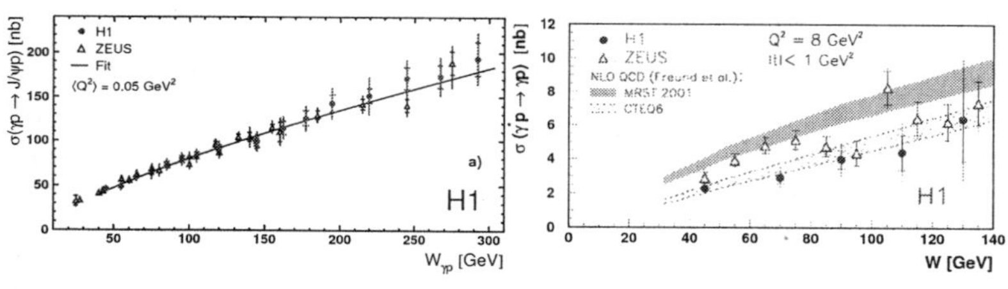

Figure 4. Elastic diffractive J/Ψ production (a) and (b) DVCS

Figure 4a shows the cross section for diffractive elastic J/Ψ photoproduction[6] as a function of the photon-proton centre of mass energy, W. The steep rise for all W, seen at both high and low Q^2, shows the importance of the vector meson mass as a hard scale. A steep rise with W is also seen in the W dependence of the DVCS cross section[7] (Fig. 4b).

Here the only hard scale present is Q^2. The steep rises with W for both cross sections show the sensitivity of diffractive vector meson and DVCS measurements to the gluon PDFs of the proton. For instance, Fig. 4b shows the sensitivity of an NLO prediction to inputs from two generalised parton densities derived from different input PDFs.

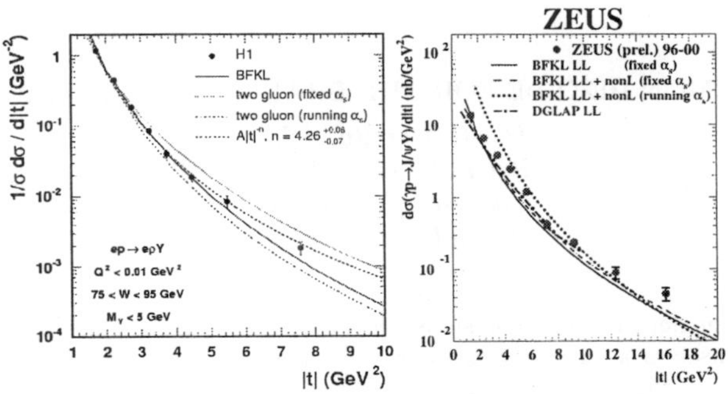

Figure 5. Diffractive photoproduction of (a) ρ and (b) J/Ψ mesons at high $|t|$.

New measurements of the diffractive photoproduction of ρ mesons[8] and J/Ψ mesons[9] at large momentum transfer (high $|t|$) confirm that $|t|$ can provide a hard scale at which to probe the diffractive exchange with vector mesons. The $|t|$ dependence of the diffractive photoproduction of high $|t|$ ρ (Fig. 5a) and J/Ψ (Fig. 5b) mesons are reasonably described by different BFKL models. However, two gluon models give a worse description of the ρ data and the J/Ψ data can only be described by the DGLAP model in the range of validity ($|t| < M_{J/\Psi}$).

References

1. P. Newman and F. P. Schilling, arXiv:hep-ex/0511032.
2. H1 Collaboration, paper 178 ICHEP04, H1Prelim-04-111
3. ZEUS Collaboration, paper 342 EPS05, ZEUS-prel-05-020
4. H1 Collaboration, paper 177 ICHEP04, H1prelim-04-113
5. ZEUS Collaboration, paper 268 LP05, ZEUS-prel-05-022
6. A. Aktas *et al.* [H1 Collaboration], Accepted by Eur.Phys.J.C, arXiv:hep-ex/0510016.
7. A. Aktas *et al.* [H1 Collaboration], Eur. Phys. J. C **44** (2005) 1 arXiv:hep-ex/0505061.
8. A. Aktas *et al.* [H1 Collaboration], Submitted to Phys. Lett. B, arXiv:hep-ex/0603038.
9. D. Szuba [ZEUS Collaboration], AIP Conf. Proc. **792** (2005) 410.

HIGH-P_T SUPPRESSION IN HEAVY ION COLLISIONS FROM THE BRAHMS EXPERIMENT AT RHIC

SELEMON BEKELE FOR THE BRAHMS COLLABORATION

The University of Kansas,
1082 Mallot Hall, 1251 Wescoe Hall Drive,
Lawrence, KS 66045, USA
E-mail: bekeleku@ku.edu

In central nucleus-nucleus collisions, deviations from binary scaling at high-p_T reveal themselves as a suppression in the transverse momentum distribution of produced particles as measured by R_{cp}, the ratio of central to peripheral data scaled by the number of binary collisions. The system size dependence of R_{cp} may help disentangle the interplay of initial (e.g. gluon saturation) and final (e.g. parton energy loss) state effects which are thought to modify the scaling behavior of the high-p_T suppression. Here we present results on R_{cp} of charged hadrons from the BRAHMS experiment comparing results from AuAu, CuCu and dAu collisions.

1. Introduction

The study of high-p_T hadron production may provide an insight into the properties of the highly excited partonic matter, often called the Quark Gluon Plasma (QGP), created in high energy heavy ion collisions. Recent data from RHIC reveal that the mid-rapidity hadron spectra at high-p_T in central AuAu collisions are strongly suppressed relative to scaled pp spectra[1].

Measurements of the nuclear modification factor are consistent with partonic energy loss in the larger interaction volume corresponding to central collisions suggesting that high-p_T suppression might be due to final state interactions. In order to make sure high-p_T suppression is indeed a final state effect, one has to look at situations where final state interactions are supposed to be less important, such as dAu collisions. The expectation is that the ratio of dAu to pp-collisions should show an enhancement over a range in momentum, a phenomenon known as the Cronin effect [2]. The argument is that energy loss in the "cold matter" of dAu collisions is quite small and therefore does not mask the Cronin effect which results from p_T

broadening from multiple scattering. On the other hand, if a suppression is observed in dAu colisions at forward rapidities instead of the Cronin effect, it may be an indication that initial state effects are important.

Transverse momentum spectra of hadrons in dAu collisions have shown a lack of high-p_T suppression at mid rapidity. This result is again consistent with the hypothesis of parton energy loss or/and parton recombination in the dense medium formed in AuAu collisions. However, a clear suppression has also been observed in measurements at forward rapidity, suggesting that not only final state effects but also initial state effects may play an important role in high-p_T particle production [1,3].

2. The BRAHMS Experiment

The BRAHMS detector system [4] consists of global detectors for event characterization, a Mid Rapidity Spectrometer(MRS) and a Forward Spectrometer(FS) covering forward rapidities. Collision centrality is characterized by a multiplicity array(MA) consisting of scintillator tiles and silicon strip detectors mounted coaxially around the beam axis. For the present studies, the forward spectrometer was positioned at 4° with respect to the beam direction, corresponding to $2.9 \leq \eta \leq 3.4$. Events within ± 20 cm of the nominal interaction vertex were considered.

We present here preliminary results on R_{cp} in CuCu Collisions at $\sqrt{s_{NN}} = 200$ GeV measured at the BRAHMS experiment. The CuCu system serves as a bridge between dAu and AuAu systems making it possible to investigate the dependence of high-p_T suppression on system size. Additionally, the data from CuCu collisions allows us to study the dependence of the spectra on the shape of the reaction zone created in the collision. For the same number of participants(N_{part}), the system created in CuCu is more spherical than the one in AuAu.

3. Overview of High-P_T suppression

The total particle production in heavy ion collisions comes from a linear combination of soft processes that scale with the number of participants, and hard processes that scale with the number of binary collisions. Suppression of high-p_T particle production is described in terms of the nuclear modification factor R_{AA}. This is the ratio of the measured hadron spectra to reference spectra from pp collisions scaled by the average number of binary nucleon-nucleon collisions N_{coll}. The value of R_{AA} is expected to be unity if a nucleus-nucleus collision were just a superposition of independent nucleon-nucleon collisions. In the absence of a good pp reference, it is also possible to characterize hadron suppression in terms of R_{cp},the ratio of central to peripheral data scaled by $\langle N_{coll} \rangle$:

$$R_{cp} = \frac{Yield^{central} / < N_{coll}^{central} >}{Yield^{peripheral} / < N_{coll}^{peripheral} >}.$$

The N_{part} (i.e., centrality) dependence offers the possibility of studying the system size dependence of the nuclear modification factor, with the system dependence (CuCu vs. AuAu) supplying additional information on the size dependence. The suppression of high-p_T yields in relativistic heavy ion collisions can have a number of causes. Theoretical models have been developed to describe certain aspects of the experimental data in terms of final state effects such as parton recombination, energy loss, or initial state effects such as CGC. Recent studies based on recombination models [5], for example, seem to give adequate explanation for the different behavior of high-p_T suppression of mesons and baryons. Another mechanism invoked to explain the suppression of high-p_T particle production is a phenomenon known as jet quenching [2,6]. The idea behind jet quenching is that high energy partons traveling through the hot and dense medium created in a heavy ion collision lose energy leading to a high-p_T suppression. A marked high-p_T suppression with increasing rapidity may also be related to the initial conditions, in particular to the possible existence of the color glass condensate(CGC)[7,8]. According to the theory of the CGC, a very high energy hadron has contributions to its wave function from gluons, quarks and anti-quarks. In terms of the momentum fraction x of the partons inside an interacting hadron, low x phenomena correspond to large rapidity. The density of low x gluons grows as energy increases leading to gluon saturation. Parton scattering centers are reduced due to gluon-gluon fusion and, as a result, there should be fewer hard scatterings leading to a reduction in the production of hadrons.

4. R_{cp} results from CuCu

Figure 1(a) shows the ratio R_{cp} of yields from CuCu collisions of a given centrality class to yields from the most peripheral collisions(40-60%), at η = 3.2, scaled by the number of binary collisions in each sample. The data for the different centrality classes are obtained from the same collider run. As a result, the ratios are largely free of run-dependent systematic errors associated with collider and detector performance. The dominant

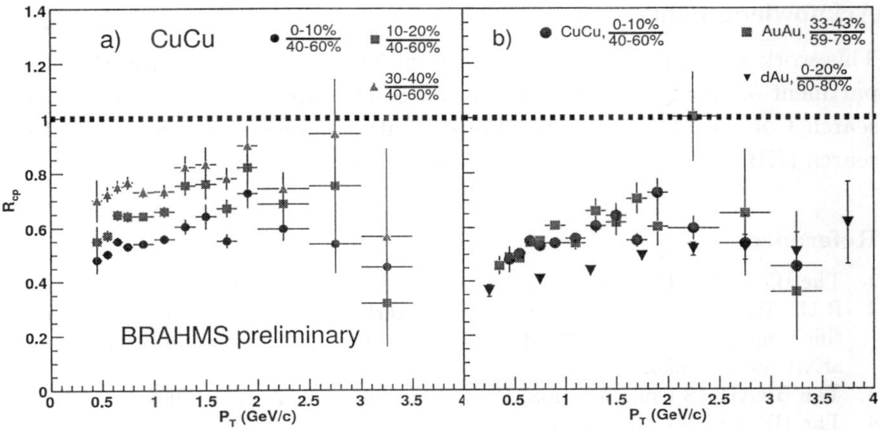

Figure 1. R_{cp} at $\eta = 3.2$ in CuCu Collisions as a function of p_T, (a) Centrality dependence, (b) comparison to results from dAu and AuAu Collisions. For CuCu, $0-10\%$ corresponds to $N_{part} \approx 96.8$ and $40-60\%$ to $N_{part} \approx 18.7$. For AuAu $33-43\%$ corresponds to $N_{part} \approx 96.3$ and $59-79\%$ to $N_{part} \approx 17.8$.

systematic error in the R_{cp} ratios come from the determination of N_{coll} in the centrality bins. One can see that there is more suppression as the collisions become more central.

The R_{cp} in CuCu in comparison with those measured in dAu and AuAu collisions at the same pseudo-rapidity is shown in Figure 1(b). It is evident that the suppression in CuCu follows a similar trend as in dAu and AuAu. Considering the case where the mean number of participants at the same collision energy are equal, one obtains R_{cp} values very similar in both CuCu and AuAu collisions.

5. Summary

In summary, we have presented preliminary results on R_{cp} in CuCu collisions at $\sqrt{s_{NN}} = 200$ GeV and $\eta = 3.2$. Our results show that one obtains similar results for R_{cp} considering the same number of participants in both CuCu and AuAu collisions. This suggests that the observed suppression depends particularly on the volume of the interaction region through which produced hadrons have to travel. To complete the system size systematics, we are currently extending these studies by also studying the correlations with respect to the reaction plane.

Acknowledgments

This work was supported by the office of Nuclear Physics of the U.S. Department of energy, the Danish Natural Science Research Council, the Research Council of Norway, the Polish State Committee for Scientific Research (KBN), and the Romanian Ministry of Research.

References

1. The BRAHMS Collaboration, Phys. Rev. Lett. **91**, 072305 (2003).
2. R.C. Hwa and X.-N. Wang, eds Quark-Gluon Plasma (World Scientific,Singapore,2004) p.123, M.Guylassy, I.Vitev, X.N.Wang, and B.W.Zhang, arXiv:nucl-th/0302077.
3. The BRAHMS Collaboration, Phys. Rev. Lett. **93**, 242303 (2004).
4. The BRAHMS Collaboration, Nuclear Instruments and Methods A**499**, 437 (2003).
5. R. J. Fries, B. Muller, and C. Nonaka, Phys. Rev. C**68**, 044902 (2003).
6. R. Baier, D. Schiff and B.G. Zakarov, Ann. Rev. Nucl. Part. Sci. **50**, 37 (2000).
7. L.V. Gribov, E. M. Levin, M.G.Ryskin, Phys. Rep. **100**, 1 (1983).
8. D. Kharzeev, E. Levin, L. McLerran, Phys. Lett. B**561**, 93 (2003).

THE SUPPRESSION OF HIGH P_T NON-PHOTONIC ELECTRONS IN AU+AU COLLISIONS AT $\sqrt{S_{NN}} = 200$ GEV AT RHIC

J.BIELCIK FOR THE STAR COLLABORATION

Department of Physics,
Yale University,
New Haven, 06511, USA
E-mail: jaroslav.bielcik@yale.edu

In these proceedings, the preliminary non-photonic electron spectra from $p+p$, $d+Au$ and $Au+Au$ collisions at $\sqrt{s_{NN}} = 200$ GeV at mid-rapidity are presented. The momentum range of reconstructed non-photonic electrons is $1.5 < p_T < 10$ GeV/c. The dominant contribution to the non-photonic electron spectrum is the semi-leptonic decay of D and B mesons. The electron nuclear modification factor (R_{AA}) shows a large suppression in central Au+Au collisions, indicating an unexpectedly large energy loss for heavy quarks in the hot and dense matter created at RHIC. Theoretical models tend to overpredict the data if the contributions from both charm and beauty decays are taken into account.

1. Introduction

The measurement of inclusive hadron yields in central Au+Au collisions at RHIC led to the discovery of suppression of hadron production at large transverse momenta (p_T) compared to p+p collisions[1,2,3]. This is generally attributed to the energy loss of light partons in the dense nuclear matter created at RHIC. The energy loss itself depends on the properties of the medium, such as gluon density and size, as well as on the properties of the probe, such as color charge and mass. Heavy quarks are believed to be mostly created in hard scattering processes in the initial phase of the collisions and thus are excellent probes of the hot and dense matter. Due to the large mass of heavy quarks, the suppression of small angle gluon radiation should reduce their energy loss and consequently, any suppression of heavy-quark mesons at high p_T is expected to be smaller than that observed for hadrons consisting of light quarks[4].

Heavy quark mesons can be studied by the direct reconstruction of their hadronic decays, such as $D^0 \to K^-\pi^+$. The direct reconstruction becomes a challenge in Au+Au collisions, because of the very large combinatorial background. Therefore, the current measurements by STAR are limited to $p_T < 3$ GeV/c. However, the combination of this direct reconstruction with low p_T electron identification[5,6] allows the $c-\bar{c}$ cross section to be measured. The total charm production follows binary scaling from d+Au to Au+Au collisions, supporting the idea that heavy quarks are mostly produced in the early stages of the collision.

reconstructed via hadronic decay ($D^0 \to K^-\pi^+$) in both dAu[5] and AuAu[6] collisions. An alternative way to infer information about heavy quark production is the study of electrons from semi-leptonic decays of D and B mesons which is the focus of this letter. This method allows STAR to study charm and beauty production up to substantially larger p_T. There are several sources of electrons that contribute to the inclusive spectra. We divide them into two groups: non-photonic electrons (signal) and photonic electrons (background). The non-photonic electrons are mainly from semi-leptonic decays of heavy mesons with a small contribution from the Drell-Yan process. The background photonic electrons are from γ conversions, and π^0, η Dalitz and light vector-meson decays.

2. Electron identification in STAR

The results presented in this paper were obtained from an analysis of data recorded with the STAR detector[7] in 2003 (p+p, d+Au) and 2004 (Au+Au). Electron identification is based on a combination of energy loss in the Time projection chamber, energy deposition in the Barrel electromagnetic calorimeter, and shower profile in the Shower maximum detector. Further details of the analysis can be found elsewhere[8]. After all analysis cuts a clean sample of electrons was obtained, with a p_T dependent residual hadron contamination that varies from 10 to 15%. The data sample for the Au+Au dataset was divided into 3 centrality bins (0-5%, 10-40%, and 40-80%). The electron reconstruction efficiency and acceptance were determined by embedding simulated electrons into real events. For the most central events, the electron reconstruction efficiency increases with p_T up to 5 GeV/c and then remains constant at 40% . The photonic electrons have been statistically identified and subtracted from the inclusive electron spectrum. The efficiency of the photonic background rejection was determined by embedding π^0, with realistic p_T distribution, into real events and is about 60% for the most central Au+Au events, decreases slightly with

p_T.

3. Non-photonic electron spectra and R_{AA}

In Figure 1a, the ratio of inclusive electrons to photonic background electrons is shown as a function of the p_T of the electrons. For $p_T > 2.0$ GeV/c, there is a clear enhancement of electrons with respect to the background. This enhancement becomes more evident at higher momentum, where most electrons come from non-photonic sources. Figure 1b shows the preliminary background subtracted non-photonic electron spectra for p+p, d+Au and Au+Au collisions. The error bars are statistical and the boxes depict the preliminary systematic uncertainties. Many NLO pQCD [9], as well as Pythia calculations, predict that in a range between 3 and 6 GeV/c, the amount of electrons coming from B meson decays becomes significant. STAR is capable of measuring non-photonic electrons in a momentum range above this transition. Although pQCD can predict the shape of the non-photonic electron spectrum in p+p collisions rather well, it fails to describe the absolute electron yield by a factor ≈6.

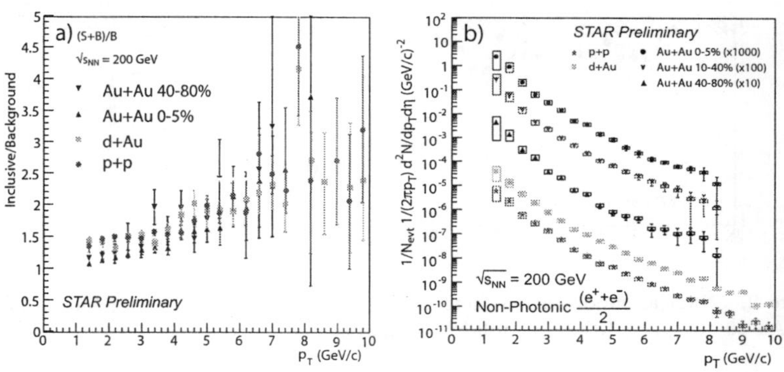

Figure 1. (a)Inclusive to photonic electrons ratio as a function of p_T. (b)Background subtracted non-photonic electron spectra for p+p, d+Au and Au+Au collisions with centralities 0-5%, 10-40% and 40-80% .

Figure 2 shows the nuclear modification factors R_{AA} and R_{dAu} for non-photonic electrons as a function of p_T. The R_{dAu} ratio (Figure 2a) seems to be systematically above unity for the entire p_T range, consistent with a small Cronin enhancement. An increased suppression from peripheral to central Au+Au events (Figure 2b-d) with respect to binary scaling is observed. Assuming that there is no other source of non-photonic electrons, this suppression indicates a strong interaction and large energy loss of heavy quarks in the medium created at RHIC. The amount of suppression in most central Au+Au collisions is similar to that of light hadrons.

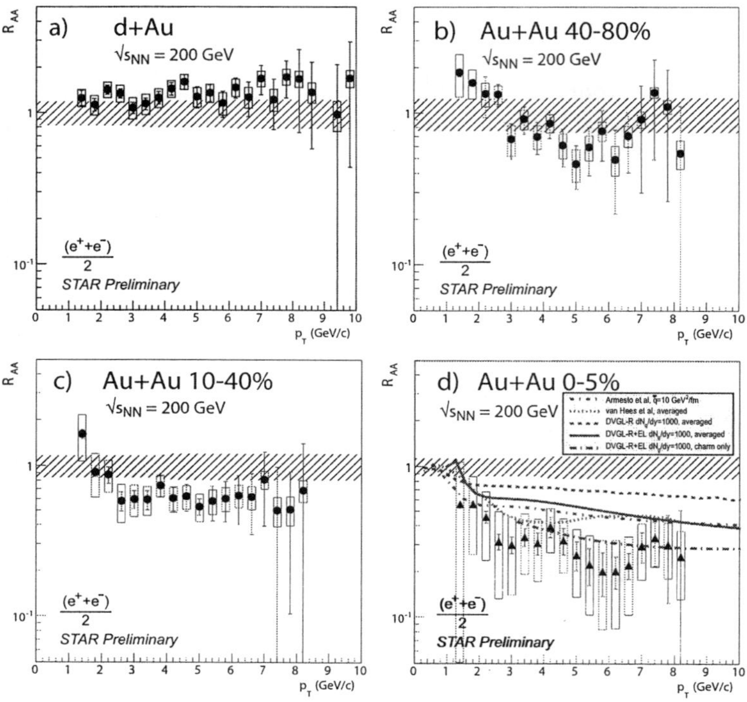

Figure 2. Nuclear modification factors R_{AA} for (a) d+Au, (b) Au+Au 40-80%, (c) Au+Au 10-40%, and (d) Au+Au 0-5% comparison with models.

Figure 2d also shows the calculations of the R_{AA} from three theoretical models. In all cases, the contribution from charm and bottom quarks were taken into account. In the first model[10] (dash-dotted curve), the medium is characterized by the time averaged BDMPS transport coefficient, $\hat{q} = 10\ GeV^2/\text{fm}$, for central Au+Au collisions. This value of $\hat{q}$ is consistent with the measurement of R_{AA} of the light hadrons. In the second model[11] (dashed curve), the DGLV theory of radiative energy loss has been applied and the created nuclear matter is characterized by a gluon density of $\frac{dN_g}{dy}$=1000, the value derived from light-quark meson suppression.

In addition, the contribution from elastic energy loss has been taken into account[12] (solid line). For the sake of comparison, a curve (dash-dot) representing the contribution only from charm sources is shown. In the third model[13] (dotted curve), the authors focus on elastic scattering of heavy quarks in the medium mediated by resonance excitations (D and B) off light quarks as well as by t-channel gluon exchange.

Although the data have large systematic and statistical errors, the data points tend to lie below the model calculations at high p_T. It is important to note that the model calculations also have large uncertainties, such as the amount of relative contribution from beauty/charm decays, that influence the final R_{AA}. Also, full understanding of the p+p collisions is necessary before we can make a final statement about heavy quark energy loss.

4. Conclusions

The non-photonic electron spectra measured by STAR for p+p, d+Au, Au+Au collisions up to $p_T \approx 8$ GeV/c were presented. An increasing suppression of non-photonic electrons with the collision centrality in Au+Au collisions is observed. This may be related to a stronger than predicted interaction between heavy quarks and the medium created at RHIC. The analysis of the full statistics from the 2004 Au+Au run will permit a more detailed study of the medium modifications for heavy quarks and allow for a better understanding of quark energy loss mechanisms.

References

1. C. Adler,*et al.* [STAR Collaboration], Phys. Rev. Lett. **89**, 202301 (2002)
2. J. Adams *et al.* [STAR Collaboration], Phys. Rev. Lett. **91**, 172302 (2003)
3. J. Adams *et al.* [STAR Collaboration], Phys. Rev. Lett. **91**, 072304 (2003)
4. Y. L. Dokshitzer and D. E. Kharzeev, Phys. Lett. B **519**, 199 (2001)
5. J. Adams *et al.* [STAR Collaboration], Phys. Rev. Lett. **94**, 062301 (2005)
6. H. Zhang [STAR Collaboration], arXiv:nucl-ex/0510063.
7. K. H. Ackermann *et al.*, Nucl. Instrum. Meth. A **499**, 624 (2003)
8. J. Bielcik [STAR Collaboration], arXiv:nucl-ex/0511005.
9. M. Cacciari, P. Nason and R. Vogt, Phys. Rev. Lett. **95**, 122001 (2005)
10. N. Armesto, M. Cacciari, A. Dainese, C. A. Salgado and U. A. Wiedemann, arXiv:hep-ph/0511257.
11. M. Djordjevic, M. Gyulassy, R. Vogt and S. Wicks, Phys. Lett. B **632**, 81 (2006)
12. S. Wicks, W. Horowitz, M. Djordjevic and M. Gyulassy, arXiv:nucl-th/0512076.
13. H. van Hees, V. Greco and R. Rapp, Phys. Rev. C **73**, 034913 (2006)

DIRECT CP VIOLATION RESULTS IN $K^{\pm} \to 3\pi^{\pm}$ DECAYS FROM NA48/2 EXPERIMENT AT CERN

C. BIINO

Istituto Nazionale di Fisica Nucleare,
via Pietro Giuria 1,
Torino, 10125, Italy
E-mail: cristina.biino@cern.ch

After firmly establishing direct CP Violation in two pions decays of neutral kaons, the NA48 experiment, during the 2003 run at CERN-SPS, has collected more than 1.6 billion of charged kaon decays into three charged pions, using a unique double beam technique which allows a high level of control on systematic effects. The measurement of the direct CP violation Dalitz plot linear slope asymmetry parameter A_g is reported. This result corresponds to more than an order of magnitude improvement in precision with respect to previous experiments and is limited by the statistics of the data sample.

1. Introduction

Forty years after its discovery, CP violation (CPV) still plays a central role in the experimental programs of high energy physics. CP violation in the neutral kaon system and in the neutral B mesons system can occur either in the mixing of two eigenstates or manifest as an asymmetry in two CP conjugate decay amplitudes (direct CP violation).

Direct CP violation is the most straightforward CP effect but it is difficult to detect experimentally and to connect to the parameters of the underlying fundamental theory, but it is a crucial window into physics beyond the Standard Model (SM).

It took more than thirty years since the discovery of CP violation in mixing of neutral kaon states in 1964 [1] to establish direct CP violation in the kaon system [2], [3]. The new world average value of :

$$Re(\epsilon'/\epsilon) = (16.7 \pm 2.6) \times 10^{-4}$$

is higher than most predictions and could be an indication of errors in the calculation of the matrix element of the decay, or might be a hint of failure of the SM. Therefore a measurement of direct CPV in other processes is quite relevant. In kaons, besides the already mentioned parameter ϵ' in $K_L \to \pi\pi$ decays, the most promising complementary observables are decay rates of GIM suppressed rare kaon decays proceeding through flavor-changing neutral currents, and the asymmetry between K^+ and K^- decays into three pions.

The usual phenomenological description of $K^\pm \to 3\pi$ decays is made in terms of the bi-dimensional Dalitz plot parameters u and v [4], related respectively to the energy sharing to the odd pion (charge opposite with respect to the other two) and among the two even pions. The matrix element is usually described in a polynomial expansion of the two Dalitz variables u and v and parametrized in terms of slopes:

$$|M(u,v)|^2 \sim 1 + gu + hu^2 + kv^2$$

where $g(\pi^\pm\pi^+\pi^-) = -0.2154 \pm 0.0035$ and $g(\pi^\pm\pi^0\pi^0) = -0.652 \pm 0.031$. Since the phase space is small, the expansion converges rapidly. The linear slope $|g|$ dominates over $|h|$ and $|k|$. The slope g can differ between K^+ and K^- only due to DCPV and the measured quantity is :

$$A_g = (g_+ - g_-)/(g_+ + g_-) = \Delta g/2g$$

The measurement of differences among the above parameters describing the decay distributions is independent on fluxes and can be performed just by comparing the Dalitz plot shapes.

Several experiments have searched for the asymmetry A_g. The precision reached so far is at the level of few 10^{-3} [5]. SM predictions vary between few 10^{-6} to few 10^{-5}. Models beyond the SM allow a wider range of A_g predicting substantial enhancements [6].

After a successful program of investigations with neutral kaons which culminated with the establishment and accurate measurement of direct CP violation in the decays of neutral kaons into two pions [3], the NA48 collaboration is now devoted [7] to measure the parameter A_g in $\pi^\pm\pi^+\pi^-$ and $\pi^\pm\pi^0\pi^0$ modes to a precision $\sim 2 - 4 \times 10^{-4}$. The above figure requires collecting very large samples and therefore intense beams, large acceptance and data acquisition bandwidth. These have to be matched with a careful control of systematics.

NA48/2 has a great potential for closing the gap between experimental and theoretical results, reaching the interesting region in which one can start to detect enhancements due to physics behind the SM.

In this paper the analysis and results based on about 1.6 billions of $K^{\pm} \rightarrow \pi^{\pm}\pi^{+}\pi^{-}$ decays taken in the first data taking period are presented.

2. Beams and Detector

NA48/2 uses the original detector [3] and a completely novel beam line system, designed and built to transport simultaneously positive and negative kaon beams, superimposed in space, focused and with narrow momentum spectra. The beam design plays an important role in enforcing large cancellations of detector acceptance effects.

The beams of charged particles are produced by 400 GeV/c primary protons from the SPS imping at zero degree angle on a beryllium target of 2 mm diameter and 40 cm length at a chosen nominal intensity of $7\times10^{11}ppp$ (with 16.8 s cycle time and 4.8 s flat-top). A schematic vertical section of the simultaneous K^{+} and K^{-} beam line is shown in Fig. 1 The two beams have an acceptance opening angle of ± 0.36 mm in both planes, defined by a common collimator.

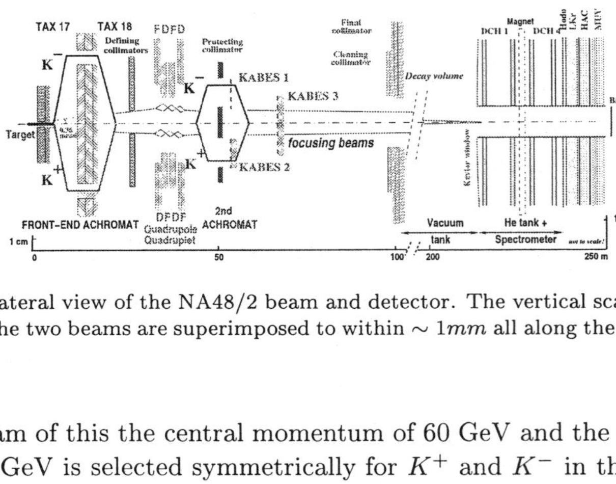

Figure 1. Lateral view of the NA48/2 beam and detector. The vertical scale is strongly enhanced. The two beams are superimposed to within $\sim 1mm$ all along the decay region.

Upstream of this the central momentum of 60 GeV and the momentum bite of ±3 GeV is selected symmetrically for K^{+} and K^{-} in the front end achromat unit, consisting of 4 dipole magnets which splits the two beams in the vertical plane and recombines them on the same axis. The two beams, containing about 6.4×10^{7} particles per pulse (12 times more pions than kaons), with a kaon charge ratio $K^{+}/K^{-} \sim 1.8$ (the analysis is independent of this ratio) pass through magnetic focusing and muon sweeping stages. A system of four quadrupoles is designed to focus particles of each sign similarly in both planes to small spot sizes ($\sim$ 5 mm r.m.s.) at the spectrometer position. The two beams, after passing the cleaning and the final collimators, remain superimposed to within 1 mm while traveling through the $\sim$ 114 m long vacuum decay region. This superposition symmetrizes the acceptance and contributes to the reduction of systematic biases. The two beams are again split and cleaned in the second achromat, that together with the KABES detector could serve also as a beam spectrometer allowing a 1% momentum measurement of the incoming kaons. Any structure remaining inside the beam spots is naturally exchanged between K^+ and K^- by regular inversions of all beam magnets fields.

The decay region is contained in a vacuum tank. Note that the decay products from charged pions in the beam, due to the limited transverse momentum available, are mostly confined to remain within such pipe, without illuminating the detector itself.

The main part of the NA48 detector used for the measurement of the CP violating asymmetry in $K^\pm \to \pi^\pm\pi^+\pi^-$ decays is the magnetic spectrometer, composed of four drift chambers and a dipole magnet (120 MeV/c horizontal kick) enclosed in a helium filled tank. The drift chambers have an octagonal shape with an area of about 4.5 m^2. Each is made of four sets of two staggered planes of sense wires oriented along four 45^0 directions. The momentum resolution is $\sigma(p)/p = 1.0\,\% \oplus 0.044\,p\,\%$ (p in GeV/c).

The interesting decays are triggered with a two-level trigger system. At the first level the rate is reduced to few hundred KHz by requiring at least two hits in a scintillator hodoscope placed behind the magnetic spectrometer. The second level trigger consists of hardware coordinates builders and a farm of asynchronous microprocessors that reconstructs tracks using data from the drift chambers. At least two tracks are required to converge within 5 cm in the decay volume.

3. Measurement Method

The measurement is based on comparing u plots of K^+ and K^- decays by projecting events in the u-v Dalitz plot on the u axis to obtain one dimensional distributions $N^\pm(u)$. In case of the $\pi^\pm\pi^+\pi^-$ final state the ratio $R(u) = N^+(u)/N^-(u)$ is proportional with sufficient precision to $(1+\Delta g u)$. $A_g = \Delta g/2g$ is extracted from a linear fit to the ratio R(u). Clearly any imperfection has to be both charge-asymmetric and non-flat in u, in order to potentially bias the measure. Charge related beam and detector differences can induce asymmetric effects, the most obvious being that due to imperfect left-right detector symmetry when coupled with the lateral deflecting effect of the spectrometer magnet. While the beam is carefully aligned along the detector axis, any local imperfection of the spectrometer can introduce an acceptance asymmetry. This effect is canceled to first order by periodically reverse the magnetic spectrometer field (once per day), therefore equalizing the time-average acceptance for K^+ and K^-. Any residual difference between "upper" and "lower" beam paths in the achromats are canceled by periodically (once per week) reversing all magnetic fields along the beam lines, so that $K^\pm$ paths are exchanged.

The $\sim$ two week cycle represents a supersample which is treated in the analysis as an independent data unit. In the 2003 period we collected four supersamples. Each one contains four $K^+ \to \pi^+\pi^+\pi^-$ and four $K^- \to \pi^-\pi^-\pi^+$ samples with different combination of magnet polarities. The actual ratio R(u) is obtained as a quadruple ratio of K^+/K^- samples:

$$R(u) = R_{US}R_{UJ}R_{DS}R_{DJ} \simeq \bar{R}(1 + 4\Delta g u)$$

where the subscripts U (D) refer to the beam line magnets polarities (a sample where K^+ travel along the upper/lower beam path in the achromats), while S (J) represent the spectrometer magnet polarity in which decay products having the same charge as the corresponding beam are deflected to the right with respect to the direction of the beam, toward Saleve mountain (to the left, toward Jura mountains). A linear fit of the above ratio results in the normalization $\bar{R}$ and Δg from which A_g is extracted. The quadrupole ratio exploits several cancellations of systematic biases:

(i) beam line differences - by comparing K^+ and K^- traveling along the same paths
(ii) detector asymmetries - by comparing K^+ and K^- illuminating the detector in the same way
(iii) global time-dependent effects - by the simultaneous detection of K^+ and K^- events

The only residual effects which can affect the asymmetry measurement are time variation of asymmetries in experimental conditions on a time scale of about one subsample. This method is independent on the relative size of the samples with different magnet configuration. The statistical uncertainty depends on the statistics of the smallest sample involved.

While, due to the superposition of the two beams, the measurement doesn't require a MonteCarlo simulation, nevertheless a GEANT-based MonteCarlo program with detailed detector simulation, including time-varying local drift chamber inefficiency and alignment maps and beam line geometry variations, has been developed to check the sensitivity of the result to various systematic effects.

4. Data Analysis

NA48/2 took data in two periods. About 4 billions of $K^{\pm} \to \pi^{\pm}\pi^{+}\pi^{-}$ and about 200 millions of $K^{\pm} \to \pi^{\pm}\pi^{0}\pi^{0}$ decays have been collected during the 50 day period in year 2003 and the 60 day period in 2004 . The total recorded data volume amounts to about 200 TB.

Several stages of compactification and filtering of data, requiring at least three reconstructed tracks in the magnetic spectrometer and loose acceptance and quality cuts as well as at least one good reconstructed three tracks vertex, were necessary in order to reduce the data to a size suitable for the final analysis.

The measurement of the pion momenta is based on the knowledge of the magnetic field in the spectrometer magnet and on the tracking information from the drift chambers. Track reconstruction combines hits from all four drift chambers using measured magnetic field map rescaled to the recorded current in the spectrometer analyzing magnet and correcting for the small magnetic field due to the vacuum tank magnetization and the Earth's field, which were measured before the run. This stray-field correction reduces the azimuthal variation of the reconstructed invariant mass of $\sim 1MeV/c^2$ by an order of magnitude. The three tracks vertex is reconstructed using track segments from the first half of the spectrometer and the vertex constrained track parameters are used to compute the three pion invariant mass, with resolution $\simeq 1.7MeV/c$. Small non-Gaussian invariant mass tails arise from kink tracks in which a charged pion decayed. To avoid introducing instrumental asymmetries, no muon rejection is applied, which is possible since the sample is practically background free. The Monte-Carlo simulation shows that those tails are highly symmetric between K^+ and K^-. Only far tails are rejected by the cut $|m_{\pi\pi\pi} - m_K| < 9\,MeV/c^2$, where m_K is the PDG value. The systematic uncertainty due to pion decay, limited by the precision of generated MonteCarlo, is $\delta(\Delta g) = 0.4 \times 10^{-4}$.

The spectrometer internal alignment was calibrated using data from special runs in which muons were recorded with magnetic spectrometer off.

A fine control of the spectrometer internal alignment is obtained by continuously monitoring the difference in the reconstructed three pion invariant masses for K^+ and K^-, which can be induced by a residual horizontal misalignment between chambers before and after the spectrometer magnet, at the level of $\sim 1.5\,KeV/\mu m$. Tiny relative drifts of chambers positions (as small as a few μm per day up to 200 μm) were detected in this way and the reconstructed momenta corrected accordingly to $p' = p(1+q\beta p)$, where q is the sign of the charge, p is the track momentum (in GeV/c) and β a parameter of order $10^{-5}GeV^{-1}$ related to the measured mass difference for positive and negative events.

Each spectrometer magnetic field reversal was preceded by a full degaussing procedure, but the reproducibility of the absolute magnitude of the field integral, and its equality for both polarities, can be controlled on-line only to within $\sim 5\times10^{-4}$. Smaller variations are continuously corrected offline by forcing the mean reconstructed three-pion invariant mass to the nominal PDG kaon mass with $\sim 10^{-5}$ precision. This is done by scaling the measured track momenta symmetrically for positively and negatively charged tracks. This effect is charge symmetric and by collecting simultaneously K^+ and K^- it cancels in the ratio R(u).

Other small effects are canceled by the fact that geometric acceptance cuts were defined with respect to the average beam positions (both before and after magnetic spectrometer).

The beam tube traversing the center of the detector is determining the acceptance. The largest instrumental effect on the event density in the Dalitz plot occurs at large u values, where there is a steep drop in acceptance due to the "odd" pion being lost into the central beam pipe hole while similar "cuts" at large $(u\pm v)/2$ values, where one of the even pions is lost, are mapped onto a wide u region, after projection. To avoid inducing a u asymmetry in the way the acceptance-defining central hole is seen by the two beams, software acceptance cuts are applied which are centered on the effective beam axis, independently for K^+ and K^-. All three tracks are required to traverse the first drift chambers at least 11.5 cm from the beam center and the last drift chambers at least 13.5 cm (the latter cut takes into account the $\sim$ 2 cm lateral beam displacement due to the spectrometer magnet). These cuts are related to the beam center rather then the detector axis, the reason is that the beam optics can control on-line the beam position to only $\pm$ 1 mm. The actual beam position is continuously monitored to a better precision by calculating the momentum weighted center of gravity of the three pions, independently for K^+ and K^-. In addition to the time variation of the beam position, also the dependence from the kaon momentum variation is taken into account ($\sim \pm1$ mm both in horizontal and vertical direction). In doing this K^+ and K^- acceptance variations cancel completely and no MonteCarlo correction is needed. A conservative limit on residual systematic uncertainty, $\delta(\Delta g) = 0.3 \times 10^{-4}$ was determined by studying the sensitivity to various acceptance definitions.

The trigger is a potential source of systematic bias and is studied by using downscaled control samples from low bias triggers collected along with the main triggers. Since the beams are simultaneous, rate-dependent inefficiencies are charge symmetric. The inefficiency of the first level trigger was small, 9×10^{-4}, and stable in time. No correction is applied and an uncertainty of $\delta(\Delta g) = 0.4 \times 10^{-4}$, limited by the statistic of the control sample, is estimated. For the second level trigger inefficiencies are larger (0.4 to 1.5%) and change in time, being related to local drift chamber inefficiencies which are more important in the trigger than in the reconstruction due to reduced redundancy. No significant charge asymmetry or u dependence was observed but, since the size of the control sample is not sufficient to completely exclude such effects, a conservative approach is adopted by correcting each sample by the measured u-dependent trigger efficiency and therefore introducing a significant statistical error due to the statistical power of the control sample.

Table 1. Summary of limits on systematic and trigger uncertainties on $\Delta g = A_g/2g$, in units of 10^{-4}.

Acceptance, beam geometry	0.3
Spectrometer alignment	0.1
Spectrometer magnet	0.1
Pion decay	0.4
u computation and fitting	0.2
Accidental activity	0.2
Total systematic uncertainty	0.6
Trigger efficiency: level 1	0.4
Trigger efficiency: level 2	0.5

Other possible sources of systematic effects were studied and evaluated such as the dependence on the way the u variable is calculated or the fitting limits, effects due to uncertainty on the knowledge of the stray magnetic fields, pile-up effects, inhomogeneities in the spectrometer alignment, accuracy of time-tracking for various changes in the beam geometry and to charge asymmetric pion interactions. Additional studies were performed excluding various border parts of the Dalitz plot. This result is obtained from the fit restricted in the interval $-1 < u < 1$. Values of systematic uncertainties are shown in Table 1.

5. Result and Conclusions

The result presented is obtained as the average of three independent analysis, all of them giving consistent results within uncorrelated uncertainties. The result is calculated for each of the four supersamples of 2003 data taking and then combined, after trigger efficiency corrections, taking into account correlated systematic uncertainties.

The results from the four supersamples are statistically consistent with each other ($\chi^2/ndf = 0.4/3$). The measurement stability as a function of the supersample is shown in the left part of Fig. 2.

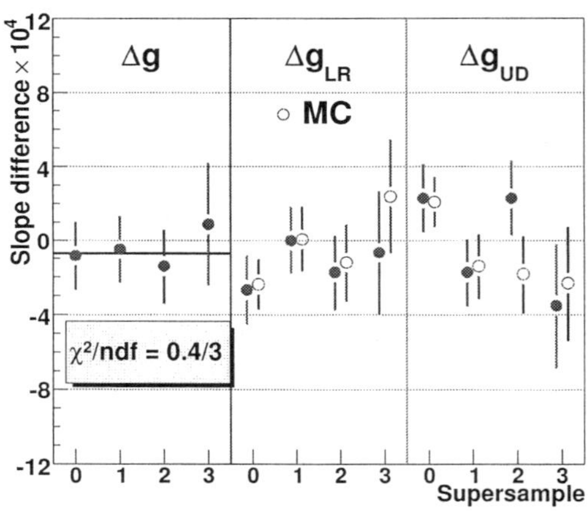

Figure 2. Left plot: stability of the preliminary result as a function of supersample. Two plots on the right side: stability of "null" asymmetries and their comparison to MonteCarlo simulation.

The result stability was checked with respect to several variables such as kaon energy and decay position, without finding any significant dependence. As a systematic check, "null" ratios were computed by building ratios of events of the same charge, deflected in opposite directions in the spectrometer magnet or distinguished only by the upper or lower path of kaons along the beam line. In this case the result is expected to be equal to zero and any asymmetry in such ratios reflects instrumental biases coupled to time variations. Such effects are at the 10^{-4} level and therefore second order effects are negligible. Moreover these results are fully reproduced by the MonteCarlo simulation as due to time variation of the detector inefficiencies and beam optics.

The difference in the linear slope parameter of the Dalitz plot for $3\pi^{\pm}$

decays of $K^\pm$, measured with the 2003 data sample, is found to be :

$$\Delta g = g^+ - g^- = (-0.7 \pm 0.9_{stat} \pm 0.6_{stat(trig)} \pm 0.6_{syst}) \times 10^{-4}$$

Converted to the direct CP violation charge asymmetry in $K^\pm \to 3\pi^\pm$ decays using the PDG value of the Dalitz plot slope $g = -0.2154 \pm 0.0035$, this leads to

$$A_g = (1.7 \pm 2.1_{stat} \pm 1.4_{stat(trig)} \pm 1.4_{syst}) \times 10^{-4} = (1.7 \pm 2.9) \times 10^{-4}$$

The precision obtained is limited mainly by the available statistics. This result is consistent with no CP violation and its precision is one order of magnitude better than earlier measurements.

Na48/2 has also collected about 200 millions of $K^\pm \to \pi^\pm\pi^0\pi^0$ decays from which a measurement of the corresponding slope asymmetry is extracted. This decay mode is disfavored statistically due to the lower branching ratio and acceptance but its more favorable Dalitz plot population leads to expected statistical uncertainty on A_g comparable to that of the charged decay mode. Systematic uncertainties are different in this case, but it is interesting to notice that for this decay the asymmetry is extracted using only information from the electromagnetic calorimeter, therefore leading to a rather complementary CP-violation measurement.

The preliminary result, presented in November 2005, is:

$$A_g^0 = (1.7 \pm 1.7_{stat} \pm 1.2_{stat(trig)} \pm 1.3_{syst} \pm 0.2_{ext}) \times 10^{-4} = (1.7 \pm 2.4) \times 10^{-4}$$

References

1. J. H. Christenson et al.,*Phys. Rev. Lett.* **13**, 138 (1964).
2. G. Barr et al., *Physics Letters* B **317**, 233 (1993); A. Alavi-Harati et al., *Phys. Rev. Lett.* **83**, 22 (1999); A. Alavi-Harati et al., *Phys. Rev.* D **67**, 012005 (2003); Erratum: *Phys. Rev.* D **70**, 079904 (2004).
3. V. Fanti et al., *Phys. Lett.* B **465**, 335 (1999); A. Lai et al., *Eur.Phys.J.* C **22**, 231 (2001); J.R. Batley et al., *Physics Letters* B **544**, 97 (2002).
4. S. Eidelman et al., (PDG), *Physics Letters* B **592** (2004).
5. W. T. Ford et al., *Phys. Rev. Lett.* **25**,1370 (1970); K. M. Smith et al., *Nucl. Phys.* B **91**, 45 (1975); W-S. Choong, Ph.D. thesis, Berkeley (2000) LBNL-47014; G. A. Akopdzhanov et al., *Eur. Phys. J.* C **40**, 343 (2005).
6. L. Maiani et al., The Second DAΦNE Physics Handbook, Vol.I, 51 (1995); A. A. Bel'kov et al., hep-ph/0311209; G. D'Ambrosio et al., *Int. J Mod. Phys.* A **13**, 1 (1998); E. P. Shabalin, *Phys. Atom. Nucl.* **68**, 88 (2005); E. Gamiz et al., *JHEP* **0310**, 042 (2003). E. P. Shabalin, ITEP preprint 8-98 (1998); G. D'Ambrosio et al., *Phys. Lett.* B **480**, 164 (2000);
7. R. Batley et al., Addendum III to Proposal P253, CERN/SPSC 2000-03, CERN, January 2000.

NEUTRINO ASTRONOMY AT THE SOUTH POLE

D.J. BOERSMA *

The IceCube Project,
222 West Washington Avenue,
Madison, Wisconsin, USA
E-mail: boersma@icecube.wisc.edu

A brief overview of AMANDA and IceCube is presented, with an introduction into the basic physics and detection principles and a small selection of the results.

1. Introduction

More than a hundred years after cosmic rays were discovered by Victor Hess their origin is still a mystery. Their energy spectrum extends to 10^{11}GeV, many orders of magnitude higher than can be reached with the current particle accelerators. There are various models and ideas about how cosmic rays acquire such high energies, such as Fermi acceleration in the jets of so-called Active Galactic Nuclei or in the shells of matter blasted away during a gamma-ray burst (GRB), in a Supernova remnant or in some other explosive astrophysical event. There are also models involving topological defects[1] such as magnetic monopoles created during the Big Bang. All these hypotheses may turn out to be wrong or insufficient. Whatever the outcome, the cosmic ray puzzle will teach us something fundamentally new about the Universe.

In the quest to solve this puzzle it is important to collect as many independent observations as possible. Measurements of, or constraints on, the flux of high energy neutrinos complement the existing and ongoing measurements of hadronic and electromagnetic cosmic rays. This is the main purpose of the AMANDA and IceCube experiments.

In many of the models for cosmic ray acceleration, hadrons (typically protons) are accelerated, which may collide with stellar or interstellar ma-

*On behalf of the icecube collaboration. Full author list: http://icecube.wisc.edu/science/publications/llwi2006.html

terial, thereby producing pions. The neutral pions decay into photons (high energy gamma rays), while the decay of each charged pion produces in total three neutrinos (ν_μ, $\bar{\nu}_\mu$ and $\nu_e/\bar{\nu}_e$), which may carry a significant fraction of the energy of the initial accelerated particle. This scenario is called the beam dump model.

There is a theoretical upper bound on the energy of cosmic rays – the so-called GZK-cutoff[2] – due to the presence of 3K cosmic microwave background (CMB): e.g. a proton with more than 50EeV can be excited to a Δ resonance by a CMB photon. The decay of a (charged) Δ resonance will result in (extremely high energy) neutrinos.

A particularly interesting aspect of neutrinos is that they are unlikely to get deflected or absorbed on their long journey from their generation to our detector, so that with their direction they point back to their origin and we can indeed speak about "neutrino astronomy".

2. Detection of high energy neutrinos in ice

When a high energy muon neutrino undergoes a charged current interaction in ice it converts into a muon. This muon keeps most of the neutrino's energy and radiates Cherenkov light. By populating the ice with photomultiplier tubes (PMTs) and measuring arrival times of the Cherenkov photons, the direction of the long muon tracks ($\sim$ 1km at E = 200GeV) can be reconstructed to an accuracy of 2°–3° in AMANDA (expected better than 1° in IceCube). For E_ν = 1TeV the angle between the muon and the original neutrino is of the order of 1°. There is a high rate of down-going muons coming from cosmic ray showers in the atmosphere directly above the South Pole; this rate is 10^6 times higher than the expected neutrino rate, so except for very high energies neutrinos can only be identified if they were coming through the Earth (up-going).

A charged current interaction by an electron or tau neutrino leads to an electromagnetic cascade. A neutral current interaction by any neutrino leads to a hadronic cascade. Cascades are relatively short (typically several tens of meters), so the angular resolution is much worse than for muons (30°–40° in AMANDA), but the energy resolution is better.

An important role in the reconstruction is played by optical properties of the ice[4]. These vary with the concentration of insoluble dust, which tracks climatological variations in the last 70,000 years. For the wavelengths around 400nm the absorption length λ_a ranges from 50 to 130 meter (average 110m) and the effective scattering length λ_{eff} ranges from 6 to 30m (average 20m).

3. Detector hardware

AMANDA

The Antarctic Muon And Neutrino Detector Array (AMANDA) consists of 19 strings with 20-42 optical modules (OMs) per string, frozen in ice at depths of 1500–2000m. The strings are distributed over a roughly circular area with a diameter of $\sim$200m. Each of the 677 OMs consists of an 8 inch photomultiplier tube enclosed in a pressure resistant glass sphere and is connected to the surface electronics by a twisted pair, coaxial and/or optical cable.

IceCube

IceCube, when completed, will have 70 strings, each with 60 Digital Optical Modules (DOMs) spaced by 17m. The strings are arranged in a hexagonal pattern spanning an area of about 1km^2. The distance between neighboring strings is 120m. AMANDA, which has inter-string distances of 30–50m, will be enclosed by IceCube and will be kept in operation, serving as a low energy sub-array. The signals of the PMTs are read out and digitized inside each DOM itself, thus avoiding signal distortion in the long cables to the surface. A full description of the design can be found in Ref. 5.

During the austral summer of 2004–2005 the first IceCube string was deployed and commissioned, demonstrating that the hot-water drill, the deployment hardware and all detector components perform according to specifications[6]. During the 2005-2006 season 8 more strings were deployed, so currently IceCube has 9 strings with in total 540 DOMs. In the following years the deployment rate should be 12-16 strings per season.

4. Research topics

Within the scope of this work only a selection of the topics being studied with AMANDA and IceCube can be presented.

The sky map as obtained from up-going tracks is given in figure 1. It is dominated by the roughly isotropically distributed muon tracks from neutrinos which originate from interactions of high energy cosmic rays in the Earth atmosphere. These so-called "atmospheric neutrinos" can be used to study systematics of the detector and to test models of cosmic ray interactions. In principle, they could also be used to study neutrino oscillation with a very long baseline, but the expected accuracy does not exceed that of the experiments dedicated to study neutrino oscillation.

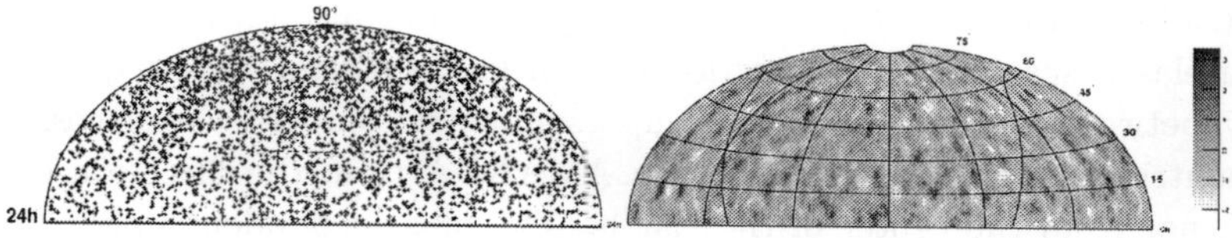

Figure 1. Left panel: directions of up-going muons with good reconstruction quality, from data taken in 2000-2003. Right panel: significance plot (gray-scale levels correspond to number of standard deviations above estimated background).

The sky map is analyzed for statistically significant excesses above this smooth atmospheric background. The total number of upward going tracks collected in 2000-2003 (807 days live-time) is 3329. The highest fluctuation was 3.4 sigma; the probability to get such an excess or greater in any angular bin in the sky map is 92%. When restricting the search to the directions of a limited number of known celestial objects, the largest excess is 10 tracks pointing back to the Crab nebula, on an expected background of 5.4; the chance probability to find such an excess or greater for any of the 33 source candidates is 64%. From these results upper limits can be derived for the neutrino fluxes from these sources[7,8].

Another way to identify extraterrestrial neutrinos is by their energy, rather than their direction. The differential flux of atmospheric neutrinos measured in AMANDA[3] is $cos\theta \frac{dE}{dE} \propto E^{-3.54\pm0.11}$, whereas several models for extraterrestrial sources predict harder spectra ($\propto E^{-2}$ or harder), which can be probed by restricting the search to higher energies at which we do expect very few atmospheric neutrinos. This is done by requiring a minimum number of PMTs with a signal. No excess over the expected background is observed yet.

If Nature is supersymmetric and R-parity is conserved, then the lightest supersymmetric particles (from now on "neutralinos") cannot decay, and we should expect a sizeable number of neutralinos to propagate freely through the Universe. They may lose energy through elastic scattering with ordinary matter and get gravitationally trapped inside a celestial body, for instance the Sun or the Earth. After losing more energy in subsequent elastic collisions they end up in the center of a celestial body, where the density of neutralinos increases until they have a finite chance of colliding with each other. They annihilate predominantly into W^+W^-, $b\bar{b}$ and $\tau\bar{\tau}$

which decay further into neutrinos and other particles. The rate at which AMANDA (and later IceCube) will detect such neutrinos from neutralino annihilation[9] depends not only on the density and velocity distribution of neutralinos in the vicinity of the solar system, but also on their mass and their elastic scattering cross section with ordinary matter. These quantities vary between various versions of supersymmetric models. No excess rate of neutrinos has been observed with AMANDA from the direction of the Sun, nor from the center of the Earth. The mass dependent upper limits which we may derive from this[10] on the neutralino density are 2–10 times higher than the limits from direct searches for cold dark matter[11]. However, the indirect and direct methods are complementary since they apply to different parts of the neutralino velocity distribution.

5. Conclusions

AMANDA has successfully demonstrated the feasibility of neutrino detection with an array of optical modules in the Antarctic ice, and produced promising first results. The IceCube experiment is not only a larger scale detector, it also involves a major technological upgrade. IceCube, now partly constructed, is already the largest neutrino telescope in the world and will grow quickly in the next few years.

References

1. S. Yoshida *et al.*, *Astrophys. J.* **479** (1997) 547.
2. K. Greisen, *Phys. Rev. Lett.* **16** (1966) 748; G.T.. Zatsepin and V.A. Kuzmin *JETP Lett.* **4** (1966) 178.
3. H. Geenen, *Energy reconstruction and spectral unfolding of atmospheric leptons with the AMANDA detector*, thesis Universität Wuppertal
4. M. Ackermann *et al.*, *Optical properties of deep glacial ice at the South Pole* in *J. Geophys. Res.*, in press, doi:10.1029/2005JD006687.
5. *IceCube preliminary design report*, see http://icecube.wisc.edu.
6. A. Achterberg *et al.*, *First Year Performance of the IceCube Neutrino Telescope*, submitted to *Astroparticle Physics*.
7. M. Ackermann *et al.*, *Search for high energy neutrino point sources in the northern hemisphere with the AMANDA-II neutrino telescope*, in *Proc. of the 29th Int. Cosmic Ray Conf.* (ICRC 2005), see also astro-ph/0509330.
8. M. Ackermann *et al.*, *Phys. Rev.* **D71** (2005) 077102.
9. W.H. Press and D.N. Spergel, *Astrophys. J.* **296** (1985) 679; J. Silk, K. Olive and M. Srednicki, *Phys. Rev. Lett.* **55** (1985) 257; T.K. Gaisser, G. Steigman and S. Tilav *Phys. Rev.* **D34** (1986) 2206.
10. M. Ackermann *et al.*, *Astropart. Phys.* **24** (2006) 459.
11. D.S. Akerib *et al.*, astro-ph 0509269.

DIRECT CP VIOLATION IN B DECAYS AT BELLE

M. BRAČKO *
University of Maribor, Smetanova ulica 17, SI-2000 Maribor, Slovenia
and
Jožef Stefan Institute, P.O.B. 3000, SI-1001 Ljubljana, Slovenia
E-mail: Marko.Bracko@ijs.si

A review of recent searches for direct CP violation in charmless B decays is given. Reported analyses were performed on a data sample containing about 386×10^6 $B\overline{B}$ pairs collected with the Belle detector operating at the KEKB e^+e^- collider.

1. Introduction

Direct CP violation (DCPV) in B decays occurs when the decay amplitudes for a B meson decay to a final state f, $B \to f$, and for its charge-conjugated mode[a], $\overline{B} \to \overline{f}$, are different. This can be observed as the CP asymmetry:

$$A_{CP} \equiv \frac{\Gamma(\overline{B} \to \overline{f}) - \Gamma(B \to f)}{\Gamma(\overline{B} \to \overline{f}) + \Gamma(B \to f)} = \frac{|\overline{\mathcal{A}}_{\overline{f}}|^2 - |\mathcal{A}_f|^2}{|\overline{\mathcal{A}}_{\overline{f}}|^2 + |\mathcal{A}_f|^2} \,. \tag{1}$$

This asymmetry of partial decay rates differs from zero, when the decay amplitudes $\mathcal{A}_f$ and $\overline{\mathcal{A}}_{\overline{f}}$ contain at least two terms with different phases. With *exactly two* terms, the amplitudes are $\mathcal{A}_f = |\mathcal{A}_1| e^{i(\phi_1+\delta_1)} + |\mathcal{A}_2| e^{i(\phi_2+\delta_2)}$ and $\overline{\mathcal{A}}_{\overline{f}} = |\mathcal{A}_1| e^{i(-\phi_1+\delta_1)} + |\mathcal{A}_2| e^{i(-\phi_2+\delta_2)}$, so that the A_{CP} from Eq. (1) reads:

$$A_{CP} = \frac{2\,|\mathcal{A}_1/\mathcal{A}_2|\,\sin(\phi_1-\phi_2)\,\sin(\delta_1-\delta_2)}{1 + |\mathcal{A}_1/\mathcal{A}_2|^2 + 2\,|\mathcal{A}_1/\mathcal{A}_2|\,\cos(\phi_1-\phi_2)\,\cos(\delta_1-\delta_2)} \,. \tag{2}$$

Eq. (2) shows that for large DCPV to be observed, both terms in the decay amplitudes need to be of comparable sizes, $|\mathcal{A}_1| \sim |\mathcal{A}_2|$, and have non-vanishing weak and strong phase differences: $(\phi_1-\phi_2), (\delta_1-\delta_2) \neq 0$. This could happen in some charmless B meson decays, for which the decay amplitudes include contributions from the *penguin* diagrams at a competing

*Representing the Belle Collaboration.
[a] Throughout the paper, charge-conjugated states are implied unless stated otherwise.

level with the *tree-level* diagrams, when the latter are suppressed in the Standard Model (SM). Since the existence of strong phase δ complicates the interpretation of results, measurements of DCPV for various decay modes are needed to test the Kobayashi-Maskawa (KM) picture[1] of CP violation in the SM. However, it is also interesting to search for DCPV in modes where it is not expected, since possible contributions from physics beyond the SM could modify the decay amplitudes through loop diagrams.

2. Recent Direct CP violation measurements at Belle

All reported analyses use a data sample containing about 386×10^6 $B\overline{B}$ pairs collected with the Belle detector[2] at the KEKB asymmetric-energy e^+e^- collider[3] that operates at the $\Upsilon(4S)$ resonance ($\sqrt{s}$=10.58 GeV).

2.1. *Common Analyses Tools*

B meson candidates are identified with two kinematic variables: $M_{\rm bc} \equiv \sqrt{E^2_{\rm beam}/c^4 - p^2_B/c^2}$ and $\Delta E \equiv E_B - E_{\rm beam}$. Here, E_B and p_B are the reconstructed energy and momentum of the B candidate, and $E_{\rm beam}$ is the beam energy, all expressed in the centre-of-mass (CM) frame. The dominant background in charmless B decay modes originates from continuum $e^+e^-\to q\bar{q}$ ($q = u, d, s, c$) production. To separate the jet-like $q\bar{q}$ events from the spherical $\Upsilon(4S)\to B\overline{B}$ processes, we use event shape variables[4]. Additional suppression of continuum is in some analyses achieved by using the b-flavour tagging parameter[5] as a measure of the confidence that particles in the event—other than the ones from the $B \to f$ decay—originated from a flavour specific B decay. Finally, signal yields and asymmetries are obtained from unbinned maximum-likelihood fits to the ΔE-$M_{\rm bc}$ distributions or Dalitz plots for selected candidate events.

2.2. *Results for $B \to K^+\pi^-$, $K^+\pi^0$ and $\pi^+\pi^0$ Decays*

The first experimental evidence for DCPV was shown by the Belle collaboration for the mode $B^0 \to \pi^+\pi^-$, followed by the evidence for DCPV in the decay $B^0 \to K^+\pi^-$, reported by both Belle and BaBar collaborations. In the updated $B \to K\pi/\pi\pi$ analysis[6], DCPV in $B^0 \to K^+\pi^-$ mode is confirmed with the 5σ significance, including systematic uncertainties, while no asymmetry is observed in charged B meson decays, as shown in Table 1 (see Fig. 1 for fit results). No observed asymmetry for $B^+ \to \pi^+\pi^0$ mode is consistent with expectations. In contrast to this, the difference between

Table 1. Measured signal yields and asymmetries in $B \to K\pi/\pi\pi$ decay modes.

Mode	Signal yield	A_{CP}	Background A_{CP}
$\overline{B}^0/B^0 \to K^{\mp}\pi^{\pm}$	3026 ± 63	$-0.113 \pm 0.022 \pm 0.008$	-0.001 ± 0.004
$B^{\mp} \to K^{\mp}\pi^0$	1084 ± 45	$+0.04 \pm 0.04 \pm 0.02$	-0.02 ± 0.01
$B^{\mp} \to \pi^{\mp}\pi^0$	454 ± 36	$+0.02 \pm 0.08 \pm 0.01$	-0.01 ± 0.01

$A_{CP}(K^+\pi^-)$ and $A_{CP}(K^+\pi^0)$ is 3.1σ away from zero, as expected in naïve factorisation. This suggests possible contributions from the electroweak penguin process in $B^+ \to K^+\pi^0$ decay or other mechanisms.

2.3. *Results for $B^+ \to \eta h^+$ and $\eta' h^+$ Decays*

Searches for DCPV can also be performed in quasi-two-body decay modes, such as $B^+ \to \eta h^+$ and $B^+ \to \eta' h^+$ ($h = K, \pi$). CP violation could be large in decay modes $B^+ \to \eta\pi^+$, ηK^+ and $\eta'\pi^+$, due to penguin-tree interference. For $B^+ \to \eta' K^+$ decays, which are within the SM believed to proceed dominantly via gluonic penguin processes, any non-zero CP asymmetry may require explanation with new physics. In the $B^+ \to \eta h$ analysis[7], $\gamma\gamma$ and $\pi^+\pi^-\pi^0$ final states are used for the η meson reconstruction, while in the $B^+ \to \eta' h^+$ analysis[8], η' mesons are reconstructed via two decay chains: $\eta' \to \eta\pi^+\pi^-$ (with $\eta \to \gamma\gamma$) and $\eta' \to \rho^0\gamma$ (with $\rho^0 \to \pi^+\pi^-$). A fit to the ΔE-$M_{\rm bc}$ distribution in both analyses includes the signal and similar

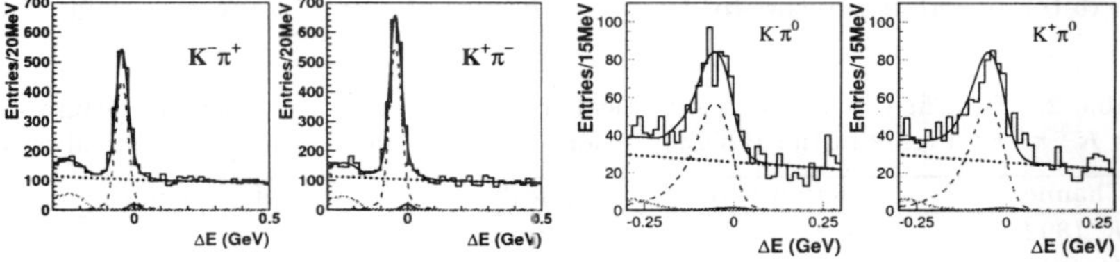

Figure 1. $M_{\rm bc}$ (top) and ΔE (bottom) distributions for $\overline{B}^0 \to K^-\pi^+$, $B^0 \to K^+\pi^-$, $B^- \to K^-\pi^0$ and $B^+ \to K^+\pi^0$ (left to right). Histograms represent data, curves display the fitted signal (dashed); backgrounds from $q\bar{q}$ (dotted), three-body B decays (dash-dotted) and K/π misidentification (hatched); and the sum of all components (solid).

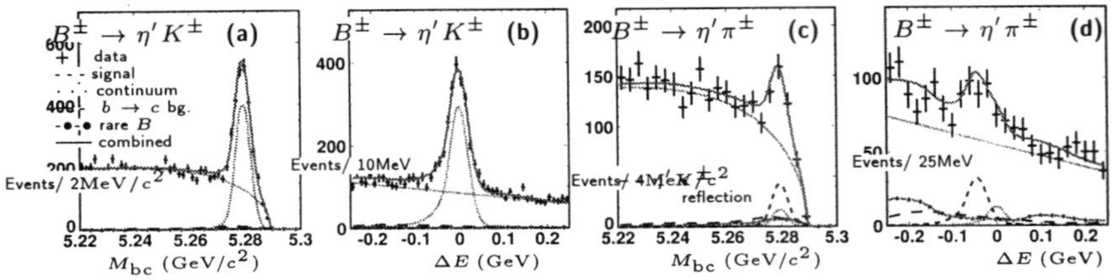

Figure 2. $M_{\rm bc}$ and ΔE distributions for: (a), (b) $B^\pm \to \eta' K^\pm$; (c), (d) $B^\pm \to \eta'\pi^\pm$.

background components: $q\bar{q}$ continuum, background from other charmless B decays (*e.g.* $B^+ \to \eta K^{*+}(\eta' K^{*+})$ for $B^+ \to \eta K^+(\eta' K^+)$) and reflections due to K-π misidentification (*e.g.* $B^+ \to \eta' K^+$ for $B^+ \to \eta'\pi^+$). For $B^+ \to \eta' h^+$ decays, $b \to c$ background is also included (see Fig. 2 for the illustration of the $B^\pm \to \eta' K^\pm, \eta'\pi^\pm$ fit results). The measured $A_{CP}(\eta\pi^+) = -0.10 \pm 0.11 \pm 0.02$ is consistent with zero, but an intriguing 2.9σ effect for the $B^+ \to \eta K^+$ asymmetry, $A_{CP}(\eta K^+) = -0.55 \pm 0.19^{+0.04}_{-0.03}$, awaits even larger data samples. No DCPV is observed in $B^+ \to \eta' h^+$ decays: $A_{CP}(\eta'\pi^+) = 0.20^{+0.37}_{-0.36} \pm 0.04$ and $A_{CP}(\eta' K^+) = 0.028 \pm 0.028 \pm 0.021$. Especially the latter result strongly supports the KM explanation within the SM and requires no additional contributions from new physics.

2.4. $B^+ \to K^+\pi^+\pi^-$ *Dalitz Analysis Results*

In decays of B mesons to two-body final states, DCPV can be observed as a difference in B and $\overline{B}$ decay rates. In B decays to three-body final states—often dominated by quasi-two-body channels—DCPV can also manifest itself as a difference in relative phase between two quasi-two-body amplitudes that can be measured via amplitude (Dalitz) analysis. In the three-body $B^+ \to K^+\pi^+\pi^-$ Dalitz analysis[9], the signal, $q\bar{q}$ and $B\overline{B}$ background yields are determined from the fit to the ΔE distribution, shown in Fig. 3(a).

Table 2. Best fit results (with statistical errors and model dependent uncertainties) for $K^\pm\pi^\pm\pi^\mp$ events in the B signal region. The A_{CP} significance is statistical only.

Channel	Fraction (%)	A_{CP} (%)	Asymmetry significance (σ)
$K^*(892)\pi^\pm$	$13.0 \pm 0.8^{+0.5}_{-0.7}$	$-14.9 \pm 6.4^{+0.8}_{-0.8}$	2.6
$K_0^*(1430)\pi^\pm$	$65.5 \pm 1.5^{+2.2}_{-3.9}$	$+7.6 \pm 3.8^{+2.0}_{-0.9}$	2.7
$\rho(770)^0 K^\pm$	$7.85 \pm 0.93^{+0.64}_{-0.59}$	$+30 \pm 11^{+11}_{-4}$	3.9
$f_0(980)K^\pm$	$17.7 \pm 1.6^{+1.1}_{-3.3}$	$-7.7 \pm 6.5^{+4.1}_{-1.6}$	1.6
$f_2(1270)K^\pm$	$1.52 \pm 0.35^{+0.22}_{-0.37}$	$-59 \pm 22^{+3}_{-3}$	2.7
$\chi_{c0}K^\pm$	$1.12 \pm 0.12^{+0.24}_{-0.08}$	$-6.5 \pm 20^{+2.9}_{-1.4}$	0.7

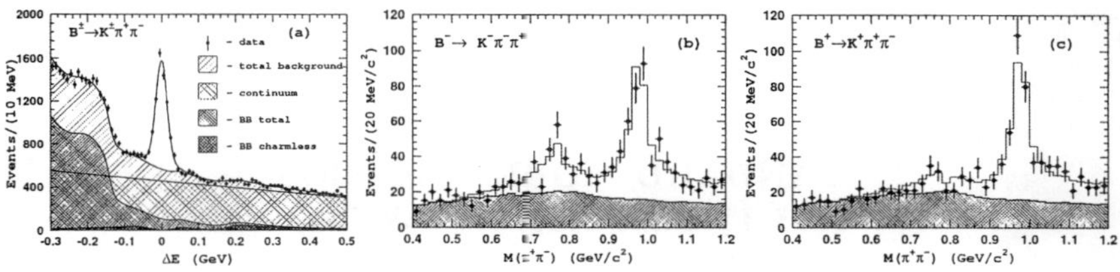

Figure 3. (a) ΔE distribution for $B^{\pm} \to K^{\pm}\pi^{\pm}\pi^{\mp}$; (b), (c) $m_{\pi^+\pi^-}$ mass spectra for B^-, B^+ events. Points with error bars are data; the open histogram is the fit result; the hatched areas are various background components.

Events in the elliptical B signal region defined around the $M_{\rm bc}$ and ΔE mean values are selected for the amplitude analysis, the background shape is determined from the $M_{\rm bc}$-ΔE sidebands. The $B^+ \to K^+\pi^+\pi^-$ signal is described by a coherent sum of $K^*(892)^0\pi^+$, $K_0^*(1430)^0\pi^+$, $\rho(770)^0K^+$, $f_0(980)K^+$, $f_X(1300)K^+$ and $\chi_{c0}K^+$ quasi-two-body channels and a non-resonant amplitude. First evidence for direct CP violation in the charged meson decay is found for the $B^+ \to \rho(770)^0K^+$ decay mode with the $\mathbf{3.9}\sigma$ statistical significance, as shown in Table 2. For illustration, compare also Figs. 3(b) and 3(c) showing the $\pi^+\pi^-$ invariant mass distributions around $\rho(770)^0$ mass region separately for B^- and B^+ events.

3. Summary and Conclusions

Four DCPV analyses of charmless B decays from the Belle collaboration are reviewed in this paper. Measured values for two-body decays[6,7,8] are still preliminary, while results of the three-body $B^+ \to K^+\pi^+\pi^-$ Dalitz analysis[9], which provides the first evidence for direct CP violation in the charged-meson decay, were already accepted for publication.

References

1. M. Kobayashi and T. Maskawa, Prog. Theor. Phys. **49**, 652 (1973).
2. A. Abashian *et al.* (Belle), Nucl. Instr. and Meth. A **479**, 117 (2002).
3. S. Kurokawa and E. Kikutani, Nucl. Instr. and Meth. A **499**, 1 (2003).
4. The method used by Belle is described in K. Abe *et al.* (Belle), Phys. Rev. Lett. **87**, 101801 (2001); K. Abe *et al.* (Belle), Phys. Lett. B **511**, 151 (2001).
5. H. Kakuno *et al.*, Nucl. Instr. and Meth. A **533**, 516 (2004).
6. K. Abe *et al.* (Belle), BELLE-CONF-0523, hep-ex/0507045.
7. K. Abe *et al.* (Belle), BELLE-CONF-0525, hep-ex/0508030.
8. J. Schümann *et al.* (Belle), hep-ex/0603001, submitted to Phys. Rev. Lett.
9. A. Garmash *et al.* (Belle), hep-ex/0512066, accepted by Phys. Rev. Lett.

STANDARD MODEL PHYSICS WITH THE ATLAS EARLY DATA

P. BRÜCKMAN DE RENSTROM*

University of Oxford,
DWB, Keble Road, Oxford OX1 3RH, UK
and
Institute of Nuclear Physics PAN, ul. Rdzikowskiego 152, Kraków 31-342, PL
EE-mail: p.bruckman1@physics.ox.ac.uk

The Standard Model, despite its open questions, has proved its consistency and predictive power to very high accuracy within the currently available energy reach. LHC, with its high CM energy and luminosity, will give us insight into new processes, possibly showing evidence of "new physics". Excellent understanding of the SM processes will also be a key to discriminate against any new phenomena. Prospects of selected SM measurements with the ATLAS detector using early LHC luminosity are presented.

1. Introduction

The LHC collider will collide protons with the design instantaneous luminosity of 10^{34} cm^{-2}s^{-1}. In 2007 we expect only a short pilot run which might, at best, deliver an integrated luminosity of 100 pb^{-1}. In 2008 LHC is foreseen to operate at, so called, "low instantaneous luminosity" of 10^{33} cm^{-2}s^{-1} (1 nb^{-1}s^{-1}) delivering around 10 fb^{-1} of data during 100 days of effective physics running. This first year of operation should already provide highly valuable physics output. In the following we discuss selected Standard Model (SM) physics analyzes planned on this initial data with ATLAS.

The ATLAS detector is a general purpose spectrometer[1,2]. Its high granularity tracking system embedded in in the two Tesla solenoidal magnetic field covers region of pseudorapidity $|\eta| < 2.5$. It consists of a silicon pixel system, silicon strip detectors and straw tube proportional chambers, altogether providing pattern recognition, momentum determination

*On behalf of the ATLAS Collaboration.

and vertex reconstruction for charged tracks. The tracking system is complemented by muon chambers combined with air-core toroids for efficient muon identification and momentum reconstruction. Muon chambers cover the range of $|\eta| < 2.7$. Electromagnetic and hadronic calorimeters, reaching down to $|\eta| < 5$, provide gamma and neutral hadron reconstruction, electron identification, energy flow and inclusive jet reconstruction.

2. SM program with the ATLAS early data

LHC contribution to SM measurements[a] will be particularly relevant in the top quark sector where measurements to date come solely from the Tevatron[3]. Measurements in the top sector offer not only stringent tests of the consistency of the SM providing an indirect limit on the mass of the SM Higgs particle but also allows probing for new physics. For similar reasons W mass measurement is important but due to difficult systematic errors involved in the analysis we expect results from LEP and Tevatron to dominate the scene at least in the first years of LHC running.

In order to do precision physics, however, we need to reduce current uncertainties on the Parton Density Functions (PDF) of the proton which give rise to large errors on certain theoretical predictions.

2.1. *$W^{\pm}$ production and the Parton Density Functions*

Because of scaling violation PDF's are in general dependent on Bjorkein x and Q^2, fourmomentum transfer involved in the hard process $f(x, Q^2)$. The x dependence cannot be calculated from first principles. It is parameterized and fitted to data. The Q^2 dependence can be predicted from the DGLAP QCD evolution equations. Both the fit and the Q^2 evolution give rise to systematic uncertainties.

Single $W^{\pm}$ production partonic crossections are known at NNLO with accuracy $< 1\%$[4]. W's can be studied e.g. in their leptonic decays $W^{\pm} \to e^{\pm}\nu(\bar{\nu})$. A simple selection requiring an isolated electron with $p_T > 25$ GeV and $|\eta| < 2.4$, $E_{\mathrm{T}}^{\mathrm{miss}} > 25$ GeV, no jets with $E_{\mathrm{T}} > 30$ GeV within $|\eta| < 2.5$ and $p_{\mathrm{T}}^{\mathrm{recoil}} < 20$ GeV leads to a clean W sample with less than 1% background contamination mainly from $W/Z \to \tau, Z \to e^+e^-$ and QCD. The total PDF uncertainty on the rapidity distributions of the electrons in the central region ($|\eta| < 2.5$) is $\sim 8\%$[5] and is retained after the selection cuts. Inclusion of the above electron rapidity distributions, resulting from

[a] Apart from eventual discovery of the SM Higgs particle, of course.

just 100 pb^{-1}, to the PDF fit can provide useful constraints on the low-x gluon shape parameter λ ($xg(x) \sim x^{-\lambda}$) [5,6]. The work is in progress.

Finally, it should be noted that due to clean signal and accurate theoretical predictions single $W^{\pm}$ and single Z production with subsequent leptonic decays may prove to be among the most reliable luminosity monitors. Recall that the observed event rate (OR) is given by:

$$\mathrm{OR} = \epsilon_{\mathrm{exp.}} \times \int dx_1 dx_2 \sigma_{ab\to X}(x_1 x_2 S = Q^2) f_a(x_1, Q^2) f_b(x_2, Q^2) \times \int L_{\mathrm{p}} dt.$$

Normalising observed event rates to the rate of the candle processes (e.g. $W^{\pm}$) would result in a great deal of cancellation. Proton luminosity drops completely and the PDF uncertainties might be substantially reduced.

2.2. $t\bar{t}$ *production and* m_t *measurement*

LHC will be a top factory. The crossection for $t\bar{t}$ production of 825 pb$\pm$5%[7] is two orders of magnitude larger than at the Tevatron ($\sim$ 6.7 pb). Even at the low luminosity (10^{33}) nearly one $t\bar{t}$ pair will be produced every second.

Higgs and quark virtual loops contribute to the W effective mass. At one loop correction to the W mass is given by:

$$m_W^2 = \frac{\frac{\pi\alpha}{\sqrt{2}G_F}}{s_W^2(1-\Delta r)} \quad \text{with} \quad s_W^2 \equiv 1 - \frac{m_W^2}{m_Z^2},$$

$$(\Delta r)_{\mathrm{top}} = -\frac{3G_F m_t^2}{8\sqrt{2}\pi^2}\frac{c_W^2}{s_W^2}, \qquad (\Delta r)_{\mathrm{Higgs}} = \frac{11 G_F m_Z^2 c_W^2}{24\sqrt{2}\pi^2}\ln\frac{m_h^2}{m_Z^2}.$$

The current world average values of for the top and W mass ($m_t = 172.7 \pm 2.9$ GeV, $m_W = 80.410 \pm 0.032$ GeV) can be converted into the 95% CL limit on the SM Higgs mass: $m_H < 186$ GeV [8]. LHC goal is to reduce δm_t to 1 GeV and ultimately δm_W to $\sim$15 MeV. While the latter involves difficult systematics, precise top mass measurement has been demonstrated feasible already with the limited statistics expected in 2008[9].

Top mass is best studied in so called gold plated channel where one of the tops decays semileptonically producing a b-jet and an electron or a muon while the other one decays hadronically to two light and one b jet. Branching for this channel is 30%. It provides a clean trigger from the high p_T isolated lepton and allows for in situ light jet energy scale calibration using W to two jet decays. B-tagging plays an important role in the reconstruction as it cleans the sample and helps jet assignment. The selection requires an isolated lepton (e or μ) with $p_T > 20$ GeV and four jest with $E_{\mathrm{T}} > 40$ GeV within $|\eta| < 2.5$, $E_{\mathrm{T}}^{\mathrm{miss}} > 20$ GeV. More than one hard jet has to be b-tagged ($\epsilon_b \approx 50\%$, $\epsilon_u ds \approx 10^{-3}$, $\epsilon_c \approx 10^{-2}$). Background (<2%) consists of mainly W/Z+jets and $WW/ZZ/WZ$ events. With 10 fb^{-1} statistical error on top mass is small (0.050 GeV). Main contribution to the systematic error comes from the b-jet energy scale (1% uncertainty gives $\delta m_t = 0.6$ GeV) and the Final State Radiation (1 GeV). The FSR uncertainty can be reduced by half when a kinematic fit is used to combine information from both hadronic and leptonic top decay. In the latter case the total systematic error has been estimated to be 0.9 GeV. In situ calibration of light jets ($m_t =< m_{jjb} - m_{jj} > +m_W$(PDG)) reduces light jet energy scale error contribution to 0.2 GeV.

Top quark decays in $\approx 100\%$ to an on-shell W boson and a b quark. SM gives a rigid prediction for the polarization of the W boson from the decay. A longitudinal W^+ is produced with probability $m_t^2/(m_t^2 + 2m_W^2)$ (0.703 for the 175 GeV top) while the remaining 30% are left-handed W^+ (right-handed W^- for a decaying $\bar{t}$). NLO calculation results in O(1%) correction to this prediction. Polarisation of the W can be studied in it's leptonic decays to e or μ. The angular distribution of the lepton is given by:

$$\frac{1}{N}\frac{dN}{d\cos\Theta} = \frac{3}{2}\left\{F_0\left(\frac{\sin\Theta}{\sqrt{2}}\right)^2 + F_L\left(\frac{1-\cos\Theta}{\sqrt{2}}\right)^2 + F_R\left(\frac{1+\cos\Theta}{\sqrt{2}}\right)^2\right\}$$

where Θ is the angle between the charged lepton and the W direction of flight in the W rest frame. With 10 fb^{-1} of data statistical error becomes negligible while the systematic uncertainty on F_0 and F_R have been estimated to be 0.016 and 0.012 respectively. The measurement may be interpreted in terms of constraints on the effective Lagrangian, putting limits on the right-handed current and dimension five left and right-handed terms[10].

Tops are produced unpolarized but their spins are correlated. At the production threshold gg fusion leads to same helicity $t\bar{t}$ pairs (90% @LHC) while $q\bar{q}$ annihilation (10%) produces opposite helicity ones. Therefore for the $t\bar{t}$ pairs near the threshold SM predicts an asymmetry between $LL+RR$ and $LR + RL$ helicity state production[11]. The helicity correlation can be studied using angular distributions of daughter lepton in top semileptonic decays. The analysis selects di-lepton top events with $m_{t\bar{t}} < 550$ GeV and fits the asymmetry parameter, A, of the double differential angular distribution:

$$\frac{1}{N}\frac{d^2N}{d(\cos\theta_1)d(\cos\theta_2)} = \frac{1}{4}(1 - A\cos\theta_1\cos\theta_2)$$

where θ's are the lepton angles in the respective top helicity frames. Using the fast simulation of the ATLAS detector it has been shown that an accuracy of around 5% on the asymmetry parameter can be achieved with 10 fb^{-1}[10]. Any deviation from the SM prediction might sign the presence of $t\bar{t}$ production through a new mechanism, e.g. a heavy scalar (Higgs?) a new vector particle (Z'?) or a spin-2 object (KK graviton?).

Finally, let us mention the rich physics program for the single top production. At LHC single tops will be produced via three mechanisms: W gluon fusion (t-channel), Wt associated production and $t\bar{b}$ weak production (s-channel). The total crossesction exceeds 300 pb. Single top production through the weak charged current is proportional to the $|V_{tb}|^2$ and allows for direct measurement of the latter[12,13]. Similarely, an excess production would indicate anomalous contributions like FCNC, charged Higgs boson (2HDM), extra gauge bosons, W', extra dimensional KK, etc[14]. V-A nature of the tWb vertex can be probed as the emerging top's are expected to be $\approx$ 100% polarised.

Acknowledgments

We would like to thank the ATLAS Standard Model Group and the Top Group for the wide choice of the analysis material. Special thanks go to: M. Cooper-Sarkar, I. Dawson, A.-I. Etienvre, C. Gwenlan, A. Lucotte, M. Moulders, D. Pallin, B. Rosende, V. Simak, K. Sliwa and A. Tricoli whose work directly contributed to this paper.

References

1. ATLAS Collaboration, ATLAS Detector and Physics Performance TDR, ATLAS TDR 15, CERN/LHCC/99-15, 25 May 1999.
2. A. Sfyrla, "ATLAS Construction: A Status Report", this proceedings.
3. `http://tevewwg.fnal.gov/top/`.
4. A. Tricoli, hep-ex/0511020.
5. A. Tricoli, A. Cooper-Sarkar, C. Gwenlan, hep-ex/0509002.
6. A. Cooper-Sarkar, hep-ph/0512228.
7. D. Chakraborty, et al. hep-ph/0303092.
8. LEP Electroweak Working Group, hep-ex/0511027.
9. I. Borjanowic, et al. hep-ex/0403021.
10. F. Hubaut, et al. hep-ex/0508061.
11. G. Mahlon, S. Parke, hep-ph/9512264.
12. B. Gonzalez-Pineiro, et al. ATL-PHYS-2000-017.
13. D. O'Neil et al. *J.Phys* **G28**, 2657-67 (2002).
14. O. Cakir, S.A. Cetin, SN-ATLAS-2004-046.

LEPTONIC B DECAYS

M.-C. CHANG
Research Center for Neutrino Science,
Tohoku university,
Sendai, Japan
E-mail: mcchang@awa.tohoku.ac.jp

Among the measured leptonic B decays in the world so far, there is no signal found except the decay $B^- \to \tau^- \bar{\nu}_\tau$. We review the data analysis methods in each experiment and compare the experimental upper limits of the branching fraction with the Standard Model predictions.

1. $B^0 \to \ell^+\ell^-$

The decays $B^0 \to \ell^+\ell^-$ are highly suppressed since they involve a $b \to d$ transition and require an internal quark annihilation. In addition, the decays are helicity suppressed by a factor of $(\mathrm{m}_\ell/\mathrm{m}_B)^2$, where m_ℓ is the mass of the lepton ℓ and m_B is the mass of the B meson.

1.1. $B^0 \to e^+e^-$ *and* $B^0 \to \mu^+\mu^-$

In the B factories [1,2,3], people do their analysis in a similar way. The $e^\pm$ and $\mu^\pm$ tracks are clearly identified by the particle identification system first. In order to reduce Bhabha and two-photon backgrounds, people require the total number of tracks to be large. Then, they use some event shape variables to separate the signal from the continuum background ($e^+e^- \to q\bar{q}$, $q = u, d, s, c$). After optimizing of the signal-to-background ratio, people open their signal box and find no candidates inside. Finally, an upper limit of the branching fraction is calculated.

At the Tevatron collider (Tevatron) [4,5], their analysis method is different. They use a special muon trigger to identify $\mu^\pm$ tracks. To reduce the combinatorial background, people refit the muon tracks. Then, people combine some variables to form a likelihood ratio function to enhance the signal. Since there are some candidates inside the signal box after the event selections, a control sample $B^+ \to J/\psi K^+ \to \mu^+\mu^- K^+$ is used to scale the number of events. In CDF, they can separate the mass of B_s and B_d due to the good mass resolution. An upper limit of $B_s \to \mu^+\mu^-$ is obtained as well.

So far, the measured limits are too far away from the SM predictions [6], except the $B_s \to \mu^+\mu^-$ mode.

1.2. $B^0 \to \tau^+\tau^-$

The first limit has been set by the BaBar group [7]. They first fully reconstruct the other B. Second, they reduce the main background $b \to c$ transitions. Then, multiplicities of e, μ, and π^0 in the recoil system must be consistent with each τ decaying channels. It is hard to say how far this upper limit is away from the SM prediction since the uncentainties of the prediction arise from two main sources: Errors on the B decay constant f_B and CKM matrix element V_{td} are still large (Eq. 1).

$$Br^{SM}(B^0 \to \tau^+\tau^-) = 1.2 \times 10^{-7} \times (\frac{f_B}{200\ MeV})^2 \times (\frac{|V_{td}|}{0.007})^2 . \quad (1)$$

2. $B^\pm \to \ell^\pm \bar{\nu_\ell}$

In the SM, the predicted branching fraction is given as in Eq. 2.

$$Br^{SM}(B^- \to \ell^- \bar{\nu}) = (\frac{G_F^2 m_B m_l^2}{8\pi})(1 - \frac{m_\ell^2}{m_B^2})^2 f_B^2 |V_{ub}|^2 \tau_B , \quad (2)$$

where G_F is the Fermi coupling constant, m_B and m_ℓ are the mass of B and ℓ, respectively, f_B is the B meson decay constant, $|V_{ub}|$ is the CKM matrix element, and τ_B is the B lifetime. The branching fraction measurement will provide the direct B meson decay constant f_B.

Meanwhile, in the two-Higgs doublet model, it provides an equation to constrain the Higgs mass (Eq. 3).

$$Br(B^- \to \ell^- \bar{\nu}) = Br^{SM}(B^- \to \ell^- \bar{\nu}) \times (1 - (\frac{\tan\beta}{m_H} m_B)^2)^2 . \quad (3)$$

2.1. $B^\pm \to \mu^\pm \bar{\nu_\mu}$ *and* $B^\pm \to \tau^\pm \bar{\nu_\tau}$

The analysis method for the two modes in the B factories are similar. People fully reconstruct one of the B mesons in the event, and compare the properties of remaining particles. The method allows people to suppress the combinatorial background from both $B\bar{B}$ and continuum events. In order to avoid experimental bias, the signal region in data is blinded until the event selection criterion are finalized.

Among the measurements [8,9,10], the best $B^{\pm} \to \mu^{\pm}\bar{\nu_{\mu}}$ upper limit of the branching fraction by Belle is still one order of magnitude higher than the SM prediction [11].

For the measurements of $B^{\pm} \to \tau^{\pm}\bar{\nu_{\tau}}$ [8,12,13,14,15], the first evidence from Belle [15] reaches 4.2 standard deviation. Fig. 1, shows the obtained E_{ECL} distribution with all τ decay modes combined, where the E_{ECL} means the energy deposit in the ECL detector. Fig. 2, shows the exclusion boundaries in the $[M_{H^+}, \tan\beta]$ plane obtained from the observed upper limit on the branching fraction of $B^- \to \tau^- \nu$ [14].

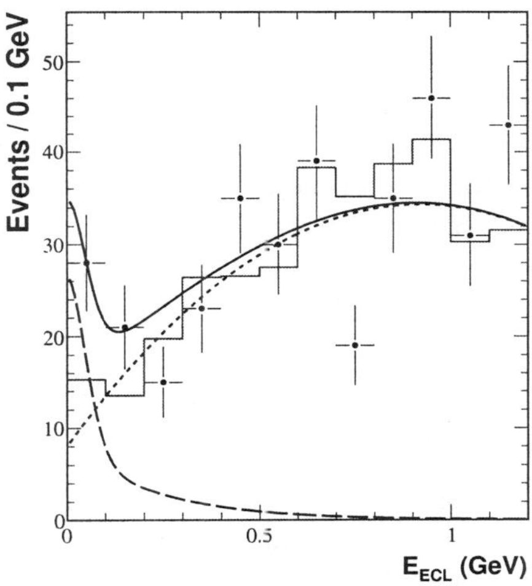

Figure 1. E_{ECL} distributions in the data after all selection requirements except the one on E_{ECL} have been applied. The data and background Monte Carlo samples are represented by the points with error bars and the solid histogram, respectively. The solid curve shows the result of the fit with the sum of the signal shape (dashed) and background shape (dotted).

3. $B^0 \to$ invisible and $B^0 \to \nu\bar{\nu}\gamma$

The first limit was given by BaBar [16]. A search for the B^0 decay contains no observable final state particles or plus a single photon. They use the

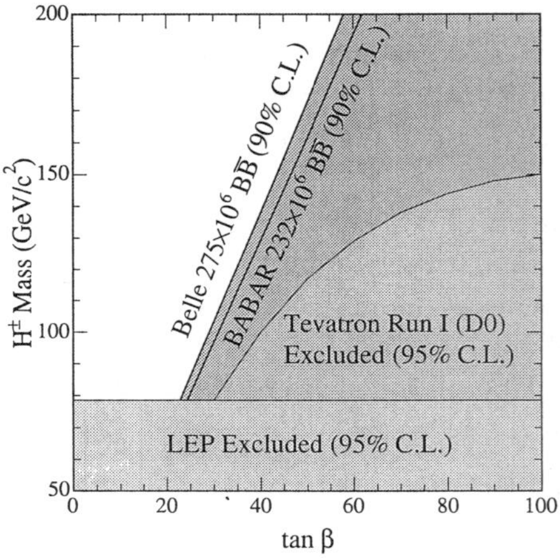

Figure 2. The 90% C.L. exclusion boundaries in the $[M_{H^+}, \tan\beta]$ plane obtained from the observed upper limit on the branching fraction of $B^- \to \tau^- \nu$.

same way as in the $B^- \to \ell^- \nu$ mode, because the B mesons are created in pairs, they fully reconstruct one of the B meson in an event. They improve the purity of the signal B by rejecting non-$B\bar{B}$ events using event shape variables. Then, they enhance the signal by a requirement on likelihood ratio function. The results are still far away from the theory prediction [17].

4. Summary

There are no signals found in any of the leptonic B decays, except the 4.2σ evidence of $B^- \to \tau^- \nu$. For all of the limits, none of them violates the SM predictions but do rule out a number of new physics scenarios.

In the current B factories, it is possible to observe the $B^- \to \tau^- \nu$ and $B^0 \to \mu^- \nu$ decays at SM levels. These results will constrain the charged Higgs in the $[m_H, \tan\beta]$ plane and provide a direct measurement of the B decay constant f_B.

For future higher luminosity experiments, Super-B and LHC, it is believed that people will continue to probe the $B^0_{d,s} \to \mu^+ \mu^-$ and will provide information for SUSY models.

References

1. CLEO Collaboration, T. Bergfeld *et al.*, Phys. Rev. D **62**, 091102 (2000).

Table 1. Best limits of the branching fraction and the SM predictions.

	Best limit	SM prediction
$B^0 \to e^+e^-$	6.1×10^{-8}	$(2.34 \pm 0.33) \times 10^{-15}$
$B^0 \to \mu^+\mu^-$	3.9×10^{-8}	$(1.00 \pm 0.14) \times 10^{-10}$
$B_s^0 \to \mu^+\mu^-$	1.5×10^{-7}	3.42×10^{-9}
$B^0 \to \tau^+\tau^-$	3.2×10^{-3}	1.2×10^{-7}
$B^- \to \mu^-\bar{\nu}$	2.0×10^{-6}	$\sim 10^{-7}$
$B^- \to \tau^-\bar{\nu}$	1.06×10^{-4}	1.59×10^{-4}
$B^- \to \nu\bar{\nu}\gamma$	4.7×10^{-5}	$\sim 10^{-9}$

2. Belle Collaboration, M.-C. Chang *et al.*, Phys. Rev. D **68**, 111101 (2003).
3. BaBar Collaboration, B. Aubert *et al.*, Phys. Rev. Lett. **94**, 221803 (2005).
4. CDF Collaboration, A. Abulencia *et al.*, Phys. Rev. Lett. **95**, 221805 (2005).
5. D0 Collaboration, V.M. Abazov *et al.*, Phys. Rev. Lett. **94**, 071802 (2005).
6. A.J. Buras, Phys. Lett. B **566**, 115 (2003).
7. BaBar Collaboration, B. Aubert *et al.*, hep-ex/0511015, submitted to Phys. Rev. Lett.
8. CLEO Collaboration, M. Artuso *et al.*, Phys. Rev. Lett. **75**, 785 (1995).
9. BaBar Collaboration, B. Aubert *et al.*, Phys. Rev. Lett. **92**, 221803 (2004).
10. Belle Collaboration, K. Abe *et al.*, hep-ex/0408132.
11. CKMfitter Collaboration, EPJC **41**, 1 (2005).
12. L3 Collaboration, M. Acciarri *et al.*, Phys. Lett. B **396**, 327 (1997).
13. BaBar Collaboration, B. Aubert *et al.*, Phys. Rev. Lett. **95**, 041804 (2005).
14. Belle Collaboration, K. Abe *et al.*, hep-ex/0507034.
15. Belle Collaboration, K. Ikado *et al.*, hep-ex/0604018, submitted to Phys. Rev. Lett.
16. BaBar Collaboration, B. Aubert *et al.*, Phys. Rev. Lett. **93**, 091802 (2004).
17. C.D. Lu and D.X. Zhang, Phys. Lett. B **381**, 348 (1996).

SEARCHES FOR THE HIGGS BOSON IN CMS *

GEORGIOS DASKALAKIS
Imperial College, London, UK
E-mail: Georgios.Daskalakis@cern.ch

The CMS potential for the Higgs boson discovery is discussed in the framework of the Standard Model (SM) and its Minimal Supersymmetric extension (MSSM).

1. Introduction

The Large Hadron Collider (LHC) is designed to collide two counter rotating beams of protons or heavy ions. Proton-proton collisions are foreseen at an energy of 7 TeV per beam with a planned start-up in 2007. The Compact Muon Solenoid (CMS) is one of the two general purpose detectors that will be installed on the collider. One of its main challenges is the discovery of the Higgs boson. In this report, the CMS potential for the Higgs boson discovery is discussed in the framework of the Standard Model (SM) and its Minimal Supersymmetric extension (MSSM). More details can be found in [1].

2. Discovery Potential for the Standard Model Higgs Boson

The main production mechanism for the Higgs boson at 14 TeV is the gluon-gluon fusion and the WW/ZZ fusion. For low Higgs boson masses (below 130 GeV/c^2), the most promising channel for discovery is the $H \to \gamma\gamma$. Its signature is quite clean, mainly based on a well calibrated electromagnetic calorimeter. The $H \to b\bar{b}$ channel is the dominant decay but due to the huge QCD background is not useful if the Higgs boson is produced via the gluon-gluon fusion. This decay mode is interesting when the Higgs boson is produced in association with $t\bar{t}$ ($t\bar{t}H, H \to b\bar{b}$). In that case, the background can be effectively reduced via b-jet tagging. Another interesting discovery channel, at the same mass range (around 130 GeV/c^2), is the

*On behalf of the CMS collaboration.

qq → qqH, H → $\tau^+\tau^-$. It can be searched for in the lepton plus τ-jet final state. The critical factor here is the reconstructed mass resolution due to the dominant QCD and electroweak Zjj background with Z → $\tau^+\tau^-$ and the central jet veto.

If the Higgs boson has a mass between 130 GeV/c^2 and 2M$_Z$, the preferred search channels are the H → ZZ* and H → WW*, which are abundantly produced and give quite clean signatures especially when the final state involves muons and/or electrons. For heavier Higgs bosons ($M_H > 2M_Z$) the golden discovery channels are the H → ZZ → 4e, 4μ, $2e2\mu$ which will allow a very fast discovery. For very high Higgs boson masses (above 500 GeV/c^2) the cross section for the qq→qqH production process is large and the decay channels H → ZZ → $ll\nu\nu$, $lljj$ were found to yield the highest sensitivity even though the large backgrounds and the large Higgs boson width make the discovery much more difficult compared to the lower Higgs boson masses.

The statistical significance expected for 30 fb^{-1} of integrated luminosity can be seen in Figure 1 when all channels are combined.

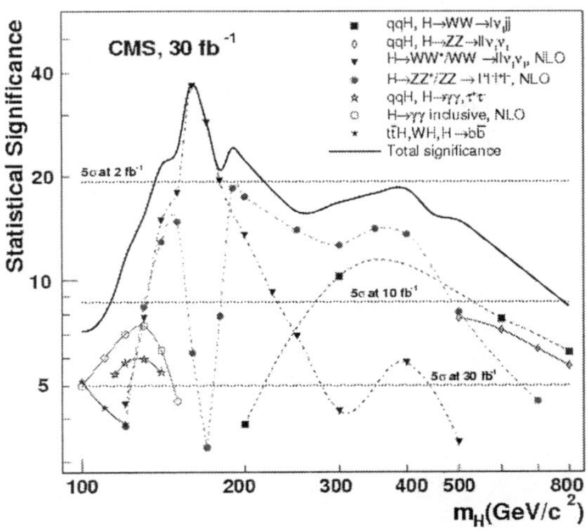

Figure 1. Expected sensitivity for the Standard Model Higgs boson observation as a function of its mass with 30 fb^{-1} integrated luminosity.

3. Discovery Potential for the MSSM Higgs Bosons

In the MSSM there are five Higgs bosons: two CP-even Higgs boson mass eigenstates h,H, a charged Higgs boson pair $H^{\pm}$ and a CP-odd neutral pseudoscalar A. At tree-level the Higgs boson sector is determined by two parameters. A common choice is the ratio of vacuum expectation values of the two doublets $\tan\beta = u_2/u_1$ and the mass of the pseudoscalar Higgs boson M_A. Radiative corrections modify the predictions of the model significantly: the mass of the lightest higgs boson at tree level is predicted to be below M_Z which is already excluded by LEP [2] but after corrections its mass may rise up to 135 GeV/c^2. Several MSSM Higgs boson scenarios have been proposed depending on different choices of the soft SUSY breaking parameters. For the results presented in this report the SUSY parameters are fixed to the values used in the LEP studies [3].

In the large M_A limit ($M_A \gg M_Z$), the so-called decoupling region, the heavy Higgs bosons (H, A, $H^{\pm}$) are almost degenerate in mass. The lighter Higgs boson h is SM-like, so its production cross sections and decay partial widths are very close to those of the SM Higgs boson. The discovery potential for the lighter scalar Higgs boson h can be seen in Figure 2 for 30 fb^{-1} of integrated luminosity.

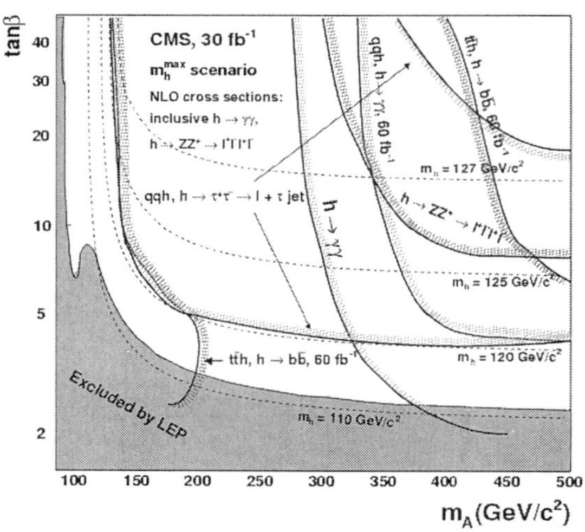

Figure 2. The 5 σ discovery contours in the (M_A , $\tan\beta$) plane for the light scalar MSSM Higgs boson with 30 fb^{-1} integrated luminosity in the M_h-max scenario.

At large tanβ the couplings of the heavy neutral Higgs bosons (H,A) to the electroweak gauge bosons are strongly suppressed while the couplings to down-type fermions are enhanced allowing a search for H,A in the H,A$\rightarrow$ $\mu^+\mu^-$ and H,A$\rightarrow$ $\tau^+\tau^-$ decay channels in the associated production gg$\rightarrow$ $b\bar{b}$H/A. In this production process, the associated b-jets can be used to suppress the Z,γ^* and QCD multi-jet backgrounds with b-tagging methods. To take advantage of the hadronic τ decays in the lepton-plus-τ-jet and two-τ-jet final states from the H/A$\rightarrow$ $\tau^+\tau^-$ decay, an efficient τ-jet (coming from τ hadronic decays) identification method is required to suppress further QCD multi-jet and W-plus-jet backgrounds. The Higgs boson mass can be reconstructed also in the H/A$\rightarrow$ $\tau^+\tau^-$ channels from the missing E$_T$ and the visible τ momenta exploiting the neutrino collinearity with the parent τ direction. The discovery potential for the H/A Higgs bosons can be seen in Figure 3 for 30 fb^{-1} of integrated luminosity.

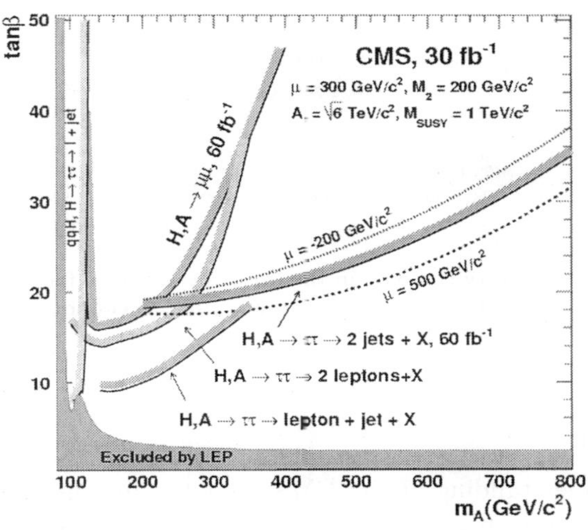

Figure 3. The 5 σ discovery contours in the (M$_A$, tanβ) plane for the heavy neutral MSSM Higgs bosons H and A with 30 fb^{-1} integrated luminosity in the M$_h$-max scenario.

The heavy charged Higgs bosons search proceeds via H$^\pm$ $\rightarrow \tau\nu_\tau, tb, Wh$ decay channels, in the associated production process gg$\rightarrow$ $btH^\pm$. The W+jet and QCD multi-jet backgrounds can be suppressed with b/τ-tagging and reconstruction of the associated top quark. The H$^\pm$ $\rightarrow \tau\nu_\tau$ decay channel is particularly interesting, when hadronic decays are required. The light charged Higgs bosons ($M_{H^\pm}$ <M_{top}) can be searched for in the $H^\pm \to \tau\nu_\tau$ decay channel in the $t\bar{t}$ production, by suppressing the backgrounds with an isolated lepton from the accompanying W decay. The discovery potential for the $H^\pm$ Higgs bosons can be seen in Figure 4 for 30 fb^{-1} of integrated luminosity.

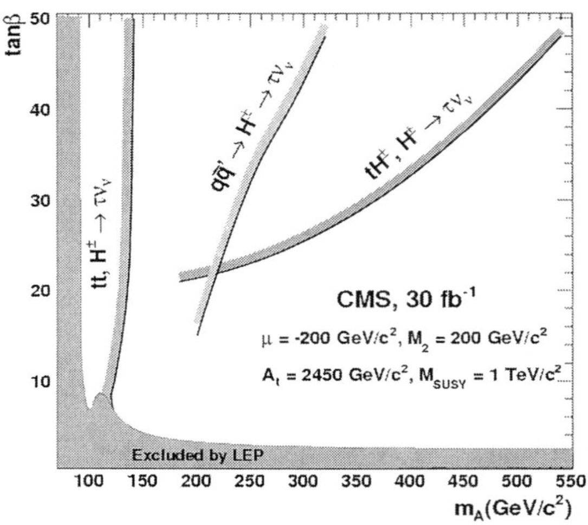

Figure 4. The 5 σ discovery contours in the (M_A , tanβ) plane for the charged Higgs bosons with 30 fb^{-1} integrated luminosity in the M_h-max scenario.

4. Conclusions

The present understanding of the CMS potential for the SM and MSSM Higgs boson discovery has been reviewed. Detailed studies are on going including systematic uncertainties and more sophisticated analysis methods.

References

1. S. Abdullin *et al.*, *CMS Note* **2003/033**.
2. LEP Higgs Working Group, *Phys. Lett.* **B565**, 61 (2003).
3. LEP Higgs Working Group, *LHWG-Note***2005-01**.

NEW PHYSICS SEARCH AT EP COLLISIONS

JEROME DEFAVEREAU

During the past years, many new physics searches have taken place at HERA. These include model-based searches, like SUSY or leptoquarks studies, but also many final-states studies. In the first case, limits could be put on many parameters of these models. In the latter, H1 saw an excess in the isolated leptons channel, which was not confirmed by ZEUS. This excess also appears in the multi-leptons search of H1.

1. Introduction

The HERA electron(positron)-proton collider allows many searches beyond the standard model (SM). The conditions are ideal to put limits on models that allow quark-lepton fusion, like R_p-violating SUSY or leptoquark models.

Many final state searches are also performed, as a general way to look for new physics. Excess in high transverse momentum (P_T) or high mass final states could obviously be the signature of the presence of massive resonnances. From that point of view high-P_T leptons which are well measured are good candidates for this kind of analysis, as the SM cross-sections are predicted to be low.

2. R-parity violating SUSY

One of the major issues of the SM is that it fails to solve the divergent mass of the Higgs boson due to radiative corrections. Supersymmetry (SUSY) is one of the most promising candidates to solve this. It is expected to appear at the TeV scale.

R-parity is a multiplicative symmetry which distinguishes between SM particles with R_p = 1 and supersymmetric particles with R_p = -1. It is defined by $(-1)^{3B+L+2S}$ where L, B and S are respectively the baryon nuber, the lepton number and the spin of the considered particle.

Even if R_p-conserving models are often favoured, the most general superpotential allows R_p-violating (RPV) terms. The most interesting of those terms for HERA physics is :

$$W_{RPV} = \lambda'_{ijk} L_i Q_j D_k$$

Where ijk denote generations. This coupling allows the resonant production of squarks at HERA for masses up to $\sqrt{s} = 318$ GeV. The decay products of this squark can be either SUSY particles like gauginos or SM particles.

2.1. *Squarks*

A search for Squark produced resonantly has been done by H1 on HERA I data. ZEUS has focused on the stop production in HERA I e^+p data. General final states for Squark decay are the following :

- electron + jet (RPV)
- neutrino + jet (RPV)
- electron + multi-jets (R_p conserving)
- neutrino + multi-jets (R_p conserving)

No evidence for Squarks was found. New limits on the λ' coupling and on the Squarks mass can be found in fig. 1. Fixing the coupling to 0.3, Squarks masses lower than 255 GeV could be excluded, and this limits extend to 270 for the stop Squark.

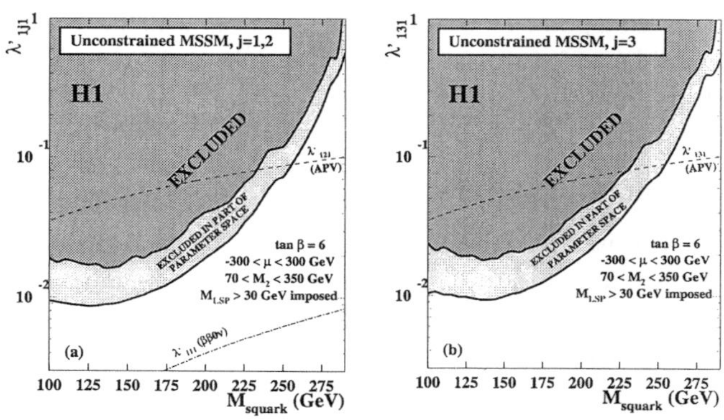

Figure 1. 95% CL limits on squarks masses and couplings

2.2. *Gauginos*

The existence of a non-zero λ' coupling allows the t-channel production of Gauginos. This production is completely independant of the squark sector but depends on the Slepton masses. The Gaugino produced decays into an electron or a neutrino and a slepton. This slepton is taken to decay into two jets. The final state is then composed of an electron or missing P_T + multi-jets.

This analysis has been performed by both H1 and ZEUS with HERA I data. The resulting 95 % limits on the μ-M_2 plane can be seen on Fig. 2. The Gaugino mass used is 103 GeV corresponding to the lower limit obtained by ALEPH and DELPHI at LEP.

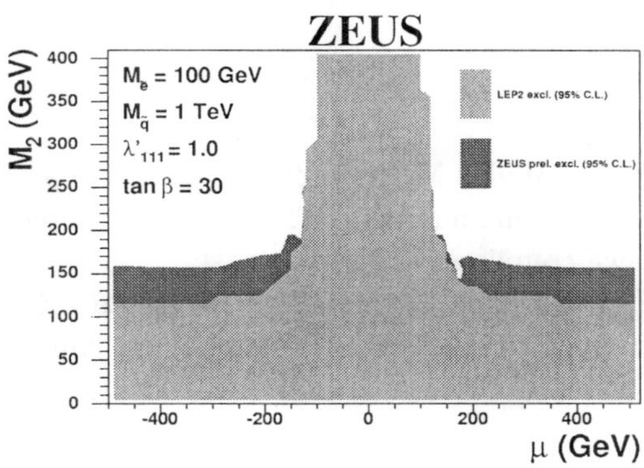

Figure 2. 95% CL limits on μ and M_2

2.3. *gravitinos*

In the Gauge Mediated Supersymmetry Breaking (GMSB) model, the gravitino is the lightest supersymmetric particle. It couples to the SM and SUSY particles via ordinary gauge couplings.

If the Gaugino, produced in the way decribed in the previous section, is the next-to-lightest particle, It decays naturally to a gravitino and a photon. The gravitino escapes detection. the final state is then composed of missing P_t, a photon and a jet. Both H1 and ZEUS could draw limits on the GMSB parameters, as can be seen on Fig. 3.

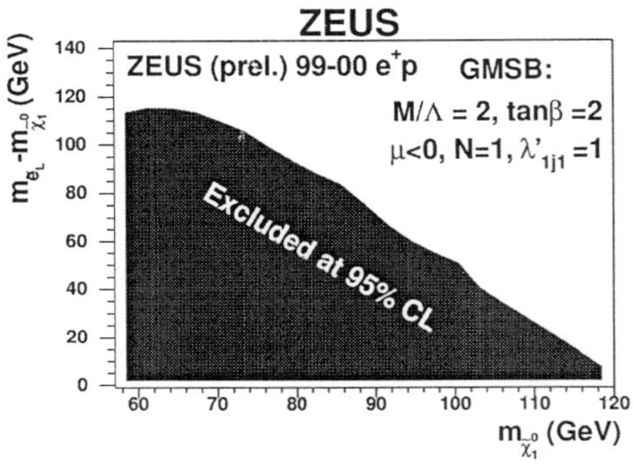

Figure 3. 95% CL limits on GMSB parameters

3. Leptoquarks

The Buchmller-Ruckl-Wyler Model predicts bosonic leptoquarks carrying both leptonic and baryonic numbers. Those bosons couple to a lepton and a quark via Yukawa couplings.

The whole HERA I data was analysed by H1 and ZEUS, showing no evidence for leptoquarks. The limits obtained by H1 can be seen on Fig. 4.

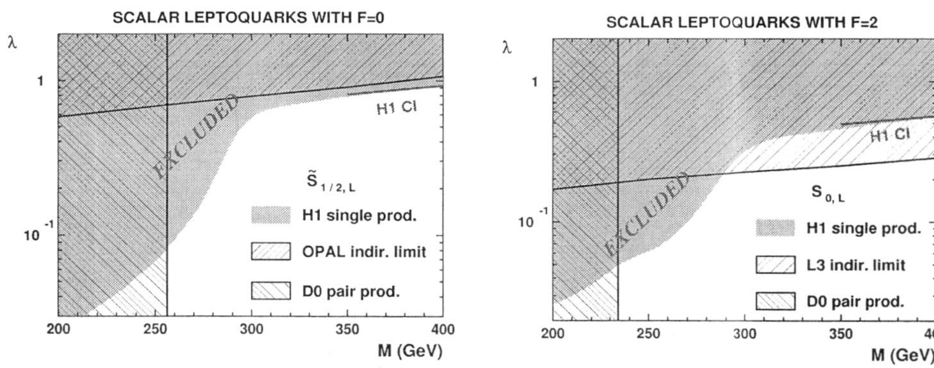

Figure 4. 95% CL limits on leptoquarks masses and couplings

4. General search for new physics

H1 performed a general search, looking for multi-object final states with two to five High-P_T (> 20 GeV) objects, such as leptons, jets or missing E_T. The whole HERA I data (115 pb^{-1}) and the early HERA II data (45 pb^{-1}) have been analysed. All channels show a good agreement with the SM.

5. Isolated leptons

In the Standard Model, high-P_T isolated leptons are mainly produced by W boson decay. Single W production is a background to many processes beyond the SM.

Both H1 and ZEUS performed an analysis on this channel, and H1 finds an excess of isolated electrons and muons at high P_T of the hadronic system, which is not confirmed by ZEUS.

Analysis by the H1 collaboration suggest that the excess occurs only in e+p collisions while data from e-p collisions remains compatible with the Standard Model (Fig. 1).

A comparison between H1 and ZEUS results with similar selections can be found in Table 2.

Table 1. Isolated leptons results from H1

candidates		electron	muon
94-05 e^+p (158 pb^{-1})	Full Sample	19 / 14.6 ± 2.0	9 / 3.9 ± 0.6
	$P_T^X > 25$ GeV	9 / 2.3 ± 0.4	6 / 2.3 ± 0.4
94-05 e^+p (121 pb^{-1})	Full Sample	11 / 12.6 ± 1.8	1 / 3.3 ± 0.5
	$P_T^X > 25$ GeV	2 / 2.4 ± 0.5	0 / 2.0 ± 0.3

Table 2. Comparison between isolated leptons results from H1 and ZEUS.

Electron candidates	12 GeV $< P_T^X <$ 25GeV	$P_t^X > 25$ GeV
99-00 e^+p (ZEUS) (prel.)	1 / 1.04 ± 0.11	1 / 0.92 ± 0.09
03-04 e^+p (ZEUS) (prel.)	0 / 0.46 ± 0.10	0 / 0.58 ± 0.08
94-00 e^+p (H1)	1 / 1.96 ± 0.27	9 / 2.30 ± 0.40

6. Multi-leptons

Multi-lepton (electrons and muons) topologies arise mainly in the Standard Model via photon-photon interactions. The H1 collaboration performed a search for events with di-lepton and tri-lepton topologies on HERA I and HERA II data (210 pb^{-1}). The results of this analysis are generally in good agreement with the Standard Model. However, requiring the scalar sum of the P_T to be greater than 100 GeV, H1 sees 4 events where only 0.61 $\pm$ 0.11 are espected from the SM.

7. Conclusion

Although no signature of physics beyond the Standard Model was found at HERA, many limits on models parameters have been improved.

The results on high P_T (multi-)lepton from H1 remains challenging since the observed excesses over the SM are not confirmed by ZEUS. Further analysis by ZEUS on the muon channel, and by both ZEUS and H1 on larger data sets will help clarify this situation.

RECENT ELECTROWEAK RESULTS FROM DØ

J. D. DEGENHARDT*†
University of Michigan

Presented are recent electroweak results from the DØ experiment.

1. Introduction

The DØ experiment is a multipurpose particle detector at the Fermilab Tevatron. The Fermilab Tevatron is a $p\bar{p}$ collider, with $\sqrt{s} = 1.96$ TeV, where many electroweak phenomena may be studied.

2. Description of the DØ Detector

The DØ detector has a central-tracking system located within a 2 T superconducting solenoidal magnet [1], with designs optimized for tracking and vertexing at pseudorapidities $|\eta| < 3$ [2]. A liquid-argon and uranium calorimeter has a central section covering pseudorapidities $|\eta|$ up to ≈ 1.1, and two end calorimeters (EC) that extend coverage to $|\eta| \approx 4.2$, with all three housed in separate cryostats [3]. A Muon system covering $|\eta| < 2$, consists of a layer of tracking detectors and scintillation trigger counters in front of 1.8 T iron toroids, followed by two similar layers after the toroids [4]. Luminosity is measured using plastic scintillator arrays placed in front of the EC cryostats. The trigger and data acquisition systems are designed to accommodate the high luminosities of Run II.

3. *W* and *Z* Cross Sections

Measurement of the production cross sections multiplied by the leptonic branching fractions ($\sigma\ B$) for W and Z bosons is useful for precision tests of

*on behalf of the DØ collaboration
†e-mail: jdegenha@umich.edu

QCD. The ratio, R, of the measured (σ B) values can be used to indirectly determine the width of the W boson. Preliminary measurements of W and Z boson production with subsequent decays to leptons are reported in Tab. 1. The leptonic decays are the purest decay channels and allow for the most precise measurements. From these measurements R was determined to be $R = 10.82 \pm 0.16$ (stat) ± 0.25 (sys) ± 0.13 (pdf), using the electron decay channels. In Tab. 1 the $Z \to \tau\tau$ cross section measurement is listed. This demonstrates the detector's ability to identify τ leptons and subsequently allow the detector to perform new phenomena searches using these leptons.

Table 1. Shown is the measured (σ B) of the weak bosons in their leptonic decays by the DØ collaboration. The uncertainties are broken into the statistical, systematic, PDF, and luminosity contributions. The luminosity uncertainty is the irreducible systematic contribution from the luminosity detector. The PDF contribution is the largest theoretical contribution to the systematic uncertainty of the measurement.

cross section	measurement (pb)	luminosity (pb)$^{-1}$
$\sigma_W B(W \to e\nu)$ [6]	2865.2 ± 8.3 (stat) ± 62.8 (sys) ± 40.4 (pdf) ± 186.2 (lumi)	177 ± 11
$\sigma_W B(W \to \mu\nu)$ [7]	2989 ± 15 (stat) ± 81 (sys) ± 194(lumi)	96 ± 6
$\sigma_Z B(Z \to ee)$ [6]	264.9 ± 3.9 (stat) ± 8.5 (sys) ± 5.1 (pdf) ± 17.2 (lumi)	177 ± 11
$\sigma_Z B(Z \to \mu^+\mu^-)$ [8]	291.3 ± 3.0 (stat) ± 6.9 (sys) ± 18.9 (lumi)	148 ± 10
$\sigma_Z B(Z \to \tau^+\tau^-)$ [5]	$237 \pm$ (stat) ± 18 (sys) ± 15 (lumi)	226 ± 15

4. Z boson rapidity

The measurement of PDFs at high momentum transfer (Q^2) and high Bjorken x at hadron colliders has traditionally been performed using high-momentum jet spectra. An alternative tool for studying the PDFs is dilepton production. The Z/γ^* process offers a complementary measurement with systematic errors almost wholly different than those in the standard analyses. The leptonic signature allows for precise measurement of the energies involved. There are significant theoretical advantages to the process as well; the Z/γ^* process rapidity distributions can be calculated with NNLO precision [9].

The measurement of the differential cross-section

$$\frac{1}{\sigma}\frac{d\sigma(p\bar{p} \to Z/\gamma^* \to e^+e^-)}{dY}, \tag{1}$$

extracted in bins of Z-boson rapidity $Y = \frac{1}{2}\ln\frac{E^Z + p_z^Z}{E^Z - p_z^Z}$, is performed [10]. At leading order, the rapidity of the Z is directly connected to the difference in Bjorken x parameters of the initiating quark and antiquark, so this measurement can provide information to constrain the parton distribution functions at $Q^2 \approx M_Z^2$. The measured distribution is show in Fig. 1.

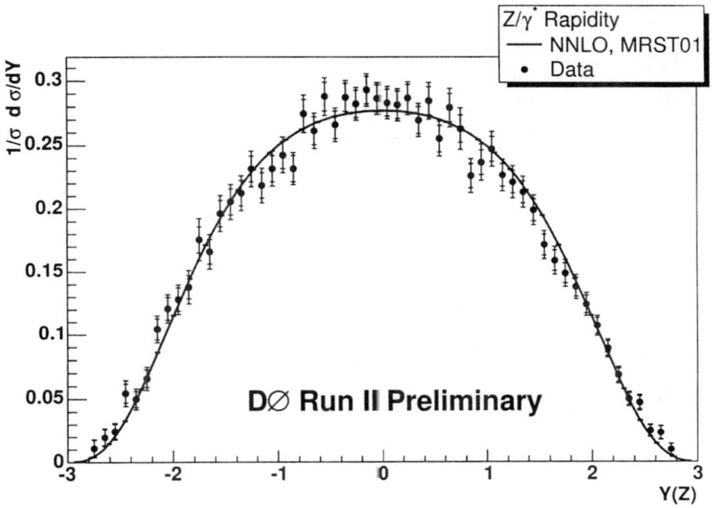

Figure 1. Measured Z boson rapidity at the DØ experiment. The outer error bars show the total error, combining statistical and systematic errors, while the inner error bars indicate the statistical error alone. The solid line shows the NNLO prediction based on the MRST 2001 PDF set.

5. *W* Boson Width

The W boson width, Γ_W, is precisely predicted to be $\Gamma_W = 2.090 \pm 0.008$ GeV in terms of the masses and coupling constants of the gauge bosons [11]. A direct measurement of Γ_W is obtained in studies of the transverse mass spectrum of $W \to e\nu$ events [12]. The W boson transverse mass is defined as $M_T = \sqrt{2E_T^e E_T^\nu (1 - \cos(\phi_e - \phi_\nu))}$, where E_T^e, E_T^ν are the transverse energies and ϕ_e, ϕ_ν are azimuthal angles of electron and neutrino respectively. The transverse mass distribution exhibits a Jacobian edge near M_W, events with $M_T > M_W$ arise from a combination of the non-zero W width and detector resolutions. Differences in Γ_W show up in the tail region, where the Breit-Wigner lineshape (width component) dominates over the Gaussian lineshape (detector resolution component).

The Γ_W is determined from a binned maximum likelihood fit to the transverse mass distribution in the region $100 < M_T < 200$ GeV. The choice $M_T > 100$ GeV minimizes the total error. The Γ_W measured by the DØ collaboration is $\Gamma_W = 2.011 \pm 0.093(\text{stat}) \pm 0.107(\text{syst})$ GeV.

6. Diboson Production

Studies of events containing pairs of vector bosons provide important tests of the standard model (SM) of electroweak interactions. In the SM, diboson production is determined by the $SU(2)_L \otimes U(1)_Y$ structure of the theory. Experimentally measuring these cross sections are a direct test of the SM and any deviations from the SM predictions would be evidence of new physics.

Diboson production is extensively studied at the DØ experiment. The production cross sections of $Z\gamma$ [13], $W\gamma$ [14], WW [15], and WZ [16] are reported in Tab. 2. These measurements were performed by studying the electron and muon final states due to the pureness of the signal. No deviations from the SM predictions have been observed to date.

Table 2. Shown is the measured diboson production cross sections with the approximate luminosity in which they are measured. Also listed are the predicted SM cross sections.

cross section	measurement (pb)	luminosity (pb^{-1})	SM prediction (pb)
$\sigma_{Z\gamma}\ B$	4.2 ± 0.4 (stat+sys)$0.3 \pm$ (lumi)	300	$3.9^{+0.1}_{-0.2}$ [17]
$\sigma_{W\gamma}$	14.8 ± 1.6 (stat) ± 1.0 (sys) ± 1.0 (lumi)	162	16.0 ± 0.4 [18]
σ_{WW}	$18.8^{4.3}_{3.8}$ (stat)$^{+1.2}_{0.9}$ (sys) ± 0.9 (lumi)	230	$12.0 - 13.5$ [19,20]
σ_{WZ}	$4.5^{+3.8}_{-2.6}$ (stat+sys)	300	3.7 ± 0.1 [20]

Note: The $Z\gamma$ was measured with the leptonic branching fraction.

7. Conclusion

The DØ experiment continues to collect data and currently has over 1.0 fb^{-1} of recorded data. The DØ collaboration is continually improving its analyses and its understanding of the detector. Expect many new results to come out of the collaboration in the near future.

Acknowledgments

We thank the staffs at Fermilab and collaborating institutions, and acknowledge support from the DOE and NSF (USA); CEA and CNRS/IN2P3 (France); FASI, Rosatom and RFBR (Russia); CAPES, CNPq, FAPERJ, FAPESP and FUNDUNESP (Brazil); DAE and DST (India); Colciencias

(Colombia); CONACyT (Mexico); KRF and KOSEF (Korea); CONICET and UBACyT (Argentina); FOM (The Netherlands); PPARC (United Kingdom); MSMT (Czech Republic); CRC Program, CFI, NSERC and WestGrid Project (Canada); BMBF and DFG (Germany); SFI (Ireland); The Swedish Research Council (Sweden); Research Corporation; Alexander von Humboldt Foundation; and the Marie Curie Program.

References

1. V. Abazov *et al.* (DØ Collaboration), "The Upgraded D Ø Detector", submitted to Nucl. Instrum. Methods Phys. Res. A, physics/0507191.
2. DØ uses a cylindrical coordinate system. In this system $\eta \equiv -\ln(\tan(\theta/2))$, where θ is the polar angle measured from the positive z direction, which is defined by the direction of the proton beam. The azimuthal angle, ϕ is the angle in the x, y plane. Transverse quantities are defined as $p_T = |p| \sin\theta$, and $E_T = E \sin\theta$.
3. S. Abachi *et al.* (DØ Collaboration), Nucl. Instrum. M ethods Phys. Res. A **338**, 185 (1994).
4. V. Abazov *et al.* (DØ Collaboration), "The Muon System of the Run II DØ Detector", physics/0503151.
5. V. Abazov *et al.* (DØ Collaboration), Phys. Rev. D **71**, 072004 (2005)
6. V. Abazov *et al.* (DØ Collaboration), DØ Note 4403-CONF (2004)
7. V. Abazov *et al.* (DØ Collaboration), DØ Note 4750-CONF (2004)
8. V. Abazov *et al.* (DØ Collaboration), DØ Note 4573-CONF (2004)
9. C. Anastasiou, L. Dixon, K. Melnikov and F. Petriello, Phys. Rev. D **69**, 094008 (2004).
10. V. Abazov *et al.* (DØ Collaboration), DØ Note 4873-CONF (2005)
11. K. Hagiwara *et al.* Phys. Rev. D **66** 010001 (2002).
12. V. Abazov *et al.* (DØ Collaboration), DØ Note 4563-CONF (2005)
13. V. Abazov *et al.* (DØ Collaboration), Phys. Rev. Lett. **95**, 051802 (2005)
14. V. Abazov *et al.* (DØ Collaboration), Phys. Rev. D **71**, 091108 (2005)
15. V. Abazov *et al.* (DØ Collaboration), Phys. Rev. Lett. **94**, 151801 (2005)
16. V. Abazov *et al.* (DØ Collaboration), Phys. Rev. Lett. **95**, 141802 (2005)
17. U. Baur, T. Han, and J. Ohnemus, Phys. Rev. D **57**, 2823 (1998).
18. U. Baur and E. L. Berger, Phys.Rev. D **41**, 1476 (1990); The prediction was derived from the Monte Carlo generator and the K-factor with the CTEQ6L PDF set. The uncertainty in the prediction is determined by the uncertainty in the PDF set.
19. J. Ohnemus, Phys. Rev. D **44**, 1403 (1991);
J. Ohnemus, Phys. Rev. D **50**, 1931 (1994).
20. J. M. Campbell and R. K. Ellis, Phys. Rev. D **60**, 113006 (1999).

SEARCH FOR EXOTIC PHYSICS WITH ATLAS

Physique des particules, Université de Montréal,
C.P. 6128, Succ. Centre-ville,
Montréal, Qc H3C 3J7 Canada

P. A. DELSART

At the LHC, the program of research in particle physics beyond the Standard Model is extremely rich. With the ATLAS detector, besides SUSY mainstream studies, many exotic theoretical models will be investigated. They range from compositeness of fundamental fermions to extra dimension scenarii through GUT models and include many variants. I shall review some selected typical studies by the ATLAS collaboration on exotic physics, highlighting the discovery prospects and the recent analyses using the latest full detector simulations.

1. Exotic Physics in ATLAS

For many well known reasons, the Standard Model (SM) of particle physics is not considered to be a final theory and many models have been proposed to extend it. Amongst them, are famous supersymmetric models which combine a lot of phenomenological advantages. But a large number of other beyond SM and non-SUSY models are viable and are studied by the ATLAS collaboration. Here, we will report on some of these "exotic" models, focusing on 3 studies performed with fully simulated ATLAS detector effects.

2. Selected topics

2.1. *New gauge boson search*

Many theoretical models predict the existence of new interactions and hence new gauge bosons. Amongst these new particles the search for a Z-like boson decaying to 2 leptons is the golden channel because of the very clean signal it yields. At the LHC, these di-lepton channels have been studied [1,6] : the typical limits reachable for the resonance mass is of a few TeV whereas the mass already excluded by other experiments are around 1 TeV.

The first study on this topics with ATLAS full simulation concentrated on the possibility of discriminating between different models of Z' decaying to electrons pairs [2]. Several variables may be used for the discrimination and in this study two of them were studied in detail : *(i)* $\sigma_{ll} \times \Gamma$, , the product of the production cross section, the branching ratio into lepton pairs *(ii)* and the width of the resonance and the forward-backward asymmetry.

For the first variable, the width is measured on the fully simulated events through a fit on the invariant mass distribution. The fitted function is the convolution of a gaussian (detector resolution) and Breit-Wigner like function. Then σ_{ll} is extracted by counting the event under the peak (*i.e.* in a $[M - 4\Gamma, M + 4\Gamma]$ window).

Table 1. discrimination variables for different models : E_6 GUT models, Sequential Standard Model and Left-Right symmetric models

	σ_{ll}^{gen}(fb)	σ_{ll}^{rec}(fb)	$\sigma_{ll}^{rec} \times \Gamma_{rec}$(fb.GeV)
SSM	78.4 ± 0.8	78.8 ±1.8	3668±138
$E_6(\xi)$	47.6 ± 0.6	48.4 ± 1.3	828 ± 45
LR	50.8 ± 0.6	51.1 ± 1.3	1515 ± 75
	A_{FB} Generation	A_{FB} observed	A_{FB} dillution corrected
SSM	0.088 ± 0.013	0.06 ± 0.022	0.108 ± 0.027
$E_6(\xi)$	-0.386 ± 0.013	-0.144 ± 0.025	-0.361 ± 0.03
LR	0.177 ± 0.016	0.10 ± 0.26	0.0186 ± 0.032

The forward-backward asymmetry for a spin 1 particle corresponds to the factor A_{FB} in the differential cross section :

$$\frac{d\sigma}{d\cos\theta^*} \propto \frac{3}{8}(1 + \cos^2\theta^*) + A_{FB}\cos\theta^* \tag{1}$$

where θ^* is the angle between the incoming quark and the outgoing electron in the Z' rest frame. In proton-proton collisions the direction of the incoming quark is unknown and is approximated by identifying it to the reconstructed Z' direction (assuming that the quark carries more momentum than the anti-quark from the sea). This approximation introduces an important bias called dilution which has to be corrected. This was done in this study by considering the probability of wrong identification which was parametrized at generator level as a function of the Z' rapidity. The asymmetry was finally computed by fitting the quantity $A_{FB}(M_{ll}) = \frac{N_+ - N_-}{N_+ + N_-}$ in different rapidity bins. Results at reconstructed level are in good agreement with values at generation level and the discrimination power of both variables is illustrated in table 1 for different test-models with 100 fb^{-1} and M=1.5 TeV.

A precise measurement of these variables will allow, by their combination, to discriminate between numerous models. It is however difficult to quantify precisely the discrimination power given the very large number of models. This is why the on-going work on this topic will make use of the CDDT parametrization [3] which unify the description of most models and allows direct comparisons.

2.2. *Models with extra-dimensions*

In ATLAS, numerous models of extra-dimension are under study. Large extra-dimension, narrow graviton resonance, black-hole production and other themes (see a list of topics in [4]). Here, we present a study based on a large extra dimension scenario. It is inspired by string theories and assumes the existence of TeV^{-1} size extra-dimension in which only the gravitons and their super-partners propagate. In ATLAS, phenomenological, fast simulation[5] and full simulation[7] studies of the production of a graviscalar through couplings to quarks and gluons have been performed. The final state arising from the processes $q + \bar{q} \rightarrow \phi + q$, $q + g \rightarrow \phi + q$ and $g + g \rightarrow \phi + g$. is a jet and missing transverse energy. At LHC, the observation of such a signal is possible only by requiring a very high missing E_T which is here chosen to be larger than 500 GeV. The main backgrounds are then $pp \rightarrow \text{jet} + W/Z$ where the W/Z decays involve neutrinos. Together with the high missing E_T cut, the analysis imposes a veto on isolated and identified leptons. The significance of the signal events surviving the cuts are then computed as a function of the dimensionless couplings g and c to quarks and gluons. Figure 1 illustrates the 5σ discovery limits in the (g, c) plane. These limits can be translated into maximum values of the funda-

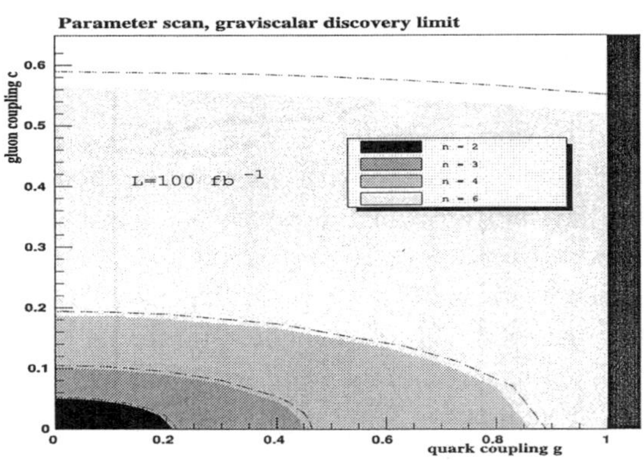

Figure 1. 5 σ discovery limits. The n parameter is the number of extra-dimension

mental Planck mass for which a signal is observable. The values depend on the relative values of g and c and on the number n of extra-dimension.

They range from ~16 TeV for n=2 to ~5 TeV for n=6. However in the latter case, the cut-off of the effective theory used for the calculation is close to this limit and hence no reliable prediction can be made for some part of the parameter space. Fast and full simulation results agree which each other and stress two important points for future studies : use of a refined missing E_T simulation and careful attention to PDF which have significant influence.

2.3. *Vector boson scattering*

In the absence of a light higgs boson, the Standard Model faces the problem of unitarity violation in the $V_L V_L \to V_L V_L$ process. Careful study of this scattering may therefore be crucial for EWSB understanding. Several models address this problem. In ATLAS two of them were studied in more detail : first is the chiral Lagrangian model [8,9] a low energy effective theory based on the flavor $SU(2)_L \times SU(2)_R$ symmetry. The second is a higgsless warped extra-dimension model [10].

Test-points were chosen in both models corresponding to resonances of respectively 1.15 TeV and 0.7 TeV for the W and Z scattering. Fast[11] and full simulation of the $qq \to qqWZ$ process (W and/or Z decaying to lepton giving 3 different final states, corresponding to cross-sections between 1.1fb and 13.4fb) were performed. The main backgrounds are a SM irreducible background where W and Z are radiated in a quark diffusion process and two reducible backgrounds with much higher cross-section, $t\bar{t}$ events and W + 4jets events. The signal is characterized by 2 opposed jets in the forward regions and isolated energy in the central region coming from the decay of the Z and W. In the analysis several cuts are necessary to ensure a good background rejection :

- One or 2 identified and isolated leptons (electrons or muons)
- Tagging of 2 opposite forward jets
- Veto on extra central jets of P_T >40GeV and veto on b-jets.
- Angular cut between reconstructed W and Z.
- Mass cuts on reconstructed W, Z and WZ resonance.

Results given in Fig. 2 show that signal discovery is possible in ATLAS within a few years of running at nominal luminosity. On the experimental point of view, these channels have very special jets characteristics (forward jets, merging of highly boosted jets from W or Z) and the ATLAS on-going effort on jet calibration should allow significant improvements of their study.

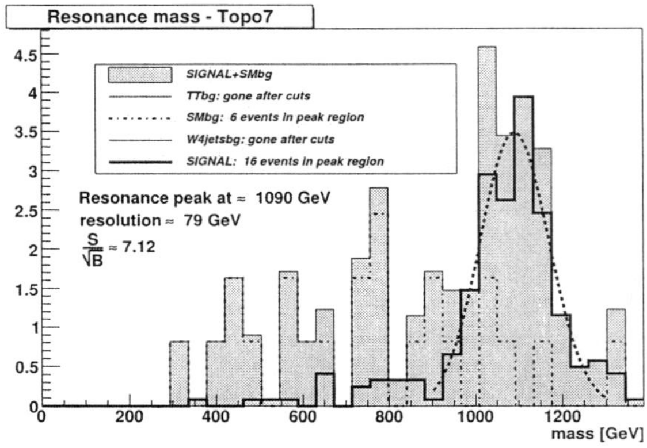

Figure 2. Mass distribution for a chiral Lagrangian model with 2 jets, 1 lepton and missing E_T in final state. This is a preliminary result (as background statistic is low) for $\mathcal{L}$=100fb^{-1}.

3. Conclusion

As presented here, recent studies with a full detector simulation have been performed in ATLAS Exotic Physics group and they confirmed the results of previous fast simulation investigations. The collaboration's tools (events simulation, reconstruction, calibration, *etc...*) are improving and are approaching their final version. They will allow very precise prediction concerning the physics potential of the first data of the LHC, to which the collaboration's next simulated data challenge is dedicated.

References

1. M. Dittmar, A.-S. Nicollerat, and A. Djouadi, *Phys. Lett.* **B583** (2004) 111–120, [hep-ph/0307020].
2. M. Schäfer, F. Ledroit, and B. Trocmé, *Atlas Note* (2005). ATL-COM-PHYS-2005-026.
3. M. Carena, A. Daleo, B. A. Dobrescu, and T. M. P. Tait, *Phys. Rev.* **D70** (2004) 093009, [hep-ph/0408098].
4. Exotic Physics web page http://atlas.web.cern.ch/Atlas/GROUPS/PHYSICS/EXOTICS/.
5. G. Azuelos, P. H. Beauchemin, and C. P. Burgess, *J. Phys.* **G31** (2005) 1–20, [hep-ph/0401125].
6. ATLAS Technical Data Report "Detector and Physics Performance"
7. O. Oye, *Atlas Note* (2005). ATL-PHYS-PUB-2005-008.
8. T. Appelquist and C. W. Bernard, *Phys. Rev.* **D22** (1980) 200.
9. A. C. Longhitano, *Nucl. Phys.* **B188** (1981) 118.
10. C. Csaki, C. Grojean, L. Pilo, and J. Terning, *Phys. Rev. Lett.* **92** (2004) 101802, [hep-ph/0308038].
11. A. Myagkov, *Atlas Note* (1999). ATL-PHYS-99-006.

GETTING READY FOR PHYSICS AT THE LHC WITH THE CMS DETECTOR

V. DROLLINGER*

Università di Padova,
Dipartimento di Fisica "Galileo Galilei",
Via F. Marzolo, 8, 35131 Padova, Italy
E-mail: drollinger@pd.infn.it

In order to get ready for physics at the LHC, the CMS experiment has to be set up for data taking. The data have to be well understood before new physics can be investigated. On the other hand, there are standard processes, well known from previous experiments and from simulation, which will help to understand the data of the detector in the early days of the LHC.

1. Goals and Needs

The main goals for physics at the LHC are searches for new physics and precision measurements. One of the main tasks at the LHC is to probe the existence of the Higgs boson or the source of electroweak symmetry breaking in general. Furthermore, many models with new physics wait to be investigated experimentally. In particular, for phenomena that are expected to show up at high energies, the LHC is reaching completely new territory. In general, precision measurements are somewhat less important than searches at hadron colliders, but there are cases where high statistics is a benefit, or new production channels are open. A prominent example is precision physics with top quarks.

Knowing the goals it is not difficult to imagine what the needs are. First of all, it is essential to have a functioning detector, which allows continuous data taking over several hours. The quality of these data has to be high in the sense that none of the detector subsystems has major failures, and that the resolutions are sufficient to carry out the triggering, reconstruction and selection of events with a high precision. In addition to a high event-reconstruction efficiency, it is necessary to have a good particle identification in order to keep the corresponding fake rates as low as possible. This is essential to avoid huge backgrounds at hadron colliders. Furthermore, the detector needs to be synchronized, aligned and calibrated in order to reach the ultimate detector performance. In particular at the

*Work supported in part by the European Community's Human Potential Programme under contract HPRN-CT-2002-00326, [V.D.].

LHC, a high performance trigger system is needed in order to reduce the event rate by about six orders of magnitude.

Since it is clearly impossible to review this complex subject in five pages, a brief overview is presented and a few examples are given.

2. The CMS Experiment

The CMS detector[1] is a general purpose detector, which is designed to meet the goals mentioned above. Figure 1 shows schematic views of the CMS detector and the structure of the trigger and data acquisition (DAQ). The CMS detector is designed like a classical collider detector with

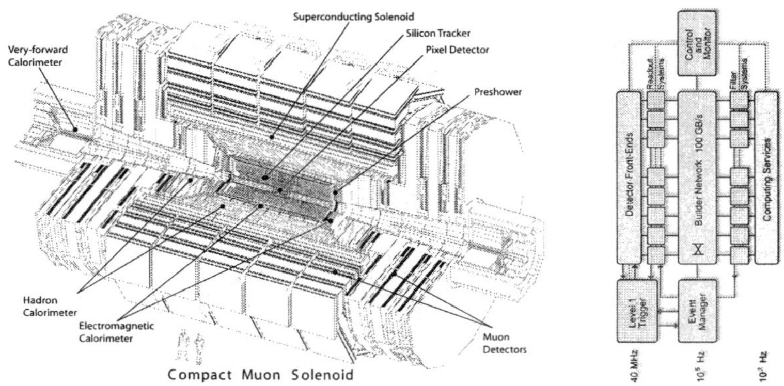

Figure 1. Drawing of the CMS detector, and of the trigger and DAQ scheme.

tracker (including pixel vertex detectors) and muon system acceptance up to $\eta = 2.5$, an electromagnetic calorimeter acceptance up to $\eta = 3$, and hadron calorimetry up to $\eta = 5$. The calorimeters are designed to be able to contain jets in the TeV range and a superconducting magnet provides a magnetic field of 4 T, which allows t+ransverse momentum measurements up to about 1 TeV. A large fraction of the CMS detectors has been built and is currently tested, as illustrated in Figs. 2 and 3.

The trigger and DAQ system is designed to cope with initial event rates of 40 MHz (100 Tbyte/s), which is reduced by the Level 1 (L1) trigger to 100 kHz (100 Gbyte/s) and further reduced to about 100 Hz (100 Mbyte/s) by the High Level Trigger (HLT). Whereas the L1 trigger consists of dedicated trigger boards, the HLT trigger is running CMS reconstruction and selection software on a computer farm.

On the software, simulation and reconstruction side a lot of progress have been made and is described in detail in Ref.[1]. The preparation for

Figure 2. Left: barrel components (front: hadron calorimeter, back: magnet and muon system) of the CMS detector. Right: forward components (muon system) of CMS.

physics analyses is going on and will be documented in an upcoming CMS Physics Technical Design Report. It is also important to setup strategies on how to perform the analyses, such as background normalization from the data itself, employing advanced analysis techniques.

3. From Cosmic Rays to first LHC Collisions

After many test beams for individual detector components have been carried out, the main focus at present is the test of groups of detectors with cosmic rays. One example of the inner tracker is shown in Fig. 3. A large scale test, including many different detector types and the entire magnet with all iron return yokes, called the "Magnet Test Cosmic Challenge" is in preparation.

In order to arrive at a fully operational experiment, many steps are necessary and the comparison of simulation and data is very helpful in understanding each step. On one hand, the simulation is employed to model the detector and to make accurate predictions of many things like trigger rates, efficiencies, resolutions, fake rates. On the other hand, the data can be used to tune the simulation to describe reality even more accurately. One example of a detailed simulation is the tracker alignment, shown in Fig. 3, where the transverse momentum momentum resolution of the CMS tracker is predicted for three alignment scenarios. The performance of the long-term alignment scenario gets close, but is not identical, to the performance of a perfectly aligned tracker.

The LHC pilot run in the year 2007 will be the first source of collision data. The numbers of events of important Standard Model processes

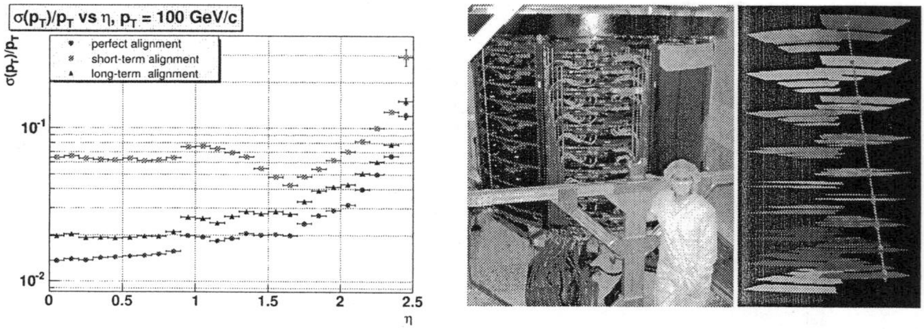

Figure 3. Left: simulation of three tracker alignment scenarios with single muons. Right: cosmic rays seen with one endcap of the inner tracker of CMS.

expected for one particular scenario[2] are summarized in Table 1. Even for a low luminosity and a short running period the number of minimum bias events and di-jet events is huge. These events are important to study the calorimeter response and to understand the general structure, i.e. the underlying event, of events at the LHC. Energetic leptons, coming from W or Z decays, are produced with sufficient statistics in order to study electron and muon triggering, reconstruction, identification and the corresponding efficiencies. Only a moderate number of top quarks, are expected in the beginning of the LHC. These are important both as a signal and as a background. Even during the pilot run, new physics can be potentially

Table 1. Expected number of events for the pilot run of the LHC with $\sqrt{s_{pp}} = 14$ TeV. One month of running with a luminosity of $L = 10^{30}$ cm^{-2} s^{-1} is assumed.

process	min.bias	jets $_{E_T > 60\ \mathrm{GeV}}$	$W^{\pm} \to \ell^{\pm}\nu$	$Z \to \ell^+\ell^-$	$t\bar{t} \to \ell^{\pm}\nu$ jets
# events	$\gtrsim 10^{12}$	$\gtrsim 10^{8}$	$\gtrsim 5 \times 10^{4}$	$\approx 5 \times 10^{3}$	$\approx 3 \times 10^{2}$

produced, but it is rather unlikely that such events could be recognized without understanding the detector in detail.

4. Help from Others

Many aspects, discussed above, are not completely new. Clearly, a lot of experience on experimental aspects is available from other (collider) experiments, and an effort has been started to propagate the know how to the LHC experiments[3]. In addition, it is very important to have a good idea of what kind of physics to expect at the LHC and collaborations with theorists have been established[4]. Theoretical predictions are required for setting up physics analyses and also for designing the detector and optimizing the reconstruction software.

One concrete example of how the LHC experiments can benefit from previous collider experiments and from theoretical work is bottom fragmentation in top ($t \to Wb$) and Higgs ($h \to b\bar{b}$) decays[5]. In order to understand bottom fragmentation in these processes, a Monte Carlo event generator is tuned to e^+e^- data by matching the B-hadron spectra x_B, the energy fraction of the B-hadron normalized to the energy of the b-quark, of the $Z \to b\bar{b}$ decays, shown in Fig. 4. In a second step, the tuned event gen-

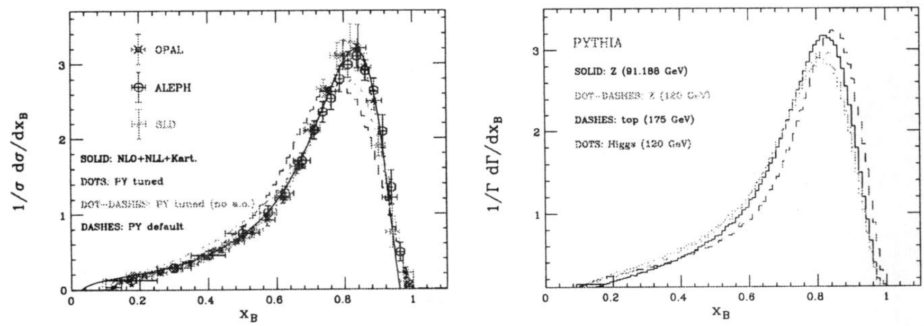

Figure 4. Bottom fragmentation in $Z \to b\bar{b}$ decays of PYTHIA tuned to e^+e^- data (left) and predictions for $t \to Wb$ and $h — b\bar{b}$ decays (right).

erator is employed to predict x_B spectra for other processes than $Z \to b\bar{b}$, namely $t \to Wb$ and $h \to b\bar{b}$ which are of interest at the LHC and at the Tevatron, too.

5. Conclusions

The main goal of studying physics at the LHC is coming closer: a big fraction of the CMS detector has been built and is currently tested with cosmic rays. In context with the CMS Physics Technical Design Reports, many aspects of the detector an physics performance have been studied in detail in preparation for the CMS physics program.

References

1. CMS Collaboration, `cmsinfo.cern.ch`, `cmsdoc.cern.ch/cms/cpt/tdr`, and references to Technical Design Reports therein.
2. G. Rolandi, private communications.
3. TeV4LHC, `conferences.fnal.gov/tev4lhc`; HERA/LHC, `www.desy.de/%7Eheralhc`.

4. Physics at TeV Colliders, wwwlapp.in2p3.fr/conferences/LesHouches.
5. G. Corcella and V. Drollinger, *Nucl. Phys.* **B730**, 82-102 (2005).

W AND Z CROSS SECTION MEASUREMENT AT CDF*

I. FEDORKO†

National and Capodistrian University of Athens
30 Panepistimiou str.
Athens 10679 Greece
and
Istituto Nazionale di Fisica Nucleare
Edificio C - Polo Fibonacci Largo B. Pontecorvo, 3
Pisa 56127 Italy E-mail: ifedorko@fnal.gov

We report on the new measurements of W and Z cross sections times leptonic branching ratios in $p\bar{p}$ collisions at the Tevatron at $\sqrt{s}$ =1.96 TeV. The measurements are based on the decays $W \to e\nu$, $Z \to \mu^+\mu^-$ and $Z \to \tau\tau$.

1. Introduction

The study of electroweak processes plays a key role in the broad physics program of CDF Run II. Precise measurements from the Tevatron detectors are complementary to those performed at LEP (e^+e^-) and are important tests of the Standard Model (SM). Any observed discrepancy with the SM prediction would be evidence of new physics.

We review the new measurements based on data collected during the years 2002-2004. The integrated luminosity of the data ranges from 223 to 349 pb^{-1}, depending on the measurement.

2. Cross section measurements

W and Z bosons are produced at the Tevatron by $q\bar{q}$ annihilation. Due to the large hadronic jet background, it is difficult to detect decays involving only quarks. Thus, leptonic channels are preferred for cleaner boson identification.

---

*On behalf of CDF collaboration.

†This work has been partially supported by EEC RTN contract HPRN-CT-00292-2002.

The main ingredients of the cross section measurements are listed in the following equation:

$$\sigma(p\overline{p} \to X) \times BF(X \to l\nu(\bar{l})) = \frac{N_{events} - N_{bkg}}{\epsilon \times Acceptance \times \int \mathcal{L}dt}, \tag{1}$$

where N_{events} (N_{bkg}) is number of signal (background) events. The efficiency ϵ includes trigger, lepton reconstruction, and lepton identification efficiencies. Geometrical and kinematical *Acceptance* is usually evaluated from Monte Carlo and detector simulation. For the reviewed cross section measuremnts the integrated luminosity $\int \mathcal{L}dt$ is the dominant source of systematic uncertainty at $\pm$ 6%. An important source of systematic uncertainty to the *Acceptance* calculation is the limited accuracy of the Parton Distribution Functions.

Previous cross section measurements[1] were based on 72 pb^{-1}. The measured values were $\sigma(p\overline{p} \to W) \times BF(W \to e\nu)$= 2780 $\pm$ 14$(stat)$ $^{+63}_{-57}(syst)$ $\pm$ 166 (lum) pb and $\sigma(p\overline{p} \to Z/\gamma^*) \times BF(Z/\gamma^* \to \mu^+\mu^-)$= 248 $\pm$ 5.9$(stat)$ $^{+8.0}_{-7.2}(syst)$ $\pm$ 15.1 (lum) pb. Taking the ratio of W to Z cross sections, an indirect measurement of Γ_W was performed with an accuracy comparable to the current world average.

3. $\sigma(p\overline{p} \to W) \times BF(W \to e\nu)$ in 1.2< $|\eta_e|$ <2.8

A new measurement of the W cross section has been performed by CDF using forward electrons. $W \to e\nu$ candidates are selected by a trigger which required a high E_T (> 20 GeV) electron detected in the forward region 1.2< $|\eta|$ <2.8[a] and high $\not{E}_T$ (> 25 GeV) as a signature of an undetected neutrino. The electron is required to be isolated and associated with a high p_T track. The tracks are measured by a combination of COT and silicon detectors, with the intermediate silicon layers[2] (ISL) playing an important role in the forward region. Using 223 pb^{-1} of data we detect 48144 candidates with 4.5% of background contamination. The measured and expected distributions for W transverse mass of our candidates is reported in Fig. 1. With an overall efficiency of 7.37% we measure the cross section to be 2815 $\pm$ 13$(stat)$ $^{+94}_{-89}(syst)$ $\pm$ 169 (lum) pb. The measured value is in good agreement with the previous measurement in the central region of the CDF detector, and with theoretical prediction $\sigma = 2687 \pm 54$ pb at NNLO [3].

The CDF capability to provide cross section measurements up to $|\eta|$ <2.8 is attractive to perform comparisons with theoretical predictions.

---

[a] $\eta = -ln(tan(\theta/2))$

A good understanding of boson rapidity and visible lepton pseudorapidity distributions are key to using the process of W production as a luminosity monitor for LHC experiments ([4] and references therein).

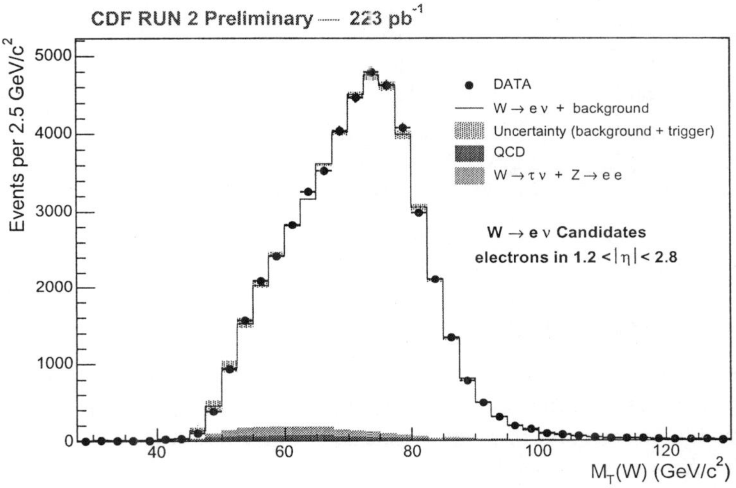

Figure 1. Transverse mass distribution of $W \to e\nu$ candidates with electron detected in forward region of CDF detector.

4. $\sigma(p\bar{p} \to Z/\gamma^*) \times BF(Z/\gamma^* \to \mu^+\mu^-)$

Identification of the Z boson is based on the reconstruction and identification of two leptons. The candidates are selected by a high-p_T muon trigger, which utilizes muon detectors surrounding the calorimeter. A muon candidate must be associated with an isolated high-p_T (>18 GeV) track extrapolated to the muon detectors. In 337 pb^{-1} we reconstructed 9620 candidates with only 8 events of background coming from the decay of the Z boson to two taus. The invariant mass distribution is reported in Fig. 2. With overall efficiency 10.91% we measure the cross section to be 261.2$\pm$ 2.7$(stat)$ $^{+5.8}_{-6.9}(syst)$ $\pm$ 15.1(lum) pb. The measured value is in good agreement with the previous measurement in central part of CDF detector and with the theoretical prediction $\sigma = 251.3 \pm 5.0$ pb based on an NNLO calculation[3].

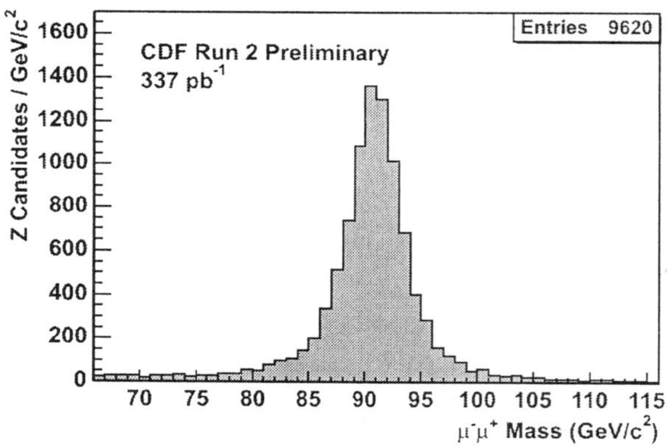

Figure 2. Invariant mass distribution of $Z \to \mu^+\mu^-$ candidates

5. $\sigma(p\bar{p} \to Z) \times BF(Z \to \tau^e\tau^h)$

Higgs and Supersymetry phenomenology predict τ-enriched signatures and τ physics plays an important role at CDF. The process $Z \to \tau\tau$ is the main irreducible background to many signatures of new physics[5].

The $Z \to \tau\tau$ cross section is measured by selecting a hadronic tau candidate and an electronic decay of the tau. The hadronic tau candidates are reconstructed by matching narrow calorimeter clusters with tracks. Around the highest p_T track an isolation cone is constructed and for signal candidates no tracks in the isolation cone are allowed. An isolated π_0 must also be reconstructed using the fine strip chamber at each decay photon's shower maximum.

To select Z candidates the *electron+track* trigger is used. Several cuts to remove conversion electrons and Drell-Yan background are applied. To increase the signal purity, selection criteria based on event topology were applied, $M_T(e, \not{E}_T) \leq 25$ GeV and $p_T(e, \not{E}_T) \geq$25GeV. The mass spectrum, defined as the invariant mass of the sum of electron, tau and $\not{E}_T$ four-momenta, is reported in Fig. 3. Using 349 pb^{-1} of data we measure the cross section to be 265 $\pm$ 20(*stat*) $\pm$ 21(*syst*) $\pm$ 15(*lum*) pb.

6. Conclusions

CDF has produced new measurements of W and Z cross sections. The extended capability for electron identification up to $|\eta| <2.8$ was used. An

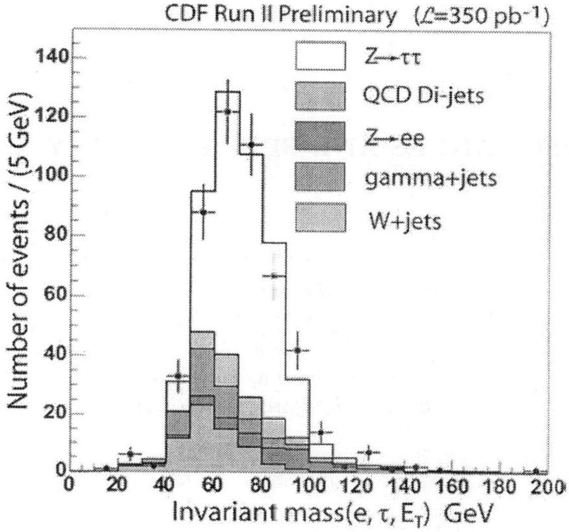

Figure 3. Invariant mass of the four-momenta of electron, hadronic tau candidate, and $\not{E}_T$ in the selected $Z \to \tau_e \tau_h$ candidate events.

low uncertainty of τ reconstruction at the level of 3% was reached. With the new results no deviation from the SM was observed.

7. Acknowledgments

I would like to thank those working hard on the CDF collaboration. I thank also the organizers of Lake Louise Winter Institute. This work has been partially supported by EEC RTN contract HPRN-CT-00292-2002.

References

1. A. Abulencia *et al.*, *Phys. Rev. Let.* **94**, 091803 (2005).
2. M. D' Onofrio *et al.*, Status report of the intermediate sillicon layers detector at CDF, *Nucl. Instrum. Meth.* **A 485**, 6 (2002).
3. P.J. Sutton *et al.*, *Phys Rev* **D45**, 2349 (1992); P.J. Rijken *et al.*, *Phys. Rev.* **D51**, 44 (1995); R. Hamberg *et al.*, *Nucl. Phys.* **B359**, 343 (1991); R.V. Harlander *et al.*, *Phys. Rev. Lett.* **88**, 201801 (2002).
4. S. Frixione, M.L. Mangano, *Nuc. Phys.* **B382**, 11 (1992).
5. A.Safonov for the CDF Collaboration, *Nucl.Phys.Proc.Suppl.* **144**, 323 (2005).

NEW RESONANCES AND SPECTROSCOPY AT BELLE

B. GOLOB*

Faculty of Mathematics and Physics,
University of Ljubljana,
Jadranska 19,
1000 Ljubljana, Slovenia
E-mail: bostjan.golob@ijs.si

The Belle experiment at the KEKB asymmetric e^+e^- collider has proven to be an excellent experimental environment for a wide variety of measurements. These include observations of several previously undiscovered particles and measurements of their properties. While some of these states are predicted in different models, other pose questions regarding their nature and represent a challenge for the description in terms of QCD. A short overview of experimental knowledge on $X(3872)$, $Z(3930)$, $Y(3940)$ and $X(3940)$ is presented.

1. Introduction

The KEKB asymmetric B meson factory[1] is operating in the National Laboratory for High Energy Accelerator Physics (KEK) in Tsukuba, Japan. Electrons and positrons are being collided at the center-of-mass (CMS) energy of about 10.6 GeV. Beside its main objective, the study of the CP violation in the system of B mesons, the Belle experiment[2] performs a large variety of other measurements. In this paper we address some intriguing discoveries of new hadronic states observed by Belle. The next four sections are in turn devoted to the determination of $X(3872)$ resonance quantum numbers (Sec.2), observation of the so called $Z(3930)$ state (Sec.3), believed to be the radially excited χ'_{c2} charmonium state, and two states of similar mass but nevertheless distinguished by other properties, $Y(3940)$ (Sec.4) and $X(3940)$ (Sec.5).

*For the Belle collaboration

2. $X(3872)$

Belle has performed a dedicated program to determine the possible quantum numbers of $X(3872)$ state, discovered in the $B^\pm \to K^\pm J/\psi\pi^+\pi^-$ decays[3]. Decays of $X(3872) \to \gamma J/\psi$ were searched for using $B \to K\gamma J/\psi$[4]. The fitted yield of B mesons reconstructed in the latter decay mode is shown in Fig.1 (left) as a function of the $\gamma J/\psi$ invariant mass. The curve represents the fit with a Gaussian signal centered at the mass of $X(3872)$ and with a width scaled from the observed χ_{c1} signal in $B \to K\chi_{c1}(\to \gamma J/\psi)$.

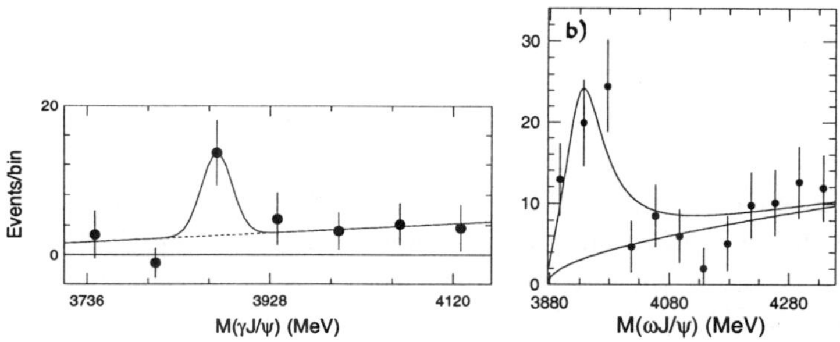

Figure 1. Yield of B mesons in $B \to K\gamma J/\psi$ as a function of $M(\gamma J/\psi)$ (left). Full curve is the result of the fit described in the text. Yield of B mesons in $B \to K\omega J/\psi$ as a function of $M(\omega J/\psi)$ (right).

The observed signal has a significance of 4σ and can be converted to $Br(X \to \gamma J/\psi)/Br(X \to \pi^+\pi^- J/\psi) = 0.14 \pm 0.05$. Observation of the radiative decay mode establishes the positive charge conjugation parity of the $X(3872)$.

Further on, several angular distributions in the more abundant $B^\pm \to K^\pm X(\to J/\psi\pi^+\pi^-)$ decay channel were studied[5]. As an example, the measured distribution of the angle between the π from $X(3872)$ and lepton from J/ψ in the J/ψ rest frame is shown in Fig.2 (left), together with expectation for the $J^{PC} = 0^{++}$ state[6]. The agreement is poor, $\chi^2/n.o.f. = 30.1/9$. The distribution of the angle between the negative B meson direction and π from $X(3872)$ in the $X(3872)$ frame, presented in Fig.2 (right), is found to agree well with the expectation for the 1^{++} state ($\chi^2/n.o.f. = 11.4/9$). In a similar manner it was demonstrated that all possible spin assignments $J = 0, 1, 2$ are disfavored, except the 1^{++} and 2^{++} states.

To distinguish between the two remaining possibilities it is important to check the decays of $X(2872) \to DD^*$; in case X would be a spin-2

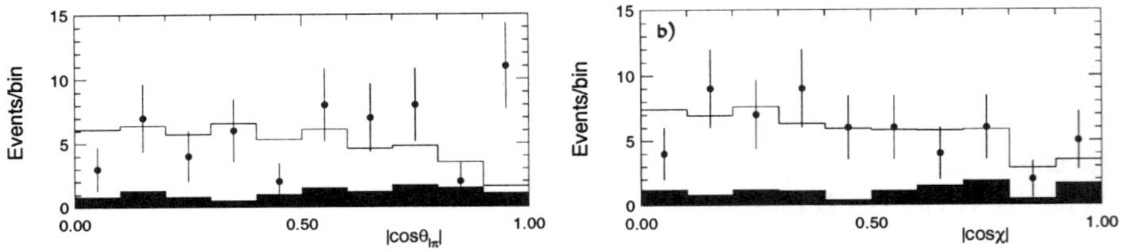

Figure 2. Distribution of angles in $X(3872)$ decays described in the text. The full histograms represent the expectation for $J^{PC} = 0^{++}$ (left) and 1^{++} (right) assignments. The hatched histograms are the contribution of background as obtained from the scaled side bands of $M(J/\psi\pi^+\pi^-)$.

state this decay would proceed through a D-wave and would thus be suppressed. A preliminary result of the measurement of $B \to KD^0\bar{D}^0\pi^0$ shows an enhancement at the $M(D^0\bar{D}^0\pi^0) = M(X(3872))$ which after correcting for the efficiency of reconstruction points to $Br(X \to D^0\bar{D}^0\pi^0)/Br(X \to J/\psi\pi^+\pi^-) \sim 10$. Such a result can hardly be accommodated by an assumption of $J^{PC} = 2^{++}$. Hence the conclusion of several measurements is a preference for the $J^{PC} = 1^{++}$ assignment for the $X(3872)$. This is in agreement with a possible molecular interpretation of this state[7].

3. $Z(3930)$

In the two photon collisions the production of $D\bar{D}$ pairs was carefully examined[8]. The two photon production of $D\bar{D}$ is selected mainly by requiring a low transverse momentum of the pair with respect to the e^+e^- axis. The invariant mass of D meson pairs is shown in Fig.3 (left). A significant (5.3σ) peak denoted as $Z(3930)$ with $M = (3929 \pm 5 \pm 3)$ MeV and $\Gamma = (29 \pm 10 \pm 2)$ MeV is observed above the background.

Efficiency corrected θ^* angular distribution, where θ^* is the angle between the D meson and the beam axis in the $\gamma\gamma$ rest frame, agrees with the spin 2 assumption for the observed resonance (Fig.3 (right)). The measured product of the two-photon width and branching fraction $\Gamma_{\gamma\gamma}Br(Z \to D\bar{D}) = (0.18 \pm 0.05 \pm 0.03)$ keV as well as all other measured properties agree well with the expectations for the radially excited charmonium state χ'_{c2}[9].

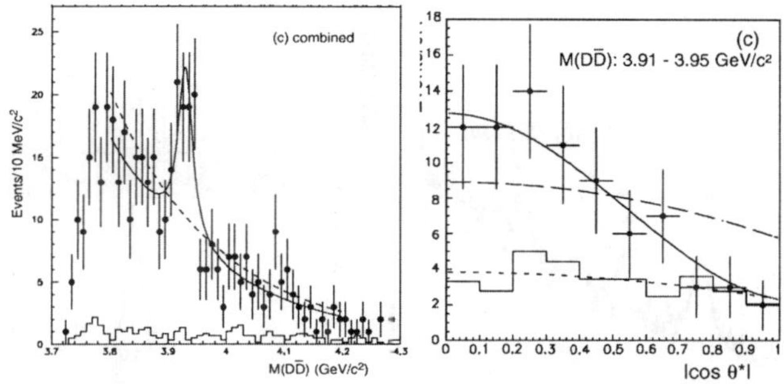

Figure 3. Invariant mass of $D\bar{D}$ produced in two photon reactions (left). Distribution of θ^* (right) with predictions for $J = 2$ (full line) and $J = 0$ (dashed line). Full histogram is the D mass side band.

4. *Y*(3940)

Following the observation of a sub-threshold decays of $X(3872) \to \omega J/\psi$[4] using $B \to KJ/\psi\pi^+\pi^-$ in a similar manner as described in Sec.2, the latter decays were further examined[10] by selecting events with $M(\pi^+\pi^-) \approx M(\omega)$. The yield of reconstructed B mesons as a function of $M(\omega J/\psi)$ is presented in Fig.1 (right) where a significant deviation from the expected phase space distribution is evident.

The signal yield of the state with the mass of $(3943 \pm 11 \pm 13)$ MeV and width of $(87 \pm 22 \pm 26)$ MeV is found to be 58 ± 11 with a statistical significance exceeding 8σ. The measured product branching fraction for this state, called $Y(3940)$, is $Br(B \to YK)Br(Y \to \omega J/\psi) = (7.1 \pm 1.3 \pm 3.1) \times 10^{-5}$. Several speculative interpretations of this state are available, ranging from a radially excited P-wave conventional charmonium (but having difficulties to accommodate a large $Br(Y \to \omega J/\psi)$) to more exotic $c\bar{c}$-gluon hybrid (being able to explain the apparent absence of $D^{(*)}D^{(*)}$ decays, however with predicted masses $\geq$ 4300 GeV)[11].

5. *X*(3940)

Another state of similar mass but of different nature than $Y(3940)$ is exposed in the studies[12] of the mass spectrum recoiling against the J/ψ in $e^+e^- \to J/\psi X$. The observed spectrum of the recoil mass $M_{rec}(J/\psi)$ is presented in Fig.4 (left). Beside the known charmonium states, η_c, χ_{c0} and $\eta_c(2S)$, another state is present at the mass of about 3940 MeV. The observed yield in this inclusive reconstruction of the state tentatively called $X(3940)$ consists of 266 ± 63 events.

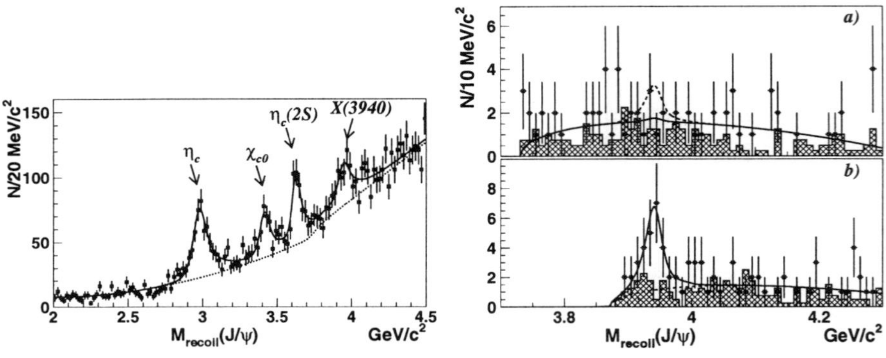

Figure 4. Spectrum of mass recoiling against the J/ψ (left). Same recoil mass for events tagged as $e^+e^- \to J/\psi D\bar{D}$ (top right) and $e^+e^- \to J/\psi D^*\bar{D}$ (bottom right). Full lines are results of the fit.

Two possible exclusive decay modes of the newly observed state were searched for: $X \to D^*\bar{D}$ and $X \to J/\psi\omega$. For the former, only a single D meson beside the J/ψ was fully reconstructed. Events with $M_{rec}(DJ/\psi)$ close to the D or D^* mass were retained for further analysis. For these, the recoil mass against the DJ/ψ system was constrained to the exact mass of $D^{(*)}$ which improved the resolution on the $M_{rec}(J/\psi)$ by almost a factor of 3 (from ~30 MeV to ~10 MeV). The resulting recoil mass against the J/ψ, which corresponds to the invariant mass of the $D^*\bar{D}$ system, is shown in Fig.4 (right).

While no significant signal at the mass of about 3940 MeV is observed for the events tagged as $e^+e^- \to J/\psi D\bar{D}$, there is a clear peak in events tagged as $e^+e^- \to J/\psi D^*\bar{D}$. The mass of $X(3940)$ resulting from the fit is $(3943 \pm 6 \pm 6)$ MeV and the upper limit on its width is 52 MeV (90% C.L.). From the ratio of the signal yields of inclusive and exclusive reconstruction the branching fraction for decays to a pair of charmed mesons is limited to $Br_{>2}(X \to D^*\bar{D}) >$ 45% (90% C.L.). The subscript > 2 denotes the measured Br including only $D^*\bar{D}$ final states with more than two charged tracks.

A search for the $X \to J/\psi\omega$ reveals no significant signal and the upper limit is placed on $Br(X \to J/\psi\omega) < 26\%$. Hence the $X(3940)$ appears not to be the same state as the $Y(3940)$ (Sec.4).

References

1. S. Kurokawa, E. Kikutani, Nucl. Instr. Meth., **A499**, 1 (2003), and other papers included in this volume.
2. Belle Coll., A. Abashian *et al.*, Nucl. Instr. Meth. **A479**, 117 (2003).
3. Belle Coll., S.-K. Choi, S. Olsen *et al.*, Phys. Rev. Lett. **91**, 262001 (2003).
4. Belle Coll., K. Abe *et al.*, hep-ex/0505037.
5. Belle Coll., K. Abe *et al.*, hep-ex/0505038.
6. J.L. Rosner, Phys. Rev. **D70**, 094023 (2004).
7. E.S.Swanson, Phys. Lett. **B588**, 189 (2004).
8. Belle Coll., S. Uehara *et al.*, Phys. Rev. Lett. **96**, 082003 (2006).
9. S. Godfrey, N. Isgur, Phys. Rev. **D32**, 189 (1985); C.R. Münz, Nucl. Phys. **A609**, 364 (1996).
10. Belle Coll., S.-K. Choi, S. Olsen *et al.*, Phys. Rev. Lett. **94**, 182002 (2005).
11. F.E. Close, P.R. Page, Nucl. Phys. **B443**, 233 (1995); C. Banner *et al.*, Phys. Rev. **D56**, 7039 (1997).
12. Belle Coll., K. Abe *et al.*, hep-ex/0507019, submitted to Phys. Rev. Lett.

COSMIC NEUTRINOS BEYOND THE STANDARD MODEL

ULRICH HARBACH AND MARCUS BLEICHER

Institut für Theoretische Physik
Johann Wolfgang Goethe-Universität
and
Frankfurt Institute for Advanced Studies
Max-von-Laue-Str. 1
60438 Frankfurt am Main, Germany
E-mail: harbach@th.physik.uni-frankfurt.de

We show that the enhancement of the ν-N cross section due to new physics is suppressed in models that include a minimal length scale, resulting in a strongly reduced potential observation rate of micro black holes at neutrino telescopes.

The string theory-motivated model of large extra dimensions[1,2,3,4] predicts, among other things, a strong increase of the ultra high energy neutrino-nucleon cross section due to new physics, e.g. black hole production. These events could be detected directly in a subsurface detector[5], and substantial rates have been predicted in realistic calculations[6]. However, effects due to a minimal length scale, which is also predicted by string theory, strongly question the validity of the semiclassical approach to the black hole production cross section at low energies and thus increase the minimal mass of black holes produced[7,8]. Furthermore these effects also substantially decrease high energy cross sections in general and thus also the black hole production cross section[9]. In this paper we examine the influence of these minimal length effects on the black hole production rates in subsurface neutrino detectors.

In the Standard Model the neutrinos only interact weakly with other particles via exchange of a W or Z boson. Within the electroweak model the charged current differential cross section for scattering of a neutrino with an isoscalar nucleon $N = (p+n)/2$ can be written in terms of the Bjorken scaling variables $x = Q^2/2M\nu$ and $y = \nu/E_\nu$ as

$$\frac{d^2\sigma}{dxdy} = \frac{2G_F^2 M_N E_\nu}{\pi}\left(\frac{M_W^2}{Q^2+M_W^2}\right)^2 \left(xq(x,Q^2) + x\bar{q}(x,Q^2)(1-y)^2\right) \quad . \quad (1)$$

Here, $G_F = 1.16632 \times 10^{-5}\mathrm{GeV}^{-2}$ is the Fermi constant, M_W and M_N are the weak boson and nucleon masses, E_ν is the neutrino energy in the nucleon rest frame, Q^2 and ν are the transferred momentum and energy, and $q(x,Q^2)$, $\bar{q}(x,Q^2)$ are the quark and antiquark distribution functions for an isoscalar nucleon. For numerical calculations we use the CTEQ6 parton distribution functions.

In models with Large Extra Dimensions[1,2] the observed weakness of gravity compared to the other fundamental forces (and thus, the hugeness of the Planck mass M_{Pl}) is only a consequence of the size of d extra spatial dimensions. The fundamental mass scale $M_f^{2+d} = M_{Pl}^2/R^d$ of gravity can be as low as the electroweak symmetry breaking scale.[a] Accordingly, at distances below the size of the extra dimensions, gravitation gains strength and black hole formation becomes possible on small scales[10] due to the enormous rise of the Schwarzschild radius in higher dimensions[11]

$$r_S(M_{BH}) = \left(\frac{8\Gamma(\frac{d+3}{2})}{\sqrt{\pi}(d+3)}\frac{M_{BH}}{M_f^{d+2}}\right)^{\frac{1}{d+1}} \quad .$$

The cross section for black hole production for two point particles can be estimated on geometrical grounds and is given by

$$\sigma_{ij\to BH}(\sqrt{s}) = \pi r_S^2(\sqrt{s})\theta(\sqrt{s} - M_{min}) \quad .$$

where $\sqrt{s}$ is the total energy of the colliding particles in the center-of-mass frame and θ is the Heaviside step function which provides a threshold M_{min} for black hole production, i.e. an energy $M_{min} \gg M_f$, above which we can trust our semiclassical picture.

For ν-N interactions, where the nucleon has to be treated as a compound object, the cross section reads

$$\sigma_{\nu N\to BH}(\sqrt{s}) = \int_0^1 \mathrm{d}x\pi r_S^2(\sqrt{xs})\theta(\sqrt{xs} - M_{min})[q(x,\mu) + \bar{q}(x,\mu)] \quad , \quad (2)$$

where $s = 2M_N E_\nu$. For the factorisation scale of the parton distribution functions, we use the canonical choice $\mu = 1/r_S$.

To include effects of a minimal length, we use the model developed in [12, 13]. It is assumed that at arbitrarily high momentum p of a particle, its wavelength is bounded by some minimal length L_{f} or, equivalently, its wave-vector k is bounded by a $M_{\mathrm{f}} = 1/L_{\mathrm{f}}$[14]. Thus, the relation between the

[a] Here, R^d is the volume of the compactified space. Note that there are various definitions of a new fundamental scale in literature, depending on the way of compactification.

momentum p and the wave vector k is no longer linear $p = k$ but a function $k = k(p)$[b] The quantisation in this scenario is straightforward and follows the usual procedure. The arising physical modifications can be traced back to an effective replacement of the usual momentum measure by a measure which is suppressed at high momenta:

$$\frac{\mathrm{d}^3p}{(2\pi)^3} \to \frac{\mathrm{d}^3p}{(2\pi)^3}\left|\frac{\partial k}{\partial p}\right| \quad ,$$

This replacement is founded by the finiteness of the integration bounds in k-space. Here the absolute value of the partial derivative denotes the Jacobian determinant of $k(p)$. For our calculations, we will use the specific relation from Ref. 13 for $k(p)$.

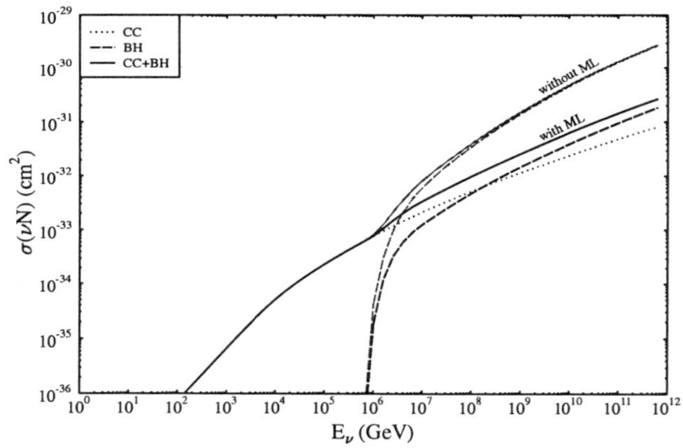

Figure 1. Total ν-N cross section as a function of the incident neutrino energy. The dotted line depicts the charged current cross section, the dashed lines depict the contribution from black hole production and the solid lines yield the respective sums. Here, $M_f = 1\mathrm{TeV}$, $d = 6$, $M_{min} = M_f$.

Fig. 1 shows the total $\nu - N$ cross section a a function of the incident neutrino energy. One clearly observes the enhancement of the $\nu - N$ cross section around 10^6 GeV if the minimal length is neglected (indicated as 'without ML')). However the inclusion of the minimal length effects results

[b]Note, that this is similar to introducing an energy dependence of Planck's constant $\hbar$.

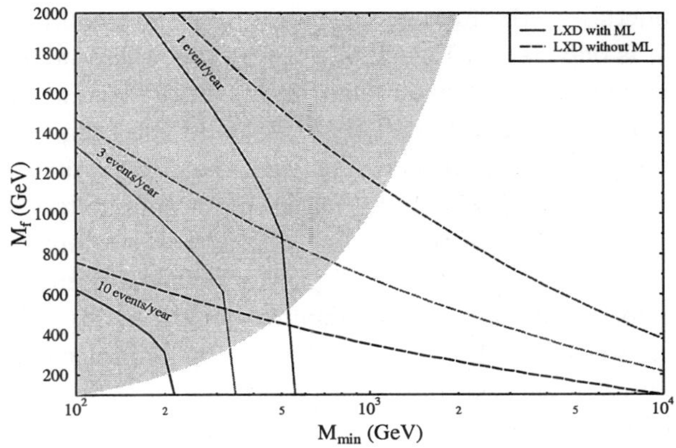

Figure 2. Black hole production rates contained in a volume of 1 km^3 at 2 km below the surface as a function of the fundamental scale and the minimal mass of the black holes. Here, a conservative estimate of the cosmogenic neutrino flux is assumed. The shaded region is excluded ($M_{min} < M_f$). Again, $d = 6$.

in a strong suppression of the cross section enhancement. A very surprising feature is that, despite the high neutrino energies far above the fundamental scale, the charged current cross section remains uninfluenced by the effects of the minimal scale. This fact holds because high momentum transfers are strongly suppressed by the boson propagator in the standard model.

To elaborate these results into realistic predictions for a subsurface neutrino detector such as IceCube, one has to make assumptions about the cosmic neutrino flux $F_\nu(E_\nu)$. For the present study we have taken the flux from Ref. 15. Further one has to take into account the geographical situation of a detector, i.e. the screening of the neutrino flux by the surrounding earth. The number of black hole events per time t and solid angle Ω with detection threshold energy E_{th} in a subsurface detector with volume V reads

$$\frac{\mathrm{d}^2 N}{\mathrm{d}t\mathrm{d}\Omega} = \frac{\rho_{det} V}{M_N} \int_{E_{th}}^{\infty} \mathrm{d}E_\nu F_\nu(E_\nu) \sigma_{\nu N\to BH}(E_\nu) \exp\left(-\sigma_{\nu N\to X} X(\theta)/M_N\right) \quad .$$

Here, ρ_{det} is the mass density of the detector material, $X(\theta)$ is the column density of material between the detector and the upper atmosphere[16] and $\sigma_{\nu N\to X}$ is the total cross section as a sum of the black hole and the charged current cross section. The result for the total number of black hole events per year as a function of the fundamental scale M_f and the minimal black hole mass M_{min} is depicted in Fig. 2. As can be seen the number of black hole events substantially decreases when taking into account effects of a minimal length scale. For a value of $M_f = 1\text{TeV}$ in the range of the electroweak symmetry breaking scale and even with the most optimistic case of $M_{min} = M_f$, there will be basically no black hole events at IceCube. This result is nearly independent of the number of extra dimensions. Changing the number of extra dimensions only slightly affects the number of black holes produced both with and without minimal length scale.

In summary we have shown that the strong enhancement of the neutrino-nucleon cross section due to black hole production is severly reduced if one includes the existence of a minimal length scale as motivated by string theory. This result is in line with the low observation rate of horizontal air showers. In addition the altered mass distribution of produced black holes results in a strong suppression of the black hole detection rate in neutrino telescopes such as IceCube.

Acknowledgments

U.H. thanks the Frankfurt Institute of Advanced Studies for financial support through a PhD scholarship. This work was supported by GSI and BMBF.

References

1. N. Arkani-Hamed, S. Dimopoulos and G. R. Dvali, Phys. Lett. B **429**, 263 (1998) [arXiv:hep-ph/9803315].
2. I. Antoniadis, N. Arkani-Hamed, S. Dimopoulos and G. R. Dvali, Phys. Lett. B **436**, 257 (1998) [arXiv:hep-ph/9804398].
3. D. Cremades, L. E. Ibanez and F. Marchesano, Nucl. Phys. B **643**, 93 (2002) [arXiv:hep-th/0205074].
4. C. Kokorelis, Nucl. Phys. B **677**, 115 (2004) [arXiv:hep-th/0207234].
5. Y. Uehara, Prog. Theor. Phys. **107**, 621 (2002) [arXiv:hep-ph/0110382].
6. M. Kowalski, A. Ringwald and H. Tu, Phys. Lett. B **529**, 1 (2002) [arXiv:hep-ph/0201139].
7. M. Cavaglia, S. Das and R. Maartens, Class. Quant. Grav. **20**, L205 (2003) [arXiv:hep-ph/0305223].
8. M. Cavaglia and S. Das, Class. Quant. Grav. **21**, 4511 (2004) [arXiv:hep-th/0404050].
9. S. Hossenfelder, Phys. Lett. B **598**, 92 (2004) [arXiv:hep-th/0404232].
10. P. C. Argyres, S. Dimopoulos and J. March-Russell, Phys. Lett. B **441**, 96 (1998) [arXiv:hep-th/9808138].
11. R. C. Myers and M. J. Perry, Annals Phys. **172**, 304 (1986).

12. S. Hossenfelder, M. Bleicher, S. Hofmann, J. Ruppert, S. Scherer and H. Stoecker, Phys. Lett. B **575**, 85 (2003) [arXiv:hep-th/0305262].
13. S. Hossenfelder, Phys. Rev. D **70**, 105003 (2004) [arXiv:hep-ph/0405127].
14. D. V. Ahluwalia, Phys. Lett. A **275**, 31 (2000) [arXiv:gr-qc/0002005].
15. R. J. Protheroe and P. A. Johnson, Astropart. Phys. **4**, 253 (1996) [arXiv:astro-ph/9506119].
16. D. A. Morris and R. Rosenfeld, Phys. Rev. D **44**, 3530 (1991).

LEPTON FLAVOR VIOLATING τ DECAYS AT *BABAR*

C. HAST

Stanford Linear Accelerator Center,
2575 Sand Hill Road,
Menlo Park, CA 94025, USA
E-mail: hast@slac.stanford.edu

We report the results of three searches for lepton-flavor and lepton-number violations in the decay of the tau lepton. The analyses have been performed with around 2×10^8 $e^+e^- \to \tau^+\tau^-$ events collected by the BABAR detector at the PEP-II storage ring at a center-of-mass energy near 10.58 GeV. A search for the nonconservation of lepton flavor number in the decay $\tau^\pm \to \mu^\pm\gamma$ yielded no evidence for a signal and we set an upper limit on the branching ratio of $B(\tau^\pm \to \mu^\pm\gamma) < 6.8 \times 10^{-8}$ at 90 % confidence level. In the search for the non-conservation of lepton flavor in the decay $\tau^\pm \to \mathrm{e}^\pm\gamma$ we find no evidence for a signal and set an upper limit on the branching ratio of $B(\tau^\pm \to \mathrm{e}^\pm\gamma) < 1.1 \times 10^{-7}$ at 90% confidence level. A search for lepton-flavor and lepton-number violation in the decay of the tau lepton into one charged lepton and two charged hadrons was performed in 14 different decay modes. The observed data are compatible with background expectations, and upper limits are set in the range $B(\tau \to \ell hh) < (0.7 - 4.8) \times 10^{-7}$ at 90% confidence level. The results of these three analyses have been published in [1 2 3].

1. Introduction

Lepton flavor conservation differs from other conservation laws in the Standard Model (SM) because it is not associated with an underlying conserved current symmetry. Consequently, new theories attempting to describe nature beyond the SM often include lepton flavor violating processes such as the neutrino-less decay of a μ or τ lepton, which have long been identified as unambiguous signatures of new physics. Some theoretical estimates which respect the current limit on $\mu^\pm \to e^\pm\gamma$ decays [4] allow for lepton flavor violating processes such as $\tau^\pm \to \mu^\pm\gamma$ and $\tau^\pm \to e^\pm\gamma$ up to their existing experimental bounds [5,6], which are in the 10^{-7} to 10^{-8} range. Observation of LFV in tau decays would be a clear signature of physics beyond the SM, while non-observation will provide further constraints on theoretical models.

2. The *BABAR* Detector at PEP II

The data used in these analyses were collected with the *BABAR* detector at the PEP-II asymmetric-energy e^+e^- storage ring. The data sample consists of $200\,\text{fb}^{-1}$ recorded at a luminosity-weighted center-of-mass energy $\sqrt{s} = 10.58\,\text{GeV}$. With an estimated cross section for tau pairs of $\sigma_{\tau\tau} = (0.89 \pm 0.02)$ nb [7], this data sample contains over 200 million tau decays.

Charged-particle (track) momenta are measured with a 5-layer double-sided silicon vertex tracker and a 40-layer drift chamber inside a 1.5-T superconducting solenoidal magnet. An electromagnetic calorimeter (EMC) consisting of 6580 CsI(Tl) crystals is used to identify electrons and photons, a ring-imaging Cherenkov detector (DIRC) and energy loss in the tracking system are used to identify charged hadrons, and the instrumented magnetic flux return (IFR) is used to identify muons. Further details on the *BABAR* detector are found in elsewhere [8].

3. Event Selection

Candidate signal events are required to have a 1-1 or 1-3 topology, where one tau decay yields one charged particle (1-prong), while the other tau decay yields one or three charged particles (3-prong). Two (four) well reconstructed tracks are required with zero net charge, originating from a common region consistent with $\tau\tau$ production and decay. Pairs of oppositely charged tracks, likely to be from photon conversions in the detector material, are ignored if their e^+e^- invariant mass is less than $30\,\text{MeV}/c^2$. The event is divided into hemispheres using the plane perpendicular to the thrust axis, calculated from the observed track momenta and EMC energy deposits, in the center-of-mass (CM) frame. One hemisphere must contain exactly one track while the other must contain exactly one or three. Backgrounds from $q\bar{q}$ processes, Bhabha events and other $\tau\tau$ decays are greatly reduced by cuts on the event thrust, missing mass and missing transverse momenta, and the polar angle of the missing momentum associated with the neutrinos in the event.

4. Further $\tau^\pm \to \mu^\pm\gamma$ and $\tau^\pm \to e^\pm\gamma$ Selection and Results

The signature of the signal processes $\tau^\pm \to \mu^\pm\gamma$ and $\tau^\pm \to e^\pm\gamma$ is the presence of an isolated μ or e and γ having an invariant mass consistent with that of the τ ($1.777\,\text{GeV}/c^2$ [9]) and a total energy ($E_{\mu\gamma}$, $E_{e\gamma}$) equal to $\sqrt{s}/2$ in the event center-of-mass (c.m.) frame, and properties of the other particles in the event which are consistent with a SM τ decay.

The signal-side hemisphere, defined with respect to the thrust axis, is required to contain one track with c.m. momentum less than 4.5 GeV/c and at least one γ with a c.m. energy greater than 200 MeV. The track must be identified as a μ using DCH, EMC and IFR information or an e using DCH, EMC and DIRC information, and the γ candidate is the one which gives the mass of the $\mu\gamma$ ($e\gamma$) system closest to the τ mass. This provides the correct pairing for 99.9% of selected signal events. The resolution of the $\mu\gamma$ ($e\gamma$) mass is improved by assigning the point of closest approach of the μ (e) track to the e^+e^- collision axis as the origin of the γ candidate and by using a kinematic fit with $E_{\mu\gamma}$ constrained to $\sqrt{s}/2$. This energy-constrained mass ($m_{\rm EC}$) and $\Delta E = E_{\mu\gamma}(E_{e\gamma}) - \sqrt{s}/2$ are independent variables apart from small correlations arising from initial and final state radiation.

The mean and standard deviation of the $m_{\rm EC}$ and ΔE distributions for reconstructed MC signal events are: $\langle m_{\rm EC}\rangle = 1777\,{\rm MeV}/c^2$, $\sigma(m_{\rm EC}) = 9\,{\rm MeV}/c^2$, $\langle\Delta E\rangle = -9(15)\,{\rm MeV}$, $\sigma(\Delta E) = 45(51)\,{\rm MeV}$ (numbers in () are for $e\gamma$), where the shift in $\langle\Delta E\rangle$ comes from photon energy reconstruction effects.

We blind the data events within a 3σ ellipse (signal box) centered on $\langle m_{\rm EC}\rangle$ and $\langle\Delta E\rangle$ until completing all optimization and systematic studies of the selection criteria. For the optimization of the selection criteria for the $\tau^\pm \to \mu^\pm\gamma$ analysis a neural net technique is utilized, wheras for $\tau^\pm \to e^\pm\gamma$ a cut based approach was choosen.

The $m_{\rm EC}$ versus ΔE distribution for $\tau^\pm \to \mu^\pm\gamma$ events after the final selection is presented in fig. 1 (left). The elipse shows the blinded area, the grey area shows a MC signal distribution. The projection of $m_{\rm EC}$ of a three sigma band in ΔE is show in fig. 1 (right) together with a MC generated signal of a few times 10^{-7}. No evidence for a signal is seen.

The same distributions for $\tau^\pm \to e^\pm\gamma$ events after the final selection are presented in fig. 2. The rectangle shows the blinded area. Again, no evidence for a signal is seen.

For the decay $\tau^\pm \to \mu^\pm\gamma$ we set an upper limit on the branching ratio of $B(\tau^\pm \to \mu^\pm\gamma) < 6.8 \times 10^{-8}$ [1] and for the decay $\tau^\pm \to {\rm e}^\pm\gamma$ the upper limit on the branching ratio is $B(\tau^\pm \to {\rm e}^\pm\gamma) < 1.1 \times 10^{-7}$ [2], both at 90% confidence level.

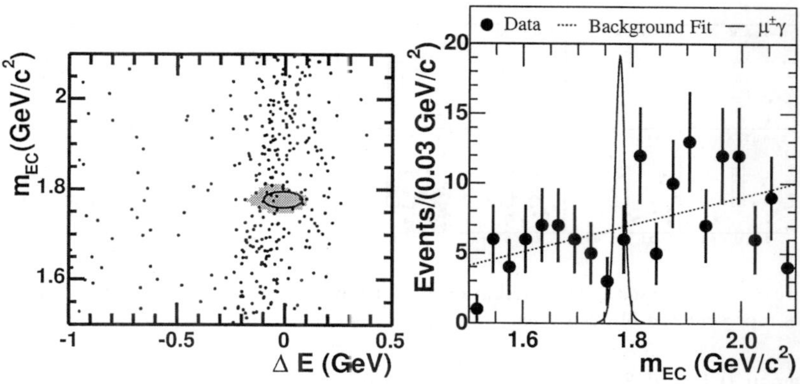

Figure 1. For $\tau^{\pm} \to \mu^{\pm}\gamma$ events $m_{\rm EC}$ versus ΔE distribution (left), and projection of $m_{\rm EC}$ of a three sigma band in ΔE.

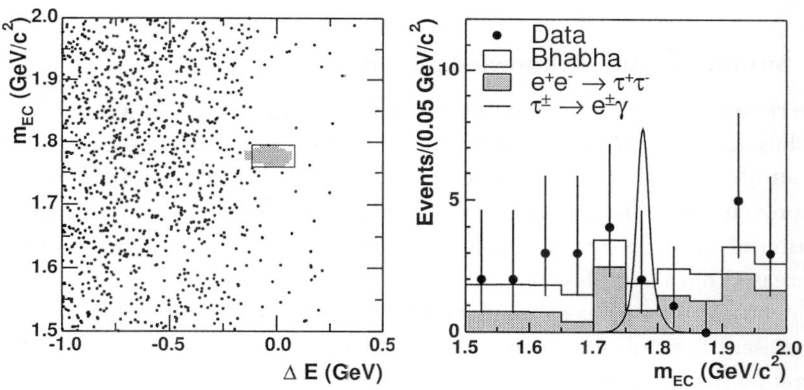

Figure 2. For $\tau^{\pm} \to e^{\pm}\gamma$ events $m_{\rm EC}$ versus ΔE distribution (left), and projection of $m_{\rm EC}$ of a three sigma band in ΔE.

5. Further $\tau \to \ell hh$ Selection and Results

One of the charged particles found in the 3-prong hemisphere must be identified as either an electron or muon candidate. Each of the other two charged particles found in the 3-prong hemisphere must be identified as either a pion or a kaon, using information from the DIRC and $\mathrm{d}E/\mathrm{d}x$. Signal candidate events are analyzed in the ΔE and $\Delta M = m_{\rm EC} - m_\tau \approx 0$ plane which is presented after the final selection in fig. 3 for the 14 analyzed channels. Backgrounds are estimated from two dimensional fits to the data. The number of expected background events in the different channels varies

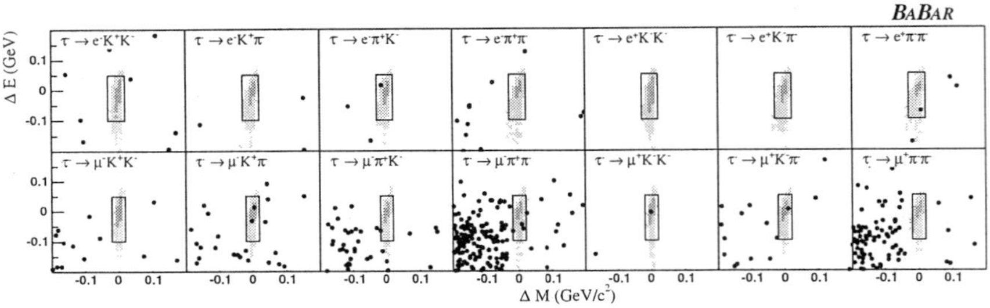

Figure 3. ΔE versus ΔM distribution for 14 different $\tau \to \ell hh$ channels. The black boxes represent the blinded area.

between 0.1 and 3.0, and the total number is 11.3 events. We observe in total 10 events in all channels combined. The observed data are compatible with background expectations, and upper limits are set in the range $B(\tau \to \ell hh) < (0.7 - 4.8) \times 10^{-7}$ at 90 % confidence level [3].

6. Summary and Acknowledgement

We report the results of three searches for lepton-flavor and lepton-number violations in the decay of the tau lepton. These *BABAR* analyses used 1/5 of an ab^{-1} of data and start to reach a sensitivity limit of 10^{-8}. Further improvement of the limits by one to two orders of magnitude, or an observation of these decays on this level, will be possible in comming years with increased data samples.

The author wishes to thank the *BABAR* collaboration for the oportunity to present their results on this conference and the organizers for their superb organization and setting of the conference venue.

References

1. B. Aubert *et al.* (*BABAR* Collab.), Phys. Rev. Lett. **95**, 041802 (2005).
2. B. Aubert *et al.* (*BABAR* Collab.), Phys. Rev. Lett. **96**, 041801 (2006).
3. B. Aubert *et al.* (*BABAR* Collab.), Phys. Rev. Lett. **95**, 191801 (2005).
4. M. L. Brooks *et al.* (MEGA Collab.), Phys. Rev. Lett. **83**, 1521 (1999). This provides the most stringent limit of $B(\mu^{\pm} \to e^{\pm}\gamma) < 1.2 \times 10^{-11}$ at 90% C.L.
5. E. Ma, Nucl. Phys. Proc. Suppl. **123**, 125 (2003).
6. J. R. Ellis, J. Hisano, M. Raidal and Y. Shimizu, Phys. Rev. D **66**, 115013 (2002).
7. B. F. Ward, S. Jadach, and Z. Was, Nucl. Phys. Proc. Suppl. **116**, 73 (2003).
8. *BABAR* Collab., B. Aubert *et al.*, Nucl. Instr. Meth. A **479**, 1 (2002).
9. J. Z. Bai *et al.* (BES Collab.), Phys. Rev. D **53**, 20 (1996).

MEASUREMENT OF $\sin 2\phi_1$ IN $b \to c$ AND $b \to s$ DECAYS AT Belle

NICHOLAS C. HASTINGS

The University of Tokyo, Japan
E-mail: hastings@hep.phys.s.u-tokyo.ac.jp

Measurements of time-dependent CP asymmetries in $B^0 \to J/\psi K^0$, ϕK^0, $K^+K^-K_S$, $\eta' K_S$, $f_0(980)K_S$ and ωK_S decays based on 386×10^6 $B\bar{B}$ pairs collected by the Belle detector are presented. With this data sample, the $J/\psi K^0$ mode provides a precision measurement of $\sin 2\phi_1$. The other modes, which proceed via $b \to s$ penguin (loop) diagrams are sensitive to new physics phases which may appear within the loop. Differing $\sin 2\phi_1$ measurements between $J/\psi K^0$ and the $b \to s$ penguin modes could be a signature of such phases.

1. Introduction

CP violation is incorporated into the Standard Model (SM) of particle physics as an irreducible complex phase in the Kobayashi and Maskawa (KM) matrix[1]. The unitarity of the matrix provides a convenient interpretation of its parameters in terms of a triangle, and hence three angles or measurable phases. One of these angles is ϕ_1 and is given by arg $\left(\frac{V_{tb}^* V_{td}}{-V_{cb}^* V_{cd}}\right)$.

In the decay chain $\Upsilon(4S) \to B^0\bar{B}^0 \to f_{CP} f_{\text{tag}}$ where one of the B mesons decays at time t_{CP} to a final state f_{CP} and the other decays at time t_{tag} to a final state f_{tag} that distinguishes between B^0 and $\bar{B}^0$, the decay rate has a time dependence given by:

$$\mathcal{P}(\Delta t) = \frac{e^{-|\Delta t|/\tau_{B^0}}}{4\tau_{B^0}}\{1 + q\,[\mathcal{S}_{f_{CP}} \sin(\Delta m_d \Delta t) + \mathcal{A}_{f_{CP}} \cos(\Delta m_d \Delta t)]\}, \quad (1)$$

where τ_{B^0} is the neutral B meson lifetime Δm_d is the mass difference between the neutral B meson mass eigenstates, q indicates the flavour of the tagging state ($q = \pm 1$ for $B^0/\bar{B}^0$), $\Delta t \equiv t_{CP} - t_{\text{tag}}$ and:

$$\mathcal{S}_{f_{CP}} \equiv \frac{2\Im(\lambda_{f_{CP}})}{|\lambda_{f_{CP}}|^2 + 1}, \quad \mathcal{A}_{f_{CP}} \equiv \frac{|\lambda_{f_{CP}}|^2 - 1}{|\lambda_{f_{CP}}|^2 + 1}, \quad \lambda_{f_{CP}} = \frac{q_d}{p_d}\frac{\bar{A}_{f_{CP}}}{A_{f_{CP}}}, \quad (2)$$

where q_d/p_d is the $B^0\bar{B}^0$ mixing parameter and $A_{f_{CP}}$ and $\bar{A}_{f_{CP}}$ are the decay amplitudes to f_{CP} for B^0 and $\bar{B}^0$ respectively. To a good approximation the SM predicts $-\xi_{f_{CP}}S_{f_{CP}} = \sin 2\phi_1$, where $\xi_{f_{CP}}$ is the CP eigenvalue of f_{CP}, and $\mathcal{A}_{f_{CP}}$ =0 for both $b \to c\bar{c}s$ and $b \to sq\bar{q}$ transitions. Therefore a comparison of CP violation parameters between these decays is an important test of the SM. The analyses presented here all use a data set of 386×10^6 $B\bar{B}$ pairs collected at the $\Upsilon(4S)$ by the belle detector[2].

2. $\sin 2\phi_1$ from $b \to c\bar{c}s$ decays

The so called gold-plated modes $B \to J/\psi K^0$ were first used to discover CP violation in the B meson system. These modes now provide precision measurements of $\sin 2\phi_1$ which can be used as a bench mark against the rarer $b \to sq\bar{q}$ modes. The final state is $J/\psi K^0$ is reconstructed in both the K_S and K_L modes, with $J/\psi \to e^+ e^-$ or $J/\psi \to \mu^+\mu^-$. Candidate K_S mesons are selected via their decays to $\pi^+\pi^-$ and $\pi^0\pi^0$. Neutral pions are reconstructed from their decays to two photons detected in the Belle calorimeter.

$B \to J/\psi K_S$ candidates are selected using energy the difference $\Delta E = E_B - E_{\text{beam}}$ and the beam-energy constrained mass $M_{\text{bc}} = \sqrt{E^2_{\text{beam}} - |\vec{p}_B|^2}$ where E_{beam} is the beam energy in the center of mass system (CMS), and E_B and $\vec{p}_B$ are the reconstructed B meson energy and momentum in the CMS. Since the energy of the K_L is not measured, $B \to J/\psi K_L$ candidates are selected by reconstructing the magnitude of the CMS B momentum which requires the K_L momentum to be inferred from its flight direction.

Very large signals of 5264 ± 73 and 4792 ± 105 with purities of 98% and 60% are selected in the $J/\psi K_S$ and $J/\psi K_L$ modes respectively. The separation in z (along the beamline) between a vertex formed using the J/ψ daughter leptons and a vertex with the remaining tracks in the event (excluding tracks from the signal side K_S and decays of other long lived particles), Δz, after applying the known boost gives the proper time, Δt between the decays of the reconstructed ($J/\psi K_S$) and tag B mesons. The flavour of the tag B meson ($q = \pm 1$) and the quality of the tag r (a continuous variable between zero and one) are determined with a likelihood method using the tracks and clusters which are not used to reconstruct f_{CP}. The Δt distributions are fit to Probability Density Functions (PDF's) (see Eq. 1) after modifications accounting for backgrounds and detector effects. The fit results are shown in Fig. 1 and yield $\sin 2\phi_1 = 0.652 \pm 0.039 \pm 0.020$ and $\mathcal{A} = 0.010 \pm 0.026 \pm 0.036$.

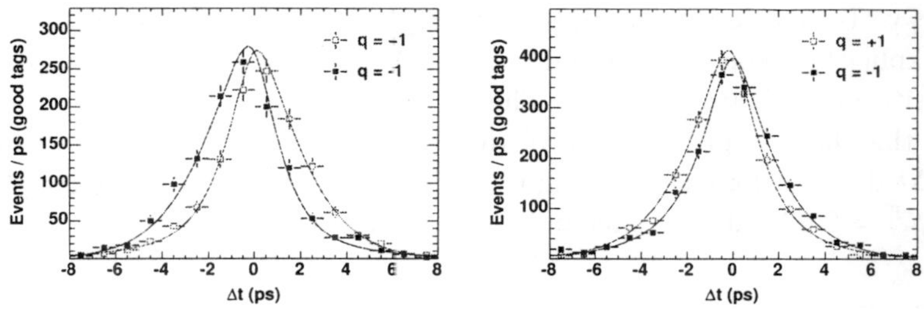

Figure 1. Fit projections for $B \to J/\psi K_S$ (left) and $B \to J/\psi K_L$ (right) with $r > 0.5$.

3. $\sin 2\phi_1$ from $b \to sq\bar{q}$ penguin decays

The $b \to s$ modes ϕK_S, $\eta' K_S$, $K_S\pi^0$ and ωK_S have CP eigenvalue, $\xi = -1$ and ϕK_L, $\eta' K_L$, $f_0(980)K_S$ and $K_SK_SK_S$ have $\xi = +1$. An angular analysis gives the CP even fraction of $K^+K^-K_S$ to be $f_+ = 0.93 \pm 0.09 \pm 0.05$[3], and we define $\xi = 2f_+ - 1 = +0.86 \pm 18 \pm 0.09$. This allows us to define an "effective $\sin 2\phi_1$" for these modes of $\sin 2\phi_1^{\text{eff}} \equiv -\xi^{-1}\mathcal{S}_f$. Unless stated otherwise, reconstruction, vertexing and flavour tagging methods in the $b \to sq\bar{q}$ modes follow those described for $B \to J/\psi K^0$.

The decay $B \to \phi K^0$ is reconstructed with $\phi \to K^+K^-$ and in both the K_S and K_L modes yielding signals of 180±16 and 78±13 and purities of 57% and 12% respectively. The $B \to \phi K^0$ vertex is determined from the charged kaon tracks from the ϕ. Fits to the Δt distributions of these samples give $\sin 2\phi_1^{\text{eff}}$ of 0.19 ± 0.32 and 1.54 ± 0.59 respectively. The combined fit shown in Fig. 2 (left) yields $\sin 2\phi_1^{\text{eff}} = 0.44 \pm 0.27 \pm 0.05$ and $\mathcal{A} = 0.14 \pm 0.17 \pm 0.07$. The plot shows the CP odd (ϕK_S) and even (ϕK_L) modes combined by ploting $-\xi\Delta t$ on the horizontal axis. The asymmetry plotted is given by $(N_+ - N_-)/(N_+ + N_-)$ where the subscript indicates the tag value $q = \pm 1$.

The mode $B \to K^+K^-K_S$ has the same final state particles as the $B \to \phi K^0$ mode, and is reconstructed in a similar fashion except that ϕ mesons are explicitly reconstructed and vetoed, and tighter constraints are applied in selecting the charged kaons. A signal of 536 ± 29 events are found with a purity of 55%. A fit to the Δt distribution (Fig. 2, centre) of the reconstructed $B \to K^+K^-K_S$ events gives $\sin 2\phi_1^{\text{eff}} = 0.60 \pm 0.18 \pm 0.04^{+0.19}_{-0.12}$ and $\mathcal{A} = -0.06 \pm 0.11 \pm 0.07$.

In the decay $B \to \eta' K_S$ the η' is reconstructed via the decay chain $\eta' \to \rho\gamma$, $\rho \to \pi^+\pi^-$ or $\eta' \to \eta\pi^+\pi^-$, $\eta \to (\gamma\gamma, \pi^+\pi^-\pi^0)$. After all event selection 830 ± 35 signal events with a purity of 61% are reconstructed. In the $B \to \eta' K_L$ mode the η' is only reconstructed via $\eta' \to \eta\pi^+\pi^-$,

$\eta \to \gamma\gamma$. To reduce backgrounds in the K_L mode, various $B \to \eta' X$ decays are explicitly reconstructed and vetoed. A signal of 187 ± 18 events with a purity of 26% are selected. The $B \to \eta' K^0$ vertices are determined from the charged pions in the η' (and ρ or η) decay. A combined fit to the Δt distributions (Fig. 2, right) for the two $B \to \eta' K^0$ modes yields $\sin 2\phi_1^{\rm eff} = 0.62 \pm 0.12 \pm 0.04$ and $\mathcal{A} = -0.04 \pm 0.08 \pm 0.06$.

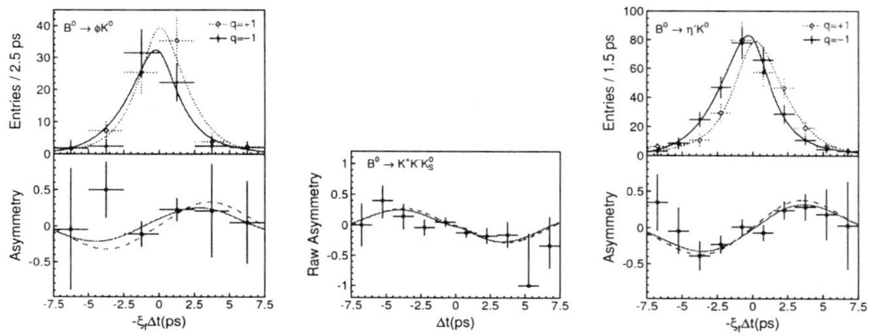

Figure 2. Fit projections ($r > 0.5$) for $B \to \phi K^0$ (left) $B \to K^+K^-K_S$ (centre) and $B \to \eta' K^0$ (right). The dashed curves on the asymmetry plots show the SM expectation.

The modes $B \to K_S K_S K_S$ and $B \to K_S \pi^0$ provide an interesting challenge in determining the B decay position. These modes have no tracks directly (spatially) from the B decay point, as such, pseudo tracks from the reconstructed K_S mesons are used. By ignoring J/ψ daughters in the vertex, $B \to J/\psi K_S$ provides a control sample for this method.

Candidate $B \to K_S K_S K_S$ decays are reconstructed with no more than one $K_S \to \pi^0\pi^0$, 105 ± 12 signal events with a purity of 64% are found. A $B \to K_S \pi^0$ signal of 344 ± 30 with purity of 25% are selected. Fits to the Δt distributions (Fig. 3) yield $\sin 2\phi_1^{\rm eff} = 0.58 \pm 0.36 \pm 0.08$, $\mathcal{A} = -0.50 \pm 0.23 \pm 0.06$ and $\sin 2\phi_1^{\rm eff} = 0.22 \pm 0.47 \pm 0.08$, $\mathcal{A} = +0.11 \pm 0.18 \pm 0.08$ for $B \to K_S K_S K_S$ and $B \to K_S \pi^0$ respectively. It is interesting to note that the statistical error for $\mathcal{A}$ in the $B \to K_S \pi^0$ mode is small compared to the $\sin 2\phi_1^{\rm eff}$ error, since although this mode has a low vertexing efficiency, events with no vertex information still provide power in $\mathcal{A}$ determination.

The $B \to f_0(980) K_S$ mode is reconstructed via $f_0(980) \to \pi^+\pi^-$ and a fit to the $m_{\pi^+\pi^-}$ determines the non resonant $B^0 \to \pi^+\pi^- K_S$ component peaking with the signal in $M_{\rm bc}$ and ΔE. A signal of 145 ± 16 events with a purity of 47% are selected. A fit to the Δt distribution (Fig. 4, left) gives

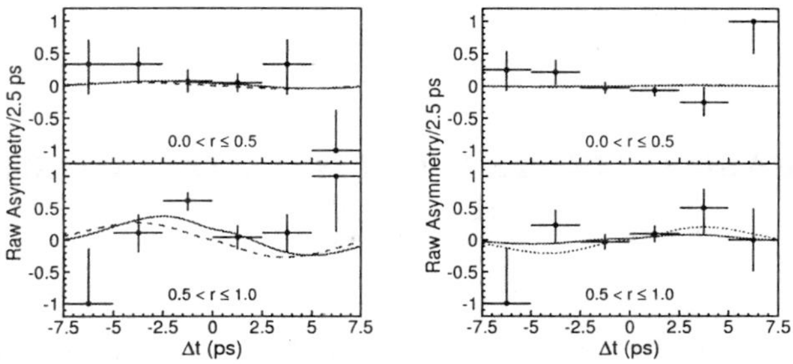

Figure 3. Fit results for $B \to K_S K_S K_S$ (left) and $B \to K_S \pi^0$ (right) in the low (upper) and high (lower) quality tag regions. The dashed line shows the SM expectation.

$\sin 2\phi_1^{\rm eff} = 0.47 \pm 0.36 \pm 0.08$ and $\mathcal{A} = 0.23 \pm 0.23 \pm 0.13$.

Selection of the $B \to \omega K_S$ mode is performed using a 3D fit to $M_{\rm bc}$ ΔE and the $\omega \to \pi^+\pi^-\pi^0$ reconstructed mass. The fit yields 68 ± 13 signal events with a purity of 17%. A fit to the Δt distribution (Fig. 4, right) gives $\sin 2\phi_1^{\rm eff} = 0.95 \pm 0.53^{+0.12}_{-0.15}$ and $\mathcal{A} = 0.19 \pm 0.39 \pm 0.13$

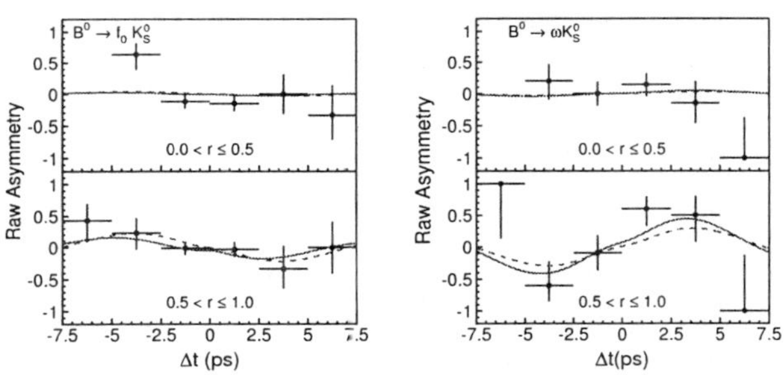

Figure 4. Fit results for $B \to f_0(980)K_S$ (left) and $B \to \omega K_S$ (right) in the low (upper) and high (lower) quality tag regions. The dashed line shows the SM expectation.

Further details of the analyses presented here can be found elsewhere[4]. Each $\sin 2\phi_1^{\rm eff}$ measurement agrees with the precision measurement of $\sin 2\phi_1$ from $B \to J/\psi K^0$ as expected from the SM.

References

1. M. Kobayshi and T. Maskawa, Prog. Theor. Phys. 49 (1973) 652.
2. S. Mori et. al., Nucl. Intrum. Meth. A479 (2002) 117.
3. Belle, A. Garmash et al., Phys. Rev. D69 (2004) 012001, hep-ex/0307082.
4. Belle, K. Abe et al., (2005), hep-ex/0507037.

DVCS MEASUREMENTS WITH NUCLEAR TARGETS AT HERMES

M. HOEK (ON BEHALF OF THE HERMES COLLABORATION)

II. Physikalisches Institut, Universität Giessen
H.-Buff-Ring 16
D-35392 Giessen, Germany
E-mail: Matthias.Hoek@physik.uni-giessen.de

For the first time, beam-spin and beam-charge azimuthal asymmetries have been measured in electroproduction of hard photons on nuclei ranging from Deuterium to Krypton. The asymmetry results from the interference between the Bethe-Heitler and deeply virtual Compton scattering processes. The data have been obtained in the period from 1998 to 2000 by the HERMES experiment at HERA/DESY by scattering the 27.6 GeV longitudinally polarized lepton beam off an internal gas target. The final running period of HERMES until 2007 is dedicated to hard exclusive reactions. A recoil detector surrounding the gas target was installed and is currently being commissioned.

1. Introduction

Lepton scattering experiments constitute an important source of information for the understanding of the structure of nucleons and nuclei. This structure is described by form factors and structure functions, which are measured in elastic and deep inelastic scattering (DIS) experiments, respectively.

About a decade ago, the new theoretical framework of Generalized Parton Distributions (GPD) was developed for the description of hard exclusive processes[1,2]. The GPD formalism implicitly includes parton distribution functions (PDF) and form factors (FF), but also embodies additional information. As an example, the total angular momentum of partons inside the nucleon can in principle be accessed through a sum rule for the moment of the sum of two GPDs[2]. There are four leading twist GPDs for each quark flavor q in the nucleon: H_q, E_q, $\widetilde{H}_q$ and $\widetilde{E}_q$. These GPDs depend upon three kinematic variables: the longitudinal momentum fraction of the parton in the initial and final hadron and the momentum transfer between these hadron states.

The cleanest presently accessible exclusive process to study GPDs is deeply virtual Compton scattering (DVCS), in which a photon γ^* with high virtuality (Q^2) interacts with a quark inside the target nucleon and produces a real photon γ. The DVCS final state is indistinguishable from that of the Bethe-Heitler (BH) process, where the photon is radiated by the incoming or outgoing lepton. The resulting interference term $\mathcal{I}$ offers the opportunity to access the DVCS amplitude which can be expressed in terms of GPDs which describe the target nucleon. The BH-DVCS interference term $\mathcal{I}$ depends on the charge (e_l) and polarization (P_l) of the incident lepton and can be extracted by measuring the beam charge and beam spin asymmetries. These asymmetries are usually expressed as a function of the azimuthal angle ϕ between the lepton scattering plane and the photon production plane[3].

2. The HERMES Experiment

The data have been collected at the HERMES experiment by scattering a 27.6 GeV longitudinally polarized electron (positron) beam stored in the HERA collider at DESY off an internal gas target. The beam polarization is measured continuously by two independent polarimeters[4].

The scattered electrons (positrons) and coincident photons are detected by the HERMES spectrometer[5] in the polar-angle range of 40 to 220 mrad. A lepton trigger was formed from a coincidence between three scintillator hodoscope planes and a lead-glass calorimeter and required an energy of more than 3.5 GeV deposited in the calorimeter. Leptons are identified with an average efficiency of 99% with a hadron contamination of less then 1%. This is accomplished by combining information from the calorimeter, a scintillator hodoscope preceded by two radiation length of lead (preshower detector), and a transition radiation detector.

The photons are identified by their energy deposition in the calorimeter and preshower detector with no associated signal in the tracking system. The calorimeter provides an angular resolution for photons of about 2 mrad.

3. The DVCS Measurements

The single photon events were selected with only one DIS particle and one photon detected in the spectrometer. In addition, the angle $\theta_{\gamma^*\gamma}$ between the real and the virtual photon was restricted to the interval of 2-45 mrad. This was necessary to ensure reasonable resolution and large acceptance in the azimuthal angle ϕ.

The recoiling target remnants are not covered by the acceptance of the HERMES spectrometer. Therefore the missing mass

$$M_x = \sqrt{m_N^2 + 2\,m_N(\nu - E_\gamma) + t} \quad , \tag{1}$$

where m_N is the nucleon mass, E_γ the photon energy and t the squared four momentum transfer, was used for the selection of exclusive events. Due to the finite resolution, which is limited by the calorimeter, the exclusive sample will not only include contributions from the exclusive coherent reaction on the nucleus and from incoherent production off protons and neutrons in the nucleus but also semi-inclusive processes where only one photon is detected within the HERMES acceptance. The semi-inclusive contribution to the exclusive sample was estimated with Monte Carlo studies and was found not to exceed 5%.

3.1. *Beam Charge Asymmetry*

The Beam Charge Asymmetry (BCA) A_C from an unpolarized target and unpolarized lepton beams can be written as:

$$A_C = \frac{d^+\sigma^{unp}(\phi) - d^-\sigma^{unp}(\phi)}{d^+\sigma^{unp}(\phi) + d^-\sigma^{unp}(\phi)} \sim \cos\phi \times \Re(I) \tag{2}$$

The BCA from a deuteron target is shown in Fig. 1 as a function of the azimuthal angle ϕ. The solid curve in the figure indicates the $\cos\phi$ behavior of A_C, in accordance with the expectations of the GPD formalism.

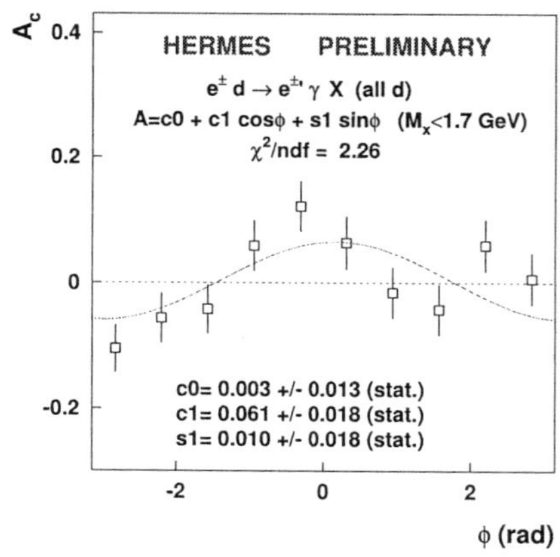

Figure 1. Beam charge asymmetry from an unpolarized Deuterium target as a function of the azimuthal angle ϕ.

3.2. *Beam Spin Asymmetry*

The Beam Spin Asymmetry (BSA) A_{LU} (here the subscripts L and U refer to longitudinally polarized beam and unpolarized target) can be written as:

$$A_{LU} = \frac{\overrightarrow{d\sigma}(\phi) - \overleftarrow{d\sigma}(\phi)}{\overrightarrow{d\sigma}(\phi) + \overleftarrow{d\sigma}(\phi)} \sim \sin\phi \times \Im(I) \tag{3}$$

The left panel of Fig. 2 shows the results obtained from an unpolarized Neon target, the solid line indicates the sin ϕ behavior of A_{LU} as expected in leading order QCD. The right panel of Fig. 2 shows the $A_{LU}^{\sin\phi}$ moment from an unpolarized Krypton target as a function of the missing mass M_x, corroborating the missing mass technique for exclusive event selection as the $A_{LU}^{\sin\phi}$ moment is vanishing for higher missing masses.

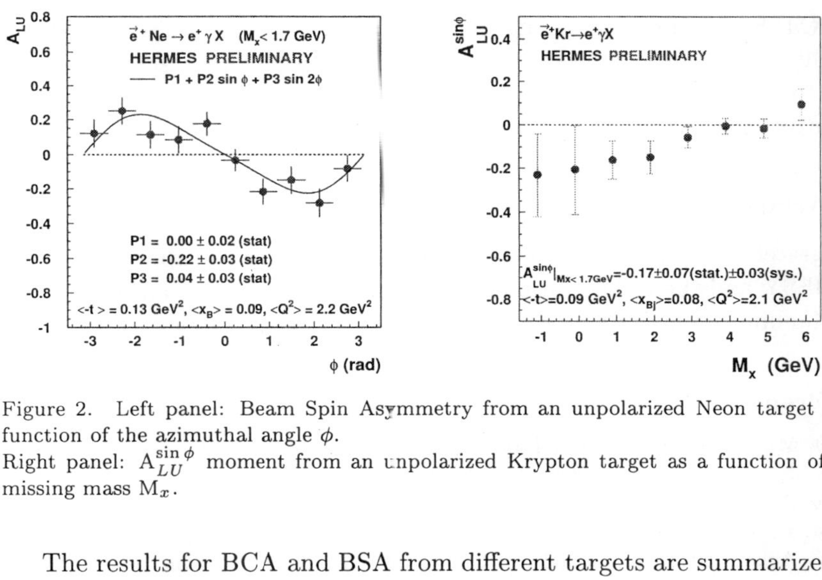

Figure 2. Left panel: Beam Spin Asymmetry from an unpolarized Neon target as a function of the azimuthal angle ϕ.
Right panel: $A_{LU}^{\sin\phi}$ moment from an unpolarized Krypton target as a function of the missing mass M_x.

The results for BCA and BSA from different targets are summarized in Tab. 1, including the results for a proton target[6].

4. Conclusion

The beam-spin and beam-charge azimuthal asymmetries associated with the DVCS-BH interference have been measured for the first time for a variety of nuclear targets ranging from Deuterium up to Krypton. Studying GPDs and their modifications in nuclear matter opens the possibil-

Table 1. $A_{LU}^{\sin\phi}$ and $A_C^{\cos\phi}$ from different targets

target	$A_{LU}^{\sin\phi}$	$A_C^{\cos\phi}$
p	$-0.18\pm0.03_{stat.}\pm0.03_{syst.}$	$0.059\pm0.028_{stat.}$
d	$-0.15\pm0.03_{stat.}\pm0.03_{syst.}$	$0.061\pm0.018_{stat.}$
Ne	$-0.22\pm0.03_{stat.}\pm0.03_{syst.}$	–
Kr	$-0.17\pm0.07_{stat.}\pm0.03_{syst.}$	–

ity of accessing spatial distributions of energy, angular momentum and shear forces inside the nuclei[7]. Theoretical calculations[8] predict the ratios $A_{LU}^{nucleus}/A_{LU}^{prot}$ and $A_C^{nucleus}/A_C^{prot}$ to approach unity when integrating over the momentum transfer t, cf. Tab. 1. In case of purely coherent reactions these ratios should be larger than unity. Future studies of the DVCS process at HERMES will be based on a separation of coherent and incoherent fractions.

Furthermore, a dedicated Recoil Detector has lately been installed in the HERMES experiment enabling the detection of recoiling target remnants. In this way excited nucleon states described by different sets of GPDs, *e.g.* Δ-resonance, can be distinguished and is thus allowing for a cleaner exclusive events sample.

5. Acknowledgments

We gratefully acknowledge the DESY management for its support, the staff at DESY and the collaborating institutions for their significant effort, and our funding agencies (BMBF, DFG, EU) for financial support.

References

1. D. Müller *et al.*, *Fortschr. Phys.* **42**, 101 (1994)
2. X. Ji, *Phys. Rev. Lett.* **78**, 610 (1997)
3. A. Belitsky *et al.*, *Nucl. Phys.* **B629**, 323 (2002)
4. M. Beckmann *et al.*, *Nucl. Instrum. Meth.* **A479**, 334 (2002)
5. K. Ackerstaff *et al.* (HERMES Collaboration), *Nucl. Instrum. Meth.* **A417**, 230 (1998)
6. A. Airapetian *et al.* (HERMES Collaboration), *Phys. Rev. Lett.* **87**, 182001 (2001)
7. M. Polyakov, *Phys. Lett.* **B555**, 57 (2003)
8. V. Guzey, M. Strikman, hep-ph/0301216

MEASUREMENT OF FORWARD-BACKWARD ASYMMETRY IN $B \to K^*\ell^+\ell^-$ AT BELLE

A. ISHIKAWA

The University of Tokyo,
Hongo 7-3-1,
Bunkyo-ku, Tokyo 113-0033, Japan
E-mail: akimasa@bpost.kek.jp

We present the first measurement of the forward-backward asymemtry and the ratios of Wilson coefficients A_9/A_7 and A_{10}/A_7 in $B \to K^*\ell^+\ell^-$, where ℓ represents an electron or a muon. The results are obtained from a data sample containing 386 million $B\bar{B}$ pairs that was collected on the $\Upsilon(4S)$ resonance with the Belle detector at the KEKB asymmetric energy e^+e^- collider.

1. Introduction

The $b \to s$ processes are sensitive to new physics effect. If new heavy particles can contribute to the decays, their amplitudes will interfere with the Standard Model (SM) amplitudes and thereby modify the decay rate as well as decay distributions.

To evaluate the new physics contributions in $b \to s$ processes, Wilson coefficients are usually used[1]. These coefficients parameterize the strength of the short distance interactions. If new physics contributes to the $b \to s$ processes, the relevant coefficients may deviate from the SM values. For electroweak penguin decays, the effective Wilson coefficients C_7^{eff}, C_9^{eff} and C_{10}^{eff} appear in the partial decay width. A next-to-next-to-leading order calculation for these effective coefficients has many correction terms[2], so leading coefficients A_7, A_9 and A_{10} are usually used for the evaluation.

Measurements of $B \to X_s\gamma$ decay, which are consistent with the SM prediction[3], strongly constrain the magunitude of the Wilson coefficient A_7[4]. Observations of exclusive $B \to K^{(*)}\ell^+\ell^-$ and inclusive $B \to X_s\ell^+\ell^-$ decays by Belle[5] allow to constrain the Wilson coefficients A_9 and A_{10}[6]. Measured branching fractions for these decay modes are consistent with the predictions within the SM and exclude the large area of the (A_9, A_{10}) plane[7]. However sign of A_7 and values of A_9 and A_{10} are not determined yet. A Measurement of the forward-backward asymmetry and differential decay rate as functions of q^2 in $B \to K^*\ell^+\ell^-$ is promising from this point of view, since the relative signs and magnitudes of these coefficients can be determined.

We report the first measurement of the forward-backward asymemtry and the ratios of Wilson coefficients[8].

2. Measurement of Forward-Backward Asymmetry in $B \to K^*\ell^+\ell^-$

The data sample corresponds to 357 fb^{-1} which contains 386 million $B\overline{B}$ pairs is used for measurement of forward-backward asymmetry in $B \to K^*\ell^+\ell^-$. Following decays of B mesons are reconstructed:$B^0 \to K^{*0}\ell^+\ell^-$ and $B^+ \to K^{*+}\ell^+\ell^-$, where ℓ represents an electron or a muon, with subdecays $K^{*0} \to K^+\pi^-$, $K^{*+} \to K^0_S\pi^+$ and $K^+\pi^0$, $K^0_S \to \pi^+\pi^-$, and $\pi^0 \to \gamma\gamma$. The charge conjuate modes are included throughout the manuscript unless otherwise specified. We observe 113.6 ± 13.0 $B \to K^*\ell^+\ell^-$ signal events with a purity of 44% (Fig 1).

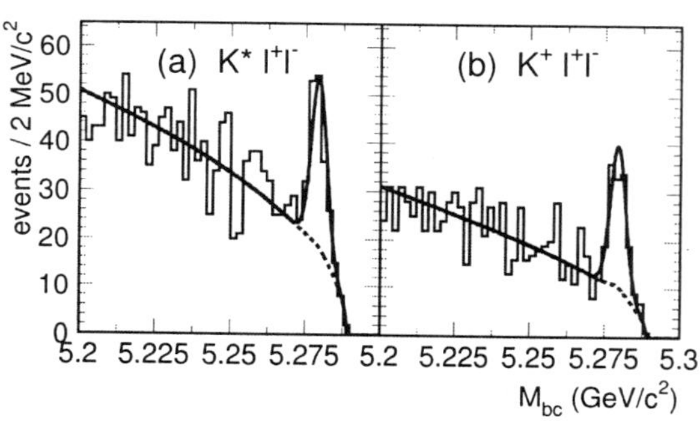

Figure 1. $M_{\rm bc}$ distributions for (a) $B \to K^*\ell^+\ell^-$ and (b) $B \to K^+\ell^+\ell^-$ samples. The solid and dashed curves are the fit results of the total and background contributions.

To extract ratios of Wilson coefficients, we perform an unbinned maximum likelihood fit with probability density function (PDF) that includes the normalized double differential decay width $(1/\Gamma)d^2\Gamma/dq^2 d\cos\theta$[9], where $\cos\theta$ is the cosine of the angle between negative (positive) charged lepton and B^0 or B^+ ($\overline{B}^0$ or B^-) meson momenta in the dilepton rest frame. The forward-backward asymmetry is defined as

$$\mathcal{A}_{\rm FB}(q^2) = \frac{\int_0^1 \frac{d^2\Gamma}{dq^2 d\cos\theta} d\cos\theta - \int_{-1}^0 \frac{d^2\Gamma}{dq^2 d\cos\theta} d\cos\theta}{\int_0^1 \frac{d^2\Gamma}{dq^2 d\cos\theta} d\cos\theta + \int_{-1}^0 \frac{d^2\Gamma}{dq^2 d\cos\theta} d\cos\theta}. \quad (1)$$

In the PDF, the signal and cross feed efficiencies and the background probability density as functions of $\cos\theta$ and q^2. The A_7 is fixed to SM value, -0.330, since the measurement of the branching fraction of $B \to X_s\gamma$ is consistent with the prediction within the SM, while the A_9/A_7 and A_{10}/A_7 are allowed to float in the fit. We measure the ratios of Wilson coefficients,

$$\begin{aligned} A_9/A_7 &= -15.3\,^{+3.4}_{-4.8} \pm 1.1, \\ A_{10}/A_7 &= 10.3\,^{+5.2}_{-3.5} \pm 1.8, \end{aligned} \quad (2)$$

which are consistent with the SM values -12.3 and 12.8, respectively. Figure 2 shows fit results projected onto the background-subtracted forward-backward asymmetry distribution in bins of q^2.

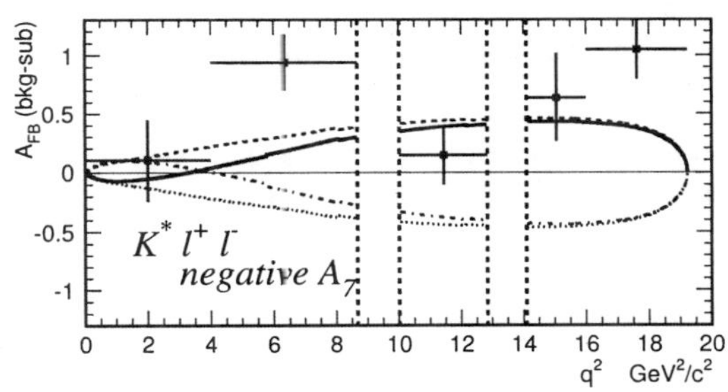

Figure 2. Fit results for the negative A_7 solution projected onto forward-backward asymmetry(solid blue) and forward-backward asymmetry curves with several input parameter including efficiency effect; A_7A_{10} sign flipped (dashed green), both A_7A_{10} and A_9A_{10} signs flipped (dash-dot red) and A_9A_{10} sign flipped(dotted magenta) to SM value. The new physics scenarios shown by the red and magenta curves are excluded.

In Fig. 3, we show confidence level (C.L.) contours in the A_9/A_7-A_{10}/A_7 based on fit likelihood smeared by systematic error, which is assumed to have a Gaussian distribution. We also calculate an interval on A_9A_{10}/A_7^2 at 95% C.L. for any allowed A_7 value,

$$-1401 < A_9A_{10}/A_7^2 < -26.4. \quad (3)$$

We determine the sign of A_9A_{10} to be negative, and exclude solutions in the first or third quadrant with more than 95% C.L. Both second and fourth quadrant solutions are allowed, so the sign of $A_7 A_{10}$ cannot be determined yet. We exclude new physics scenarios shown by the red and magenta curves in figure 2, which have positive $A_9 A_{10}$[10].

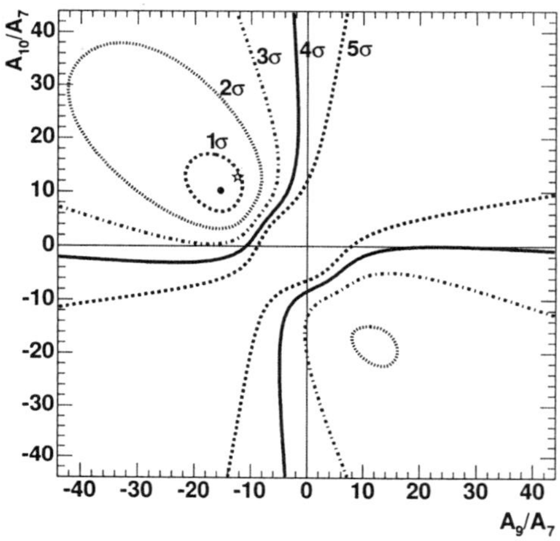

Figure 3. C.L. contours for negative A_7. Curves show 1 to 5 σ contours. Circle, star and triangle show the best-fit, the SM and the A_{10} positive cases.

3. Conclusion

We have measured the forward-backward asymmetry and the ratios of Wilson coefficients A_9/A_7 and A_{10}/A_7 in $B \to K^* \ell^+ \ell^-$. The measured Wilson coefficients are consistent with the SM prediction.

Acknowledgments

The author wish to thank the KEKB accelerator group for the excellent operation of the KEKB accelerator.

References

1. G. Buchalla, A. Buras and M. Lautenbacher, *Rev. Mod. Phys.* **68**, 1125 (1996) for example.
2. H. H. Asatryan *et al. Phys. Lett.* B **507**, 162 (2001).

3. P. Koppenburg *et al.* (Belle Collaboration), *Phys. Rev. Lett.* **93**, 061803 (2004); S. Chen *et al.* (CLEO Collaboration), *Phys. Rev. Lett.* **87**, 251807 (2001); R. Barate *et al.* (ALEPH Collaboration), *Phys. Lett.* B **429**, 169 (1998).
4. T. Besmer, C. Greub, T. Hurth, hep-ph/0105292; M. Ciuchini *et al.*, *Nucl. Phys.* B **534**, 3 (1998); C. Bobeth, M. Misiak and J. Urban, *Nucl. Phys.* B **567**, 153 (2000); F. Borzumati *et al.*, *Phys. Rev.* D **62**, 075005 (2000); T. Goto *et al.*, *Phys. Rev.* D **58**, 094006 (1998).
5. K. Abe *et al.* (Belle Collaboration), *Phys. Rev. Lett.* **88**, 021801 (2002); J. Kaneko *et al.* (Belle Collaboration), *Phys. Rev. Lett.* **90**, 021801 (2003); A. Ishikawa *et al.* (Belle Collaboration), *Phys. Rev. Lett.* **91**, 261601 (2003).
6. E. Lunghi *et al.*, *Nucl. Phys.* B **568**, 120 (2000); J. L. Hewett and J. D. Wells, *Phys. Rev.* D **55**, 5549 (1997); A. Ali, G. F. Giudice and T. Mannel, *Z. Phys.* C **67**, 417 (1995); N. G. Deshpande, K. Panose and J. Trampetić, *Phys. Lett.* B **308**, 322 (1993); A. Ali, T. Mannel and T. Morozumi, *Phys. Lett.* B **273**, 505 (1991); B. Grinstein, M. J. Savage and M. B. Wise, *Nucl. Phys.* B **319**, 271 (1991); W. S. Hou, R. S. Willey and A. Soni, *Phys. Rev. Lett.* **58**, 1608 (1987).
7. P. Gambino, U. Haisch and M. Misiak, *Phys. Rev. Lett.* **94**, 061803 (2005).
8. A. Ishikawa *et al.* (Belle Collaboration), hep-ex/0603018, submitted to *Phys. Rev. Lett.*.
9. A. Ali, P. Ball, L. T. Handoko and G. Hiller, *Phys. Rev.* D **61**, 074024 (2000).
10. E. Lunghi *et al.*, *Nucl. Phys.* B **568**, 120 (2000).

THE LATEST FROM MINOS*

HYEJOO KANG
PHYSICS DEPARTMENT
STANFORD UNIVERSITY
(FOR MINOS COLLABORATION)

382 Via Pueblo Mall
Stanford University
Stanford, CA 94305 USA
E-mail: hjkang@slac.stanford.edu

MINOS (Main Injector Neutrino Oscillation Search) is an accelerator-based long-baseline experiment designed to measure the properties of neutrino oscillations. MINOS will be able to provide a greater understanding of neutrino oscillations by making precision measurements of the neutrino mixing parameters in the atmospheric Δm^2 range. A neutrino beam generated by the NuMI facility at Fermilab is directed towards the Soudan Mine, located 735 km away in northern Minnesota. The MINOS Near Detector at Fermilab measures the neutrino beam energy spectrum, which is then extrapolated to predict the non-oscillated neutrino energy spectrum at the Far Detector located in the Soudan Mine. The ratio of the observed energy spectrum to the predicted energy spectrum at the Far Detector provides the measurement of Δm^2_{atm} and $sin^2 2\theta_{atm}$. MINOS has been taking data from the neutrinos delivered through the NuMI beamline since early 2005. In this talk, I present the physics goals, the current status of the experiment and the analysis of the first year of data.

1. Introduction

MINOS[1] studies ν_μ oscillation in the Δm^2_{atm} range with high statistics. The MINOS two detectors are magnetized tracking calorimeters. The 980 ton Near Detector at Fermilab measures the beam composition and the beam energy spectrum to predict the non-oscillated neutino energy spectrum at

*This work was supported by the U.S. Department of Energy, the U.K. Particle Physics and Astronomy Research Council, the U.S. National Science Foundation, the State and University of Minnesota, the Office of Special Accounts for Research Grants of the University of Athens, Greece, and FAPESP (Fundacao de Amparo a Pesquisa do Estado de Sao Paulo) and CNPq (Conselho Nacional de Desenvolvimento Cientifico e Tecnologico) in Brazil.

the Far Detector. The 5.4 kiloton Far Detector located deep underground in the Soudan Mine measures the oscillated energy spectrum. The Far Detector is the first large underground ν detector able to identify μ^- and μ^+, which enables MINOS to study atmospheric neutrino oscillation for μ^- and μ^+ separately.

The first physics goal of the MINOS experiment is the verification of the dominant ν_μ oscillation through a measurement of the oscillation energy dependence of ν_μ charged current (CC) events in the Far Detector. Since the MINOS baseline L is fixed in the probability $P(\nu_\mu \to \nu_\mu) = 1 - sin^2 2\theta_{atm} sin^2(1.27\Delta m^2_{atm} L/E)$, the probability can be measured by comparing the oscillated energy spectrum at the Far Detector and the un-oscillated energy spectrum predicted from the Near Detector. The goal is to make a precise measurement of Δm^2_{atm} at the 10% level. Fig.1 illustrates the ratio of the energy spectra and the allowed region in the Δm^2_{atm} and $sin^2 2\theta_{atm}$ parameter space with high statistics. The location and the depth of the signal in the ratio spectrum yield the Δm^2_{atm} and $sin^2 2\theta_{atm}$ measurements. The behavior of the oscillated energy spectrum is very different from the behaviors of other non-oscillation models such as ν decay and ν decoherence etc. Therefore it will be possible that these models can be either ruled out or confirmed with MINOS data.

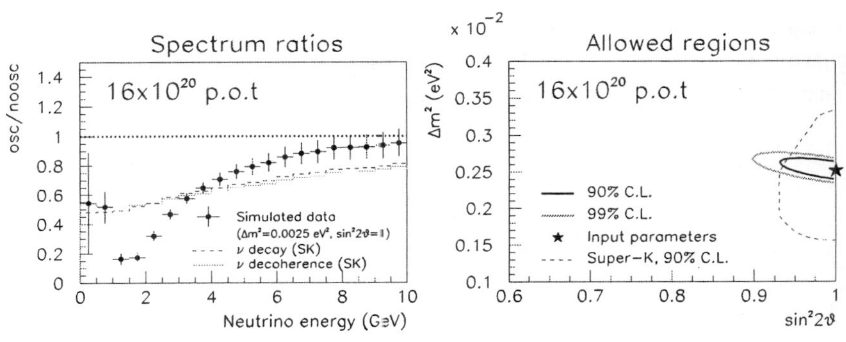

Figure 1. The ratio of the oscillated energy spectrum to the un-oscillated energy spectrum (left) of simulated data with the parameter input $\Delta m^2_{atm} = 2.5 \times 10^{-3} eV^2$ and $sin^2 2\theta_{atm} = 1$ and the spectra of non-oscillation models. The expected allowed region of MINOS in the Δm^2_{atm} and $sin^2 2\theta_{atm}$ parameter space with the allowed region from the zenith angle Super-K measurement (right)

Another physics goal is to search for $\nu_\mu \to \nu_e$ signal. The sensitivity limit on $sin^2 2\theta_{13}$ can be improved by a factor of two compared to the

CHOOZ result which is the current best limit on the parameter. Finally, the larger volume Far Detector is used to study μ^- and μ^+ oscillations of atmospheric neutrinos using charge identification[2].

2. Neutrinos at Fermilab

Neutrinos for MINOS are generated at the NuMI (Neutrinos at Main Injector)[3] beamline which was designed to deliver up to 4×10^{13} of protons every two seconds to the NuMI target. Currently, 2.2 to 2.8 $\times10^{13}$ protons are accelerated to 120 GeV at the Main Injector every two seconds. The accelerated protons hit a segmented graphite target of two interaction lengths, which results in the generation of a charged hadron beam. The hadron beam is focused with two water-cooled, parabolic aluminum magnetic horns. The focused hadrons decay in a 675 m steel vacuum decay pipe. An ionization chamber used as a hadron monitor and an aluminum beam absorber are positioned downstream of the decay pipe. To monitor muons, three ionization chambers are placed between rocks which function as muon absorbers. A special feature of the NuMI beam line is the ability to tune the neutrino energy spectrum by changing the position of the target by up to 2.5 m in the beam direction. An additional tuning can be achieved by changing the distance between the two horns by up to 40 m. Fig.2 shows the neutrino flux spectra at the Near Detector for different target positions. The experiment has been operating mainly with the neutrino beam in the "Low Energy" configuration, in which the neutrino energy gives a sensitivity of Δm^2_{atm} down to $2\times10^{-3}eV^2$.

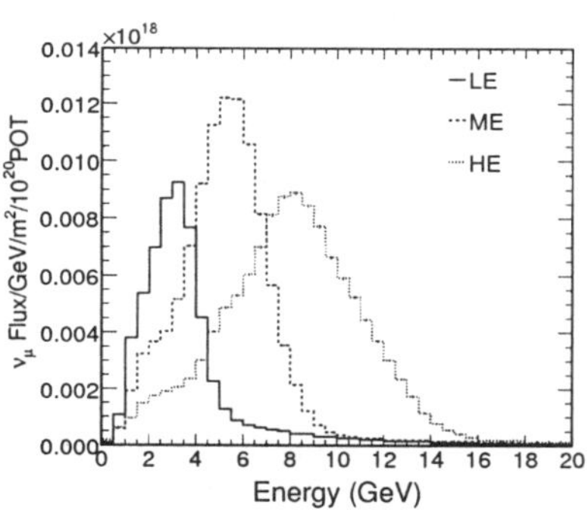

Figure 2. The neutrino energy spectra for different target positions. For the "Low Energy (LE)" configuration, the target is inside the first horn."Medium Energy (ME)" and "High Energy (HE)" configurations were used for systematics studies.

3. MINOS detectors

MINOS' two detectors are functionally identical, the detector technology and granularity being the same. The detectors are composed of layers of 2.54 cm thick steel planes and 1 cm thick and 4.1 cm wide solid scintillator strips. Scintillator strips in alternate planes are oriented at $\pm 45^o$ to the vertical, which provide event signals in two orthogonal directions. The scintillation light is collected by wavelength shifting fibers embedded in the scintillator strips and then transmitted through clear optical fibers to multi-anode photomultiplier tubes. The detectors are magnetized with a toroidal B field to an average value of 1.2 T. The electronics of the detectors read-out continuously during ~10 micro-second beam spills without any trigger necessary. The fast no-deadtime electronics of the Near Detector can read-out signals at very high event rates from the beam. The energy resolution of hadronic showers is $55\%/\sqrt{E}$. The momentum of muons can be measured to 6% by counting the number of planes that muons pass through and to 10% by observing the curvature of the tracks.

4. Charged Current (CC) analysis for ν_μ oscillation

The ν_μ oscillation analysis was performed using data taken in the "Low Energy" configuration with 0.93 $\times 10^{20}$ total accumulated protons on the target. ν^μ CC events were selected with likelihood-based particle identification (PID). The PID used three input Probability Density Functions; the event length, the fraction of event pulse heights contained in a muon track and the average track pulse height per plane. The analysis was pursued as a blind analysis for the first beam neutrino result. An unknown fraction of MINOS events were hidden based on their length and total deposited energy. The event selection and fitting procedure for the analysis were pre-defined and validated on MC before opening the box. The direct comparison of the observed energy spectrum at the Far Detector and the un-oscillated energy spectrum extrapolated from the Near Detector was performed with small corrections from MC for energy smearing and the detector acceptances. Several different methods to predict the un-oscillated Far Detector spectrum were investigated for the analysis.

The box was opened on March, 4th 2006 and the first measurement of the oscillation parameters with MINOS data was obtained. The oscillated spectrum at the Far Detector and the predicted un-oscillated spectrum are shown in Fig.3 with the measured values of Δm^2_{atm} and $sin^2 2\theta_{atm}$. The different methods to predict the un-oscillated Far Detector spectrum give consistent results. The results also consistent with the K2K and Super-K results. The size of the allowed region in the Δm^2_{atm} and $sin^2 2\theta_{atm}$ parameter space is comparable with the K2K result.

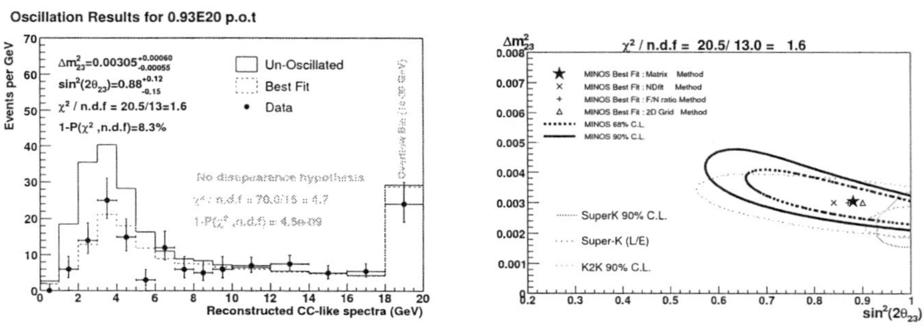

Figure 3. The oscillated spectrum at the Far Detector and the predicted un-oscillated spectrum and the best fit result. The MINOS allowed region in the Δm^2_{atm} and $sin^2 2\theta_{atm}$ parameter space as well as the K2K and Super-K results

5. Conclusion

MINOS is designed to make precision measurements of the neutrino mixing parameters in the atmospheric Δm^2 range. The first year of data taking was very stable and the detector performance was excellent. The CC analyses were pursed as blind analyses using neutrinos generated with 0.93×10^{20} protons on the MINOS target. The MINOS result is consistent with ν_μ disappearance with the parameters

$$\Delta m^2_{atm} = 3.05^{+0.60}_{-0.55}(stat) \pm 0.12(syst) \times 10^{-3} eV^2$$
$$sin^2 2\theta_{atm} = 0.88^{+0.12}_{-0.15}(stat) \pm 0.06(syst)$$

We expect to have a measurment of Δm^2_{atm} with better than 10% precision with higher statistics within five years. We also anticipate to have results on ν_e appearance and sterile neutrinos in coming years.

References

1. P. Adamson *et al.* (MINOS), MINOS Technical Design Report, NuMI-L-337 .
2. P.Adamson *et al.* (MINOS collaboration) , *Phys. Rev.* **D73**, 072002 (2006).
3. *http : //www − numi.fnal.gov/numwork/tdh/tdh_index.html*, NuMI Technical Design Handbook

BLACK HOLES AND QUASISTABLE REMNANTS AT THE LHC

BENJAMIN KOCH[1,2]*, MARCUS BLEICHER[1]† AND SABINE HOSSENFELDER [3]‡

[1] *Institut für Theoretische Physik, J. W. Goethe-Universität*
Max von Laue Strasse 1
60438 Frankfurt am Main, Germany

[2] *Frankfurt International Graduate School for Science (FIGSS)*
Max von Laue Strasse 1
60438 Frankfurt am Main, Germany

[3] *Department of Physics, University of California*
Santa Barbara, CA 93106-9530, USA

In models with large extra dimensions (LXDs) the Planck scale can be lowered to values that are within the reach of the next generation of particle colliders. Those models predict the production of TeV mass black holes and therefore a new era in short distance physics. However, the details of the black hole evaporation process especially in the final phase are unknown. We study the implications of the assumption that a quasi stable remnant is formed. Therefore the black hole particle spectrum and direct detection are taken into account.

1. LXDs, black holes and quasi stable remnants

Models with large extra dimensions provide a relatively simple framework for physics beyond the standard model (SM). At the same time they allow to make distinct predictions which are experimentally testable. They are motivated by string theory[1] and were proposed by Arkani-Hamed, Dimopoulos and Dvali[2] as a solution to the hierarchy problem.

In those models with LXDs each additional dimension is compactified on a radius R by imposing periodic boundary conditions. At the same time

*koch@th.physik.uni-frankfurt.de
†bleicher@th.physik.uni-frankfurt.de
‡sabine@physics.ucsb.edu

gravity is assumed to propagate freely into the additional space-like dimensions whereas the SM particles are explicitly bound to our 4-dimensional sub-manifold, often called the 3-brane. On distance scales much smaller than the compactification radius R gravitational laws are following the $3 + d$ dimensional description. From this, one finds by using Gauss law, the gravitational potential as a function of the radial distance from a given mass, $\phi(r) \propto \frac{1}{M_{\mathrm{f}}^{d+2}} \frac{M}{r^{d+1}}$, where the new fundamental mass-scale is given by M_{f}. Due to the compactification prescription, the gravitational potential on large distance scales follows a $1/r$ behaviour. The matching of this to Newton's law gives a relation between the observed Planck mass ($m_p \sim 10^{19}$GeV) and the true fundamental constant M_{f}

$$m_{\mathrm{p}}^2 = M_{\mathrm{f}}^{d+2} R^d \quad . \tag{1}$$

¿From this one sees that the right choice of R and d allows to reproduce a $m_p \sim 10^{19}$GeV with a M_{f} that is only in the TeV region.

Newtons law is by far not the only physics where one can study and constrain the size of the LXDs[3]. Also the equations describing black holes can be derived in the $3+d$ dimensional formulation of gravity. The solution for the higher dimensional Schwarzschild metric[4] for a black hole of mass M is found to be

$$R_H^{d+1} = \frac{2}{d+1} \frac{M}{M_{\mathrm{f}}^{d+2}} \quad . \tag{2}$$

This solution is applicable to collider physics as long as the radius of the $\sim$TeV mass black hole is much smaller than the compactification radius R. It turns out that the production of such objects might indeed be possible at the LHC and the production cross section is approximately the classical geometrical cross section

$$\sigma(M) = \pi R_H^2 \quad . \tag{3}$$

As a consistent theory of quantum gravity is still to be found, one has to treat those black holes as semi classical objects[5,6] and therefore different approaches to the production cross sections are possible[7,8,9]. The potential importance of microscopic black holes in collider signatures has already been intensively studied[10,11]. Microscopic black holes are assumed to evaporate shortly after their production. This evaporation can be categorised in three characteristic phases[12]: The balding phase, the Hawking phase and the so called Planck phase. The most critical phase is the Planck phase, which is reached as soon as the Hawking temperature reaches the black hole mass or as soon as the black hole mass approaches the Planck mass. In this phase the final fate of the black hole is determined and it is closely related to the information loss problem. In principle two scenarios are possible: Either the black hole performs a final decay or something like a quasi stable remnant is formed. There are several arguments that favour the formation of a stable or quasi stable remnant[13], like the uncertainty relation[14,15], corrections to the Lagrangian[16] or quantum hair[17]. These arguments give enough motivation to study the idea of quasi stable remnants. Before turning to the results of our numerical studies we would like to address two obvious objections against long lived black hole remnants:

- Abundance from cosmic rays: The production rate of black holes from highest energetic cosmic ray events has been studied[18] and it was found that, for the most optimistic scenario, the black hole production rate in an ice cube of volume $\sim 1\ \text{km}^3$ is around 10 black holes per year. ¿From this one can roughly estimate an upper limit of $\sim$ 510 g black hole matter that would be trapped inside our earth. A stable black hole would have a very low charge to mass ratio. Such particles have been searched for in different types of matter[19]. Ordinary mass spectrometry and accelerator mass spectrometry give upper limits on the relative abundance (X/nucleon) of such particles between 10^{-8} and 10^{-24} (depending on the mass of the charged particles), which is fully consistent with the estimated $\sim$TeV remnant concentration 9×10^{-30}.
- Naked singularities: ¿From the metric for a spherically symmetric charged black hole[4] with mass M and charge Q and from the d dimensional Newtons law one can derive a condition for finding a non imaginary black hole horizon R_H. It tells that for $\alpha Q^2 \leq \left(\frac{M}{M_{\rm f}}\right)^2$, where α is the fine structure constant, the singularity is shielded by an horizon. With $M =$ few $\times M_{\rm f}$, and Q being close to e, the left hand side is at least by a factor 100 smaller than the right hand side. So for the typical collider produced black holes, the singularity will not be naked.

2. Discriminating quasi stable remnants from a final decay

The question of observing the decay products of black holes at LHC experiments has been widely discussed in the literature[11,20]. In[13] the task of discriminating between quasi stable remnants and black holes with a final decay was discussed. It has been shown quantitatively that the formation of a remnant would lower the high p_T distribution of final state particles noticably in comparison to a final decay scenario. But one can go even further and look directly for black hole remnants: If the time of flight resolution of the detector can determine the velocity (see Figure 1, which was extracted from[13]) of one of the charged remnants, the bended path in the magnetic field would allow direct determination of the remnants mass.

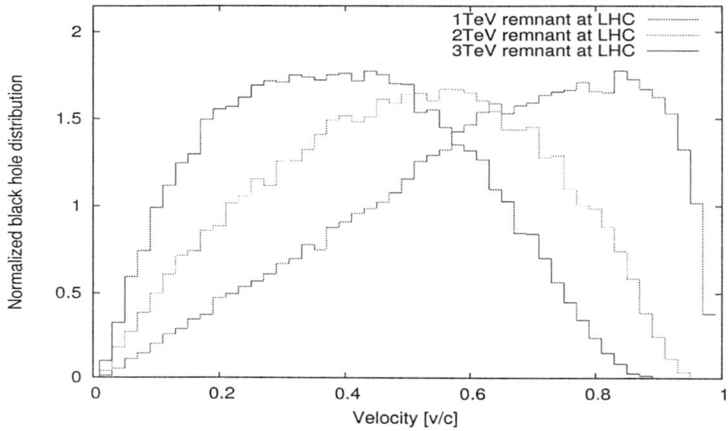

Figure 1. Remnant velocities at the LHC for the masses $M_R = 1, 2, 3$ TeV.

3. Summary

In this article we argued that the formation of quasistable remnants can not be easily excluded from naked singularities or cosmic rays and we pointed out ways for direct or indirect observation of such quasistable remnants.

References

1. I. Antoniadis, Phys. Lett. B **246**, 377 (1990);
2. N. Arkani-Hamed, S. Dimopoulos and G. R. Dvali, Phys. Lett. B **429**, 263 (1998);
3. CDF Coll. Phys. Rev. Lett. **89**, 281801 (2002); DØ Coll., Phys. Rev. Lett. **90**, 251802 (2003); K. Cheung, [arXiv:hep-ph/0409028].
4. R. C. Myers and M. J. Perry Ann. Phys. **172**, 304-347 (1986).
5. M. B. Voloshin, Phys. Lett. B **518**, 137 (2001); Phys. Lett. B **524**, 376 (2002); S. B. Giddings, eConf **C010630**, P328 (2001).
6. S. N. Solodukhin, Phys. Lett. B **533**, 153 (2002); A. Jevicki and J. Thaler, Phys. Rev. D **66**, 024041 (2002); D. M. Eardley and S. B. Giddings, Phys. Rev. **D 66**, 044011 (2002).

7. V. S. Rychkov, Phys. Rev. D **70**, 044003 (2004); K. Kang and H. Nastase, [arXiv:hep-th/0409099]; V. S. Rychkov, arXiv:hep-th/0410295; H. Yoshino and V. S. Rychkov, Phys. Rev. D **71** (2005) 104028
8. I. Ya. Yref'eva, S. B. Giddings and V. S. Rychkov, Phys. Rev. D **70**, 104026 (2004); V. S. Rychkov, [arXiv:hep-th/0410041]; O. V. Kancheli, [arXiv:hep-ph/0208021].
9. H. Yoshino and Y. Nambu, Phys. Rev. D **67**, 024009 (2003); S. N. Solodukhin, Phys. Lett. **B 533** 153-161 (2002); D. Ida, K. y. Oda and S. C. Park, Phys. Rev. D **67**, 064025 (2003).
10. S. Dimopoulos and G. Landsberg Phys. Rev. Lett. **87**, 161602 (2001); P.C. Argyres, S. Dimopoulos, and J. March-Russell, Phys. Lett. **B441**, 96 (1998).
11. L. Anchordoqui and H. Goldberg, Phys. Rev. D **67**, 064010 (2003); K. Cheung, Phys. Rev. D **66**, 036007 (2002); K. m. Cheung, Phys. Rev. Lett. **88**, 221602 (2002); S. C. Park and H. S. Song, J. Korean Phys. Soc. **43**, 30 (2003); S. Hossenfelder, S. Hofmann, M. Bleicher and H. Stocker, Phys. Rev. D **66**, 101502 (2002); M. Bleicher, S. Hofmann, S. Hossenfelder and H. Stocker, Phys. Lett. B **548**, 73 (2002); M. Cavaglia, S. Das and R. Maartens, Class. Quant. Grav. **20**, L205 (2003); M. Cavaglia and S. Das, Class. Quant. Grav. **21**, 4511 (2004); S. Hossenfelder, Phys. Lett. B **598**, 92 (2004); I. Mocioiu, Y. Nara and I. Sarcevic, Phys. Lett. B **557**, 87 (2003); A. Ringwald, Fortsch. Phys. **51**, 830 (2003); A. Chamblin and G. C. Nayak, Phys. Rev. D **66**, 091901 (2002).
12. S. B. Giddings and S. Thomas, Phys. Rev. **D 65** 056010 (2002).
13. B. Koch, M. Bleicher and S. Hossenfelder, JHEP **0510** (2005) 053; S. Hossenfelder, B. Koch and M. Bleicher, arXiv:hep-ph/0507140.
14. M. A. Markov, in: "*Proc. 2nd Seminar in Quantum Gravity*", edited by M. A. Markov and P. C. West, Plenum, New York (1984).
15. Y. B. Zel'dovich, in: "*Proc. 2nd Seminar in Quantum Gravity*", edited by M. A. Markov and P. C. West, Plenum, New York (1984).
16. J. D. Barrow, E. J. Copeland and A. R. Liddle, Phys. Rev. **D 46**, 645 (1992); B. Whitt, Phys. Rev. **D 38**, 3000 (1988); R. C. Myers and J. Z. Simon, Phys. Rev. **D 38**, 2434 (1988).
17. S. Coleman, J. Preskill and F. Wilczek, Mod. Phys. Lett. **A6** 1631 (1991).
18. A. Ringwald and H. Tu, Phys. Lett. B **525** (2002) 135; M. Kowalski, A. Ringwald and H. Tu, Phys. Lett. B **529** (2002) 1, U. Harbach and M. Bleicher, arXiv:hep-ph/0601121.
19. P. Hut and M. J. Rees, Print-83-0042 (IAS,PRINCETON); W. Busza, R. L. Jaffe, J. Sandweiss and F. Wilczek, Rev. Mod. Phys. **72** (2000) 1125; P. F. Smith et al., Nucl. Phys. B **206** (1982) 333; Z. T. Lu et al., Nucl. Phys. A **754** (2005) 361.
20. A. Barrau, J. Grain and S. O. Alexeyev, Phys. Lett. B **584** (2004) 114; C. M. Harris, P. Richardson and B. R. Webber, JHEP **0308**, 033 (2003); C. M. Harris et al., [arXiv:hep-ph/0411022]; G. L. Alberghi et al., arXiv:hep-ph/0601243; T. G. Rizzo, arXiv:hep-ph/0601029; A. Casanova and E. Spallucci, Class. Quant. Grav. **23** (2006) R45; B. Webber, arXiv:hep-ph/0511128; T. G. Rizzo, arXiv:hep-ph/0510420.

NONCOMMUTATIVE GEOMETRY AND THE PARTICLE CONTENT OF THE UNIVERSE

T.KOPF

Mathematical Institute
Silesian University at Opava
Na Rybníčku 1
74101 Opava, Czech Republic
E-mail: Tomas.Kopf@math.slu.cz

The idea of incorporating the particle content of a field theory as a mild noncommutativity in its geometry is revisited. An adjustment to the Lorentzian signature of the universe is discussed.

1. Introduction

In a desired unification of particle physics and gravity, it may turn out that gravity is a somewhat misunderstood part of particle physics or that particle physics is a somewhat misunderstood part of general relativity (geometry), not excluding both to be valid. The second view, the simultaneous description of spacetime and particle content as a geometry, though a noncommutative one, has found a remarkable expression in the works [1], [2], [3], reviewed also in [4]. See also [5], [6] for an action principle.

These are mainly concerned with a classical theory, either to be quantized later or to be understood as a partial description of the one-particle space of a quantum field theory. It is therefore not to be expected that these theories will immediately produce exact predictions for the values of coupling constants without further work on the quantization and renormalization of such models. At that stage of the development, it is important to fit the structure of the observed particle content.

The relevant noncommutative geometries, technically described as spectral triples, consisted of a commutative part describing spacetime tensored with a finite-dimensional noncommutative spectral triple responsible for the particle content. In order to have all possibilities at hand, the classification of spectral triples was worked out [7], [8].

While suitable choices led so substantial and perhaps surprising agreement with the structure of the Standard Model, there appeared some problems, in part in confrontation with recent experimental evidence. The prediction of massless neutrinos, a consequence of a form of Poincaré duality, is one of them.

Another problem was the assumption of Euclidean spacetime signature in the concept of a spectral triple, partially supported by the successes of Euclidean quantum field theory. A possible modification allowing to describe Lorentzian geometries has been given in [10] as a spectral quadruple and a commutative example, 1 + 1-dimensional de Sitter spacetime was calculated in [11]. However, these modifications affect also the structure of a possible finite noncommutative part responsible for the particle content of such a universe.

Section 2 briefly recalls some of the above mentioned ideas of noncommutative geometry, in particular spectral triples and spectral quadruples. Section 3 specifies the considered collection of mildly noncommutative theories.

2. Noncommutative geometry

The idea of noncommutative geometry is to describe spaces through algebras of coordinates on them. For classical spaces, these algebras are commutative, as the multiplication of coordinate functions (real or complex valued) gives a commutative product. In non-commutative geometry, the requirement of commutativity is dropped and as a consequence some classical concepts become less useful, like that of a point. However, important geometric concepts surprisingly survive this transition, if encoded in a suitable way [9]. A physical motivation to view such concepts as important is given by quantum theory.

To see, that the linear structure of coordinate functions is not sufficient, one may consider a classical two-point space: The coordinate functions span a 2-dimensional vector space with no information on which directions correspond to values at a given point. However, if also multiplication is supplied, the square function will map all of that linear space onto a wedge. The two half-lines forming its border determine the coordinates of the two points.

However, the algebraic structure alone is not enough to encode more advanced geometric ideas and more information on the space has to be given. The concept of a spectral triple is designed to capture the essentials of a spin manifold, its topological, differential geometric, metric and spin structure. An example is given in Section 3 while the full set of structures and axioms to be fulfilled by them can be found in [3].

A spectral quadruple builds on the spectral triple by supplying, in addition to a family of spectral triples, a time vector and allowing thus to view hyperbolic spacetimes in effect through foliations by Cauchy surfaces. For the axioms of a spectral quadruple and their discussion, see [10], [11].

3. Mildly noncommutative geometries

The classical example of a spectral triple is given by the commutative algebra of smooth functions $C^\infty(M)$ on a spin manifold M represented on sections of the spin bundle H and supplied with the Dirac operator D, the volume form γ and the charge conjugation operator J.

This is turned into a mildly noncommutative spectral triple by being tensored with a finite spectral triple. The algebra of the triple becomes then $C^\infty(M) \otimes A_F$, where A_F is decomposable into a finite sum of finite matrix algebras and has to be supplied with the structures of a spectral triple, satisfying the required axioms. It is this algebra that determines the particle content of the model.

In a similar way, spectral quadruples based on a commutative algebra provide mildly noncommutative examples by being tensored with a finite algebra provided with the necessary structures and satisfying a modified set of conditions[10].

4. Conclusion

Models of particle physics within noncommutative geometry are still at a preliminary stage. The quantization of such classical theories has to be clarified, while there are at the same time open issues in these classical (or one-particle-space) constructions. To address some of these, a classification of finite spectral quadruples, based on the results [7], [8] is desirable [12].

Acknowledgments

This work is supported by grant GAČR 202/05/2767. Additional support from Internal grant IGS SU 3/2006 of the Silesian University at Opava to present this research at the Lake Louise Winter Institute is thankfully acknowledged.

References

1. A. Connes and J. Lott, *Nucl.Phys.Proc.Suppl.* **18B**, 29 (1991).

2. A. Connes, *Journal of Math. Physics* **36**, 11 (1995).
3. A. Connes, *Commun.Math.Phys.* **182**, 155 (1996).
4. A. Connes, M. Marcolli, arXiv: math/0601054.
5. A. Chamseddine, A. Connes, *Phys.Rev.Lett.* **77** 4868 (1996).
6. A. H. Chamseddine, A. Connes arXiv: hep-th/0512169.
7. M. Paschke and A. Sitarz, *J. Math. Phys.* **39**, 6191 (1998).
8. T. Krajewski, *J. Geom. Phys.* **28**, 1 (1998).
9. A. Connes, *Noncommutative geometry* (Academic Press, San Diego, 1994).
10. Tomas Kopf and Mario Paschke: Spectral Quadruples, *Mod.Phys.Lett.* **A16**, 291 (2001).
11. T. Kopf and M. Paschke: *A spectral quadruple for de Sitter space*, *J. Math. Phys.* **43**, 818 (2002).
12. T. Kopf and M. Paschke, work in progress.

A STUDY OF MIXING IN THE $B_S^0 - \bar{B}_S^0$ SYSTEM USING THE DØ DETECTOR

D. KROP

Indiana University,
727 E. Third St.,
Bloomington, IN 47405-7105

A search for $B_s^0 - \bar{B}_s^0$ oscillations was performed with a sample of semileptonic B_s^0 decays corresponding to approximately 610 pb^{-1} of integrated luminosity accumulated with the DØ Detector in Run II at Fermilab (Tevatron). The flavor of the final state of the B_s^0 meson was determined using the muon charge from the partially reconstructed decay $B_s^0 \rightarrow \mu^+ D_s^- X$, An opposite-side tagging method was used for the initial-state flavor determination. A 95% confidence level limit on the oscillation frequency $\Delta m_s > 7.3$ ps^{-1} and sensitivity 9.5 ps^{-1} were obtained.

1. Introduction

One of the most interesting topics in B physics is B_s^0 mixing and the measurement of Δm_s. Combining Δm_s and Δm_d would allow a reduction in the theoretical uncertainty on V_{td}, and provide a critical test of the CKM formalism of the Standard Model. Currently the Tevatron is the only place in the world where B_s^0 mixing can be measured.

2. Detector Description

The following main elements of the DØ detector are essential for this analysis:

- The magnetic central-tracking system, which consists of a silicon microstrip tracker (SMT) and a central fiber tracker (CFT), both located within a 2 T superconducting solenoidal magnet;
- The muon system located beyond the calorimeter.

The SMT has $\approx 800,000$ individual strips, with typical pitch of $50-80\ \mu$m, and a design optimized for tracking and vertexing capability at $|\eta| < 3$, where $\eta = -\ln(\tan(\theta/2))$ and θ is a polar angle. The CFT has eight thin coaxial barrels, each supporting two doublets of overlapping scintillating fibers of 0.835 mm diameter, one doublet being parallel to the collision axis, and the other alternating by $\pm 3°$ relative to the axis.

The muon system consists of a layer of tracking detectors and scintillation trigger counters before 1.8 T toroids, followed by two additional layers after the toroids. Tracking at $|\eta| < 1$ relies on 10 cm wide drift tubes, while 1 cm mini-drift tubes are used at $1 < |\eta| < 2$.

3. Data sample

This analysis uses a $B \to \mu^+ D_s^- X$ data sample selected with an offline filter from all data taken from April 2002 to May 2005 with no trigger requirement. The D_s^- is reconstructed through its decay into two different channels: $D_s^- \to \phi\pi^-$, $\phi \to K^+K^-$ and $D_s^- \to K^*K^-$, $K^* \to K\pi$

Event samples were selected using cuts on the p_T of the tracks, the number of hits in the tracking detectors, the χ^2 of the vertices obtained from vertex fits, the transverse decay length significance of the D and B candidates, and the angle α_T^D between the D_s^- momentum and the direction from the primary to D_s^- vertex in the transverse plane. In addition, a likelihood ratio method was applied to further suppress backgrounds.

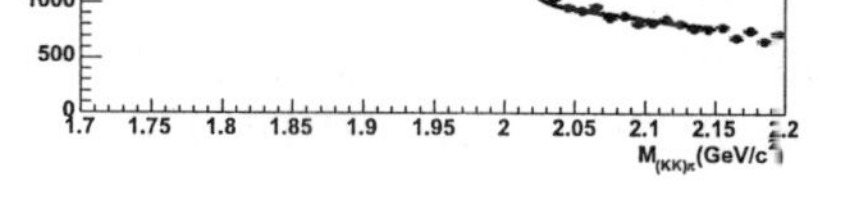

Figure 1. $(KK)\pi$ invariant mass distribution for the untagged $\phi\pi$ sample. The peaks correspond to the $\mu D^\pm$ (left) and μD_s (right) candidates. The curve represent the fitting function to this mass spectrum.

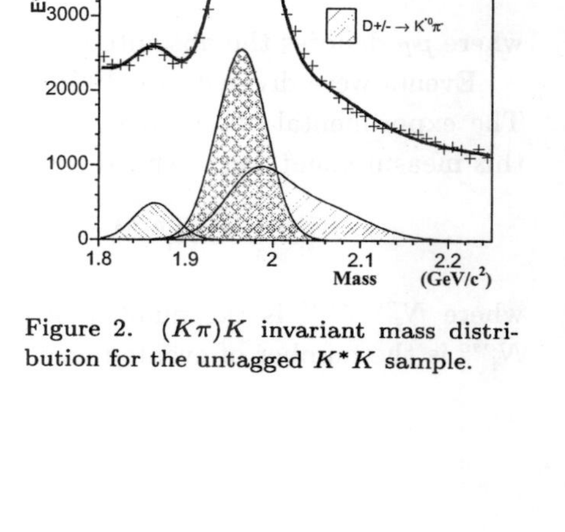

Figure 2. $(K\pi)K$ invariant mass distribution for the untagged K^*K sample.

The total number of D_s^-candidates passing the selections in the D_s mass peak is 15636 ± 193 for the $\phi\pi$ channel and is 18780 ± 782 in the K^*K channel. There are 1917 ± 66 $\phi\pi$ and 2247 ± 316 K^*K candidates which have an identified initial state flavor from the opposite-side tag.

4. Flavor Tagging

A necessary step in the B_s^0 oscillations analysis is the determination of the $B_s^0/\bar{B}_s^0$ initial and final state flavors. The presence of a muon in the B_s^0 semileptonic decay allows a determination of the final state flavor using the relations $B_s^0 \to \mu^+ X$ and $\bar{B}_s^0 \to \mu^- X$.

The inital state flavor is determined using an opposite-side tagging algorithm which is verified through a separate measurement of the B_d oscillation frequency,

$$\Delta m_d = 0.501 \pm 0.030\ (stat.) \pm 0.017\ (syst.)\ ps^{-1} \tag{1}$$

This result is in good agreement with the world average of $\Delta m_d = 0.502 \pm 0.007$ ps^{-1} [1].

5. Experimental Observables A_i^{meas}

The proper lifetime of the B_s^0 meson, $ct_{B_s^0}$, for semileptonic decays can be written as

$$ct_{B_s^0} = x^M \cdot K, \quad \text{where } x^M = (\mathbf{d_T^B} \cdot \mathbf{p_{xy}^{\mu D_s^-}})/(p_T^{\mu D_s^-})^2 \cdot M_B. \tag{2}$$

x^M is the *visible proper decay length* or VPDL. We correct for the bias due to the unreconstructed neutral particles present in the semileptonic decay by scaling the measured momentum of the B by a K-factor. The K-factor was estimated from MC simulations. For this analysis it was defined as:

$$K = p_T(\mu D_s^-)/p_T(B), \tag{3}$$

where p_T denotes the absolute value of transverse momentum.

Events were divided into 19 groups according to the measured VPDL. The experimental observables, asymmetry A_i^{meas} in each VPDL bin, for this measurement were defined as:

$$A_i^{meas} = \frac{N_i^{non-osc} - N_i^{osc}}{N_i^{non-osc} + N_i^{osc}}; \tag{4}$$

where $N_i^{non-osc}$ is the number of events tagged as "non-oscillated" and N_i^{osc} is the number of events tagged as "oscillated".

6. Amplitude method

In order to set a limit on the value of Δm_s, we chose to use a technique called the amplitude fit method [2]. This technique requires us to multiply the cosine term in the oscillation probability by a free parameter $\mathcal{A}$ as follows,

$$n_s^{non-osc/osc}(x) \propto [1 \pm (2\eta_s - 1)\cos(\Delta m_s \cdot Kx/c) \cdot \mathcal{A}], \tag{5}$$

The fitted values of $\mathcal{A}$ as a function of Δm_s were determined from the minimization of a $\chi^2(\mathcal{A})$ defined as:

$$\chi^2(\mathcal{A}) = \sum_i \frac{(A_i^{meas} - A_i^e(\mathcal{A}))^2}{\sigma^2(A_i^{meas})}. \tag{6}$$

The values of Δm_s were changed from 1 ps^{-1} to 20 ps^{-1} with a step size of 1 ps^{-1}. For each value of Δm_s the fitted value of $\mathcal{A}$ and its error were determined. To obtain a measurement of Δm_s one searches for a peak of $\mathcal{A} = 1$. If no peak is found then a limit can be set. The sensitivity of a measurement is determined by calculating the probability that $\mathcal{A} = 0$ could fluctuate to $\mathcal{A} = 1$. This occurs at 1.645σ (95% CL), where σ is the uncertainty associated with $\mathcal{A}$. The limit is determined by caclulating the probability that a fitted value of $\mathcal{A}$ could fluctuate to $\mathcal{A} = 1$. This occurs at $\mathcal{A} + 1.645\sigma < 1$.

7. Results and conclusions

Figure 3 shows the dependence of the parameter $\mathcal{A}$ and its error on Δm_s. Using a signal of 15.6k $B_s^0 \to \mu^+ \nu D_s^- X$ ($D_s \to \phi\pi$, $\phi \to KK$) and 18.8k $B_s^0 \to \mu^+ D_s^- X$ ($D_s^- \to K^{*0}K^-$, $K^{*0} \to K\pi$) decays and an opposite-side flavor tagging algorithm, we performed a search for $B_s^0 - \bar{B}_s^0$ oscillations. We obtain a 95% confidence level limit on the oscillation frequency $\Delta m_s >$ 7.3 ps^{-1} and a sensitivity of 9.5 ps^{-1}.

8. Ackowledgments

We thank the staffs at Fermilab and collaborating institutions, and acknowledge support from the DOE and NSF (USA); CEA and CNRS/IN2P3 (France); FASI, Rosatom and RFBR (Russia); CAPES, CNPq, FAPERJ, FAPESP and FUNDUNESP (Brazil); DAE and DST (India); Colciencias (Colombia); CONACyT (Mexico); KRF (Korea); CONICET and UBACyT (Argentina); FOM (The Netherlands); PPARC (United Kingdom);

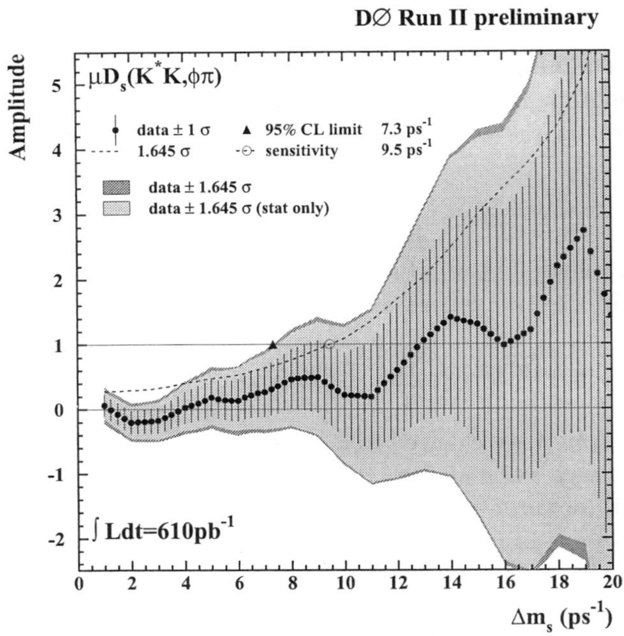

Figure 3. Combined B_s^0 oscillation amplitude.

MSMT (Czech Republic); CRC Program, CFI, NSERC and WestGrid Project (Canada); BMBF and DFG (Germany); SFI (Ireland); Research Corporation, Alexander von Humboldt Foundation, and the Marie Curie Program.

References

1. S. Eidelman et al, Phys. Lett. **B592**, 1 (2004)
2. H.G. Moser, A. Roussarie, Nucl.Instrum.Meth.A384:491–505, 1997.

EXPECTED PHYSICS PERFORMANCE OF THE LHCB EXPERIMENT

OLIVIER LEROY

Centre de Physique des Particules de Marseille
163 avenue de Luminy Case 902
13288 Marseille cedex 9, France
E-mail: Olivier.Leroy@in2p3.fr

LHCb is a dedicated B-physics experiment at the LHC. It is currently under construction, to be ready for first collisions in 2007. We give an overview of the expected physics performance: sensitivity in typical channels for study of B_s^0 mixing, CP violation and rare B decays.

1. Introduction

The LHCb experiment, its tracking and particle identification systems have been introduced elsewhere in this conference[1,2]. Here we will illustrate the LHCb physics performance with some key measurements. More detailed results can be found in[3].

CP violation is described in the Standard Model by a complex phase in the CKM matrix[4,5]. The B meson sector is a place where theoretical predictions can be precisely compared with experimental results. LHCb will measure precisely fundamental parameters of the Standard Model like the phases $\arg V_{\rm td} = -\beta$, $\arg V_{\rm ub} = -\gamma$ and $\arg V_{\rm ts} = -\phi_s/2 + \pi$. We will mesasure the B_s^0 oscillation frequency and explore many rare decays involving radiative, electroweak and gluonic penguin amplitudes. There is also a B_c^+ and b-baryon physics program and our trigger is designed to face unexpected decays. With all these measurements, we will over constrain the unitarity triangles in order to search for inconsistencies and to exhibit new physics. Indeed new physics scenarii propose new particles and new couplings which can modify the amplitudes of penguin and box diagrams.

Physics potential of LHCb is estimated using Monte Carlo simulation. Proton-proton collisions are generated using PYTHIA 6.2 including hard QCD processes, single and double diffraction. Multiple parton interactions are tuned to reproduce the track multiplicities observed at SPS and Tevatron energies. The particles are propagated through the detector using Geant simulation. Each sub-dector response, resolution, efficiency, noise and cross-talk is taken into account using test beam data. Then the offline reconstruction is done using a full pattern recognition program, including track finding and RICH reconstruction. More than 40M $b\bar{b}$ events have been generated to perform detailed background studies.

The B_s^0 proper time resolution[1] is around 43 fs. The flavour of the signal B meson is tagged using fragmentation kaon or pion as well as the opposite b-hadron decay products: kaon, leptons and vertex charge. All information is combined using a neural network. Typical effective tagging efficiencies vary between 4-5% for B^0 and 7-9% for B_s^0.

2. B_s^0 mixing, phase and $\Delta\Gamma_s$

The B_s^0 oscillation frequency, Δm_s will be measured using $B_s^0 \to D_s^- \pi^+$ decays. The statistical uncertainty will be $0.01\,\mathrm{ps}^{-1}$, with 2 fb^{-1}[a] if Δm_s is equal to $20\,\mathrm{ps}^{-1}$. We can observed B_s^0 oscillation with a significance of at least 5σ up to 68 ps^{-1}. Hence Δm_s range will be explored well beyond the Standard Model prediction.

The B_s^0 mixing phase, ϕ_s, will be measured using the CP asymmetry in $B_s^0 \to J/\psi\phi$ events. This channel is the B_s^0 counterpart of the golden mode $B^0 \to J/\psi K_S^0$. The J/ψ is reconstructed in $\mu^+\mu^-$ or e^+e^- and the ϕ in K^+K^-. An angular analysis of the decays products is necessary to separate the CP-even from the CP-odd contributions. The ϕ_s phase can also be extracted from pure CP eigenstates like $B_s^0 \to J/\psi\eta$ or $B_s^0 \to \eta_c\phi$. These channels have a much lower statistics, but do not require an angular analysis. Combining all these modes, we expected a statistical uncertainty at the level of 0.013 radians with 10 fb^{-1}. The Standard Model predicts ϕ_s around -0.036, but values order of magnitude higher can be expected in the presence of new physics.

3. The α phase

The Snyder and Quinn prescription[6] to measure α has been studied in LHCb. The decays of B^0 meson to $\pi^+\pi^-\pi^0$ is exploited, assuming that the intermediate $\rho\pi$ state is dominant. Compared to two-body decays approach, discrete ambiguities are theorically removed and an independent measurement of tree and penguin parameters is possible. The

[a] In nominal condition, we expected 2 fb^{-1} per year.

$B^0 \to \pi^+\pi^-\pi^0$ selection is based on a multivariate analysis. Both π^0 giving merged and resolved clusters in the electromagnetic calorimeter are exploited. We expect 14,000 events per year after trigger, reconstruction and offline selection. The background-over-signal ratio from inclusive $b\bar{b}$ events has been estimated to be less than 1. The full Monte Carlo parameters are used to build toy Monte Carlo experiments, in order to assess the sensitivity to α. Eleven-parameter likelihood fits are performed in time-dependent Dalitz space, on hundreds of toy Monte Carlo experiments. Assuming a flat and resonant background, we conclude that a statistical uncertainty of 10 degrees on α can be reached with 2 fb^{-1}.

4. The γ phase

The γ phase can be measured with at least six methods and only two are discussed here. The first one uses $B_s^0 \to D_s^\mp K^\pm$ events. Inteference between two tree diagrams and B_s^0 mixing in $B_s^0 \to D_s^\mp K^\pm$ allows an extraction of $\gamma + \phi_s$ from two time-dependent decay rate asymmetries. The mistag fraction can be extracted from the $B_s^0 \to D_s^- \pi^+$ sample and ϕ_s is measured with $B_s^0 \to J/\psi\phi$, as explained previously. Sensitivity have been estimated using fast Monte Carlo programs. The uncertainty on γ is 14 degrees after one year of data taking. This assumes $\Delta m_s = 20\,ps^{-1}, \Delta\Gamma_s/\Gamma_s = 0.1, \gamma$ between 55 and 105 degrees and the strong phases difference between the 2 tree diagrams to be below 20 degrees. This method has no theoretical uncertainties and is insensitive to new physics in the B_s^0 mixing.

The ADS method[7] exploits the interference between the favoured and doubly Cabibbo-suppressed decay modes of neutral D mesons to states such as $K\pi$ in charged B decays to DK. For these decays, the reversed suppression of the D decays relative to the B decays results in similar amplitude and big interference effects. No tagging nor proper time measurement is necessary, hence it is just a counting experiment. The interference depends on five parameters: γ, r_B the ratio between the colour-allowed and colour-suppressed B decays, δ_B the strong phase difference between the two diagrams, $r_D^{K\pi}$, the ratio in magnitude of the two D decay processes and $\delta_D^{K\pi}$, the strong phase difference between the two D diagrams.

Four possible B decays have to be reconstructed: $B^- \to (K^-\pi^+)_D K^-$, $B^- \to (K^+\pi^-)_D K^-$, $B^+ \to (K^+\pi^-)_D K^+$, $B^+ \to (K^-\pi^+)_D K^+$. The first and the third decays are favoured. The second and the last are suppressed, but have interference term appearing at leading order, so that they are particularly sensitive to $\delta_D^{K\pi}$ and γ. Even though they are suppressed, the expected statistics in 1 year is 50 times higher that what is currently available at the B factories. Measuring the relative rates of these 4 processes give three observables which depends on four unknowns: r_B, γ, δ_B and $\delta_D^{K\pi}$. In order to constrain the problem, it is necessary to look for the D decay into another final state such as D$\rightarrow$ K$\pi\pi\pi$.

Toy Monte Carlo experiment have been performed to estimate the sensitivity to γ. The level of background has been estimated to be around 0.5 for the two-body D decays, but is not yet known for the four-body D decays. Nevertheless, an arbitrarily B/S has been introduced in the toy Monte Carlo and the result does not depend a lot on this B/S. The expected statistical uncertainty on γ is around 5 degrees with 2 fb^{-1}.

5. Rare decays

The forward-backward asymmetry in the $\mu^+\mu^-$ rest frame for $B^0 \rightarrow \mu^+\mu^- K^{*0}$ decays is interesting because it allows to distinguish various extensions of the Standard Model[8]. The branching ratio of this decay is around 10^{-6}. We expect to select 4,400 $B^0 \rightarrow \mu^+\mu^- K^{*0}$ events per year, with a B/S ratio smaller than 2.6. With 10 fb^{-1}, we expect to locate the zero of the forward backward asymetry to $\pm$0.53 GeV, which will allow to determine the ratio of the effective Wilson coefficients C_7^{eff} and C_9^{eff} with a 13% error.

6. Conclusion

LHCb will collect unprecedented statistics of B decays. $B_s^0\overline{B}_s^0$ oscillation will be measured with a 5σ significance up to 68 ps^{-1}. Many measurements of rare decays and CP asymmetries will be performed. The statistical uncertainty on α, γ and ϕ_s after one year of data taking is expected to be 10, 5 and 2 degrees, respectively. CP phases will be determined via channels with different sensitivity to new physics, so that detailed tests of the CKM description of the quark sector will be possible. A summary of the performance is given in Table 1.

LHCb offers an excellent opportunity to spot new physics signal beyond Standard Model and to pin down its nature.

Acknowledgments

I would like to thanks the organizers of the 2006 Lake Louise Winter Institute and my LHCb colleagues for helping in preparing this talk.

Table 1. Expected statistical performance of the LHCb experiment for key measurements with 2 fb^{-1} (1 year of data taking). Untagged signal yields and background-over-signal ratios estimated from inclusive $b\bar{b}$ events are given.

	Decay channel	Annual yield	B/S	Stat. uncertainty
γ	$B^0_s \to D_s^{\mp}K^{\pm}$	5.4k	< 1	$\sigma(\gamma) \simeq 14^o$
	$B^0 \to \pi^+\pi^-$	26k	< 0.7	
	$B^0_s \to K^+K^-$	37k	0.3	$\sigma(\gamma) \simeq 6^o$
	$B^0 \to D^0(K^-\pi^+)K^{*0}$	0.5k	< 0.3	
	$B^0 \to \overline{D}^0(K^+\pi^-)K^{*0}$	2.5k	< 2	$\sigma(\gamma) \simeq 8^o$
	$B^0 \to D^0_{CP}(K^+K^-)K^{*0}$	0.6k	0.3	
	$B^+ \to D^0(K^-\pi^+)K^-$	60k	0.5	
	$B^+ \to D^0(K^+\pi^-)K^-$	2k	0.5	$\sigma(\gamma) \simeq 5^o$
α	$B^0 \to \pi^+\pi^-\pi^0$	14k	0.8	$\sigma(\alpha) \simeq 10^o$
	$B^0_s \to J/\psi\phi$	125k	0.3	
ϕ_s	$B^0_s \to J/\psi\eta$	12k	2-3	$\sigma(\phi_s) \simeq 2^o$
	$B^0_s \to \eta_c\phi$	3k	0.7	
Δm_s	$B^0_s \to D_s^-\pi^+$	80k	0.3	Δm_s up to 68 ps^{-1}
β	$B^0 \to J/\psi K^0_S$	216k	0.8	$\sin(2\beta) \simeq 0.022$
rare	$B^0 \to \mu^+\mu^-K^{*0}$	4.4k	< 2.6	NP search
decays	$B^0_s \to \mu^+\mu^-$	17	< 5.7	NP search
	$B^0 \to K^{*0}\gamma$	35k	< 0.7	$\sigma(A^{dir}_{CP}) \simeq 0.01$

References

1. J. Nardulli, "LHCb Tracking System and Its Performance", Lake Louise Winter Institute 2006 in these proceedings.
2. P. Szczypka , "LHCb Particule Identification System and Its Performance", Lake Louise Winter Institute 2006 in these proceedings.
3. LHCb Collaboration, "LHCb Reoptimized Detector and Performance TDR" , CERN/LHCC 2003-30, LHCb TDR 9, 9 Sep 2003.
4. T. Maskawa and M. Kobayashi, "CP-Violation in the Renormalizable Theory of Weak Interaction", Prog. Th. Phys. **49** (1973) 652.
5. N. Cabibbo, "Unitarity Symmetry and Leptonic Decays", Phys. Rev. Lett. **10** (1963) 531.
6. A. Snyder and H. Quinn, *Phys. Rev.* **D48**, 2139 (1993).
7. D. Atwood, I. Dunietz and A. Soni *Phys. Rev. Lett.* **78**, 3257 (1997).
8. P. Ball *et al.*, "B decays at LHC",CERN-TH 2000-101, hep-ph/0003238 (2000).

ISOSCALAR EXTRACTION OF ΔS IN THE NUCLEON AT HERMES FROM SEMI-INCLUSIVE DIS

L. A. LINDEN LEVY*

University of Illinois at Urbana Champaign
1110 West Green Street,
Urbana, IL 61801, USA
lindenle@uiuc.edu

The helicity density of the strange quark sea in the proton has been extracted from measurements of polarized semi-inclusive production of charged kaons in deep inelastic scattering of positrons from a polarized deuteron target. The isoscalar nature of the deuteron (assuming isospin symmetry) and lack of isospin for strange quarks allows the deuteron target to be used independently without relying on fragmentation models or other experimental data. In the region of measurement x >0.02 the helicity density is zero within the experimental error and agrees with a previous five flavor extraction performed by HERMES. The first moment of the axial charge in the measured region is substantially less than that inferred from LO (NLO) QCD fits that cover the entire x-range.

1. Introduction

Since the discovery of nucleon substructure, at SLAC in the 1960s, there has been a large body of experimental work done to measure the sizes of the different partonic contributions. In particular, early measurements of the polarized structure function g_1 by EMC showed that little or none of the polarization of the proton (a spin 1/2 fermion) was inherited from its quarks' spins. A second generation of polarized parton distribution function (PDF) measurements demonstrated that the initial so called "spin-crisis" was really more of a "spin-puzzle" because the quark polarizations do contribute $\sim 30\%$ of the total polarization. In turn, this has led us to search for the remaining contributions to the proton's spin which satisfy the conservation equation: $1/2 = \Delta\Sigma + L_q + J_g$.

Here $\Delta\Sigma$ describes the contribution from the quarks' spins, L_q represents quark orbital angular momentum and J_g represents the contribution

*On behalf of the HERMES collaboration.

from gluon polarization and orbital angular momentum.

One particularly successful technique for studying the flavor dependence of the quark polarizations is polarized semi-inclusive deep inelastic scattering (SIDIS). In polarized SIDIS a polarized electron interacts with a quark inside a polarized nucleon via the exchange of a virtual photon and the quark is knocked free of the nucleon, creating a jet of hadrons via fragmentation. The DIS region is characterized by a large four-momentum exchange ($Q^2 > 1$ GeV^2) between the virtual photon and the quark and by a large final-state invariant mass ($W^2 > 4$ GeV^2) which ensures that the nucleon has been broken apart. With this method, one can distinguish the quark spins that are aligned with the target nucleon spin from those anti-aligned by measuring asymmetries in target spin state production. This double spin asymmetry can be written as:

$$A_1^h(x,Q^2) = \frac{\sigma_{1/2}^h - \sigma_{3/2}^h}{\sigma_{1/2}^h + \sigma_{3/2}^h} \overset{g_2=0}{\approx} \frac{\Sigma_q e_q^2 \Delta q(x,Q^2) \int dz D_q^h(z,Q^2)}{\Sigma_{q'} e_{q'}^2 q'(x,Q^2) \int dz D_{q'}^h(z,Q^2)} \tag{1}$$

Quark flavor information can be gained by "tagging" via an observed final state hadron h, exploiting the fact that the fragmentation functions provide a link between the struck parton flavor and the final-state hadron species. A fragmentation function (D_q^h) is defined to be the number density for the final-state production of a hadron of type h when a quark of flavor q is struck.

This technique has been applied using the HERMES spectrometer, which is positioned in the east experimental hall of the HERA ring in Hamburg, Germany. The 27.6 Gev self-polarizing electrons of the HERA ring are scattered off internal polarized gas targets (^{1}H, ^{2}H) produced using an atomic beam source (ABS) [1]. The scattered beam lepton and one or more coincident final-state hadrons are detected by the spectrometer. Leptons are identified with greater than 98% efficiency at a hadron contamination of less than 1% [2]. Since 1997 the hadron species of protons, pions and kaons can be separately identified using a dual-radiator ring imaging Čerenkov detector (RICH) [3]. Before the installation of the RICH a threshold Čerenkov allowed only for the separation of pions from heavier hadrons [2].

2. 5-flavor Δq separation

In 2001, HERMES published the first 5-flavor separation of quark polarizations. The values were extracted using the so called "Purity Method" in which the double spin asymmetry (Eq. (1)) for hadron production is rewritten in terms of completely unpolarized quantities (purities) and the fractional quark polarizations:

$$A_1^h(x) = \Sigma_q P_q^h(x) \frac{\Delta q(x)}{q(x)} \tag{2}$$

The purities P_q^h represent the probability that an observed hadron h originated from the scattering off a quark of flavor q from the target, and depend only on spin-independent quantities. A total of eight asymmetries on two targets ($A_{1p(d)}$, $A_{1p(d)}^{\pi^{+(-)}}$, $A_{1d}^{K^{+(-)}}$) were measured by the HERMES collaboration. Using these asymmetries, the assumption of isospin symmetry, and some set of generated or measured purities, one can minimize a χ^2 comparison by varying the fractional quark polarizations until an acceptable stable solution is reached.

For the HERMES analysis, the purities were generated from a Monte-Carlo simulation using LEPTO [4] to produce the hard scattering event and JETSET [5], which is based on the LUND string model, to simulate the fragmentation process. These simulation tools' phenomenological parameters were tuned using a genetic search algorithm [6], such that the difference between the unpolarized hadron multiplicities from the simulation and those measured at HERMES were minimized.

The results of this measurement are presented in Figure 1. The u-quark helicity distribution is large and positive, while the d-quark distribution is small and negative, which is consistent with LO (NLO) QCD analyses of inclusive data. All of the sea quark polarizations are consistent with zero.

3. Isoscalar Analysis

The deuteron's lack of isospin makes it ideal for extracting information about the strange quark polarizations. At leading order, the inclusive asymmetry may be written

$$A_1(x) = \frac{5\Delta Q(x) + 2\Delta S(x)}{5Q(x) + 2S(x)}, \quad \text{where} \quad \begin{array}{c} \Delta Q(x) \equiv \Sigma_{q \neq s,\bar{s}}\, \Delta q(x) \\ \Delta S(x) \equiv \Delta s(x) + \Delta \bar{s}(x) \end{array} \tag{3}$$

The asymmetry for the production of K^+ and K^- combined (denoted "K") may be written

$$A_1^K(x) = \frac{\Delta Q(x)\, \mathbb{D}_Q^K + 2\Delta S(x)\, \mathbb{D}_S^K}{Q(x)\, \mathbb{D}_Q^K + 2S(x)\, \mathbb{D}_S^K}, \tag{4}$$

where charge-conjugation symmetry has been assumed for the kaon fragmentation functions: $D_q^K = D_{\bar{q}}^K$ for any flavor q of the struck quark. The

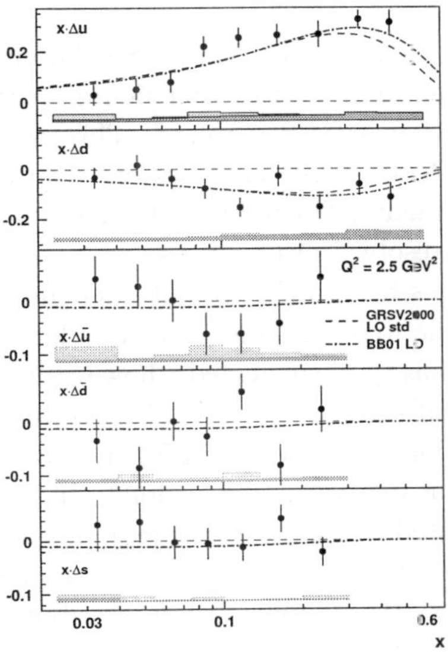

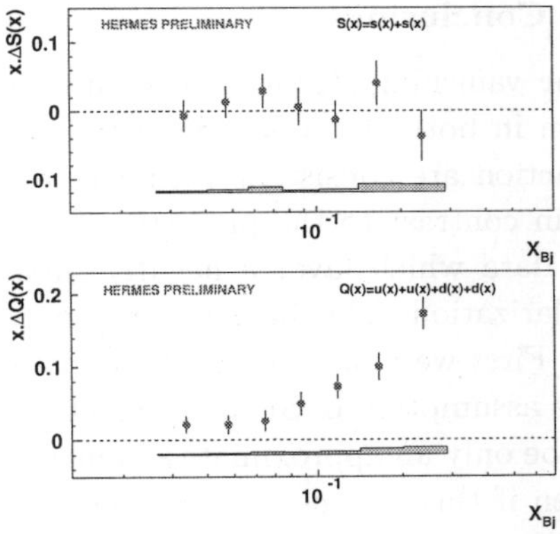

Figure 1. (left) First 5-flavor separation of the polarized quark densities in the proton ($\Delta\bar{s} \equiv 0$). Figure 2. (above) Polarized strange and non-strange quark polarizations from a second, independent analysis using only the deuteron asymmetries and no Monte Carlo input.

symbols $\mathbb{D}_Q^K$ and $\mathbb{D}_S^K$ represent integrals over combinations of the fragmentation functions (here the z dependence is suppressed for brevity):

$$\mathbb{D}_Q^K \equiv \int dz\, \left(4D_u^K + D_d^K + 4D_{\bar{u}}^K + D_{\bar{d}}^K\right) \qquad \mathbb{D}_S^K \equiv \int dz\; D_s^K + D_{\bar{s}}^K \quad (5)$$

The extraction of the strange ($\Delta S(x)$) and non-strange ($\Delta Q(x)$) PDFs becomes a simple algebraic problem if the fragmentation function integral values $\mathbb{D}_Q^K$ and $\mathbb{D}_S^K$ are available. In this analysis, these two fragmentation function integrals were extracted from HERMES unpolarized data by fitting them to the measured charged-kaon multiplicities (Eq. (6)), using the CTEQ6L [7] unpolarized PDFs to provide $Q(x)$ and $S(x)$:

$$\frac{1}{\sigma_{DIS}} \frac{d\sigma^K}{dx} = \frac{Q(x)\mathbb{D}_Q^K + 2S(x)\mathbb{D}_S^K}{5Q(x) + 2S(x)}. \quad (6)$$

The results of this extraction are shown in Figure 2. They are consistent with the previous HERMES result, and provide a simulation-independent confirmation of the previously measured s-quark polarization.

4. Conclusions

The vanishing strange polarization measured by the HERMES collaboration in both the 5-flavor purity extraction and the isoscalar deuteron extraction are consistent with zero over the measured x range. This result is in contrast to the predictions of LO (NLO) QCD analyses on inclusive g_1 data which favor a negative first moment (Δs) for the strange quark polarization. This inconsistency may be resolved in two ways.

First we must note that the predictions from inclusive data depend on the assumption of SU(3) symmetry for the hyperon octet, which is known to be only an approximate symmetry. It has been suggested [8] however that even if this symmetry is broken at the 20% level, the HERMES result is still not compatible with the global fits. The other degree of freedom is the extrapolation of the HERMES data into the unmeasured x-region [9]. In both of the analyses the moment is only calculated in the measured region, which allows for a large negative contribution from the unmeasured low-x region. The question becomes whether or not the strange polarization distribution that is required in the unmeasured region lies within the positivity bound ($|\Delta s/s| < 1$). As $s(x)$ rises dramatically at low x, a large contribution to the first moment of $\Delta s(x)$ is entirely possible. Unfortunately the mixing of two effects makes it difficult to quantify the SU(3) breaking observed at HERMES without more precise knowledge on the strange distribution for $x < 0.23$. Future measurements in this region are very valuable and will help to quantify the SU(3) symmetry-breaking magnitude (if any) that HERMES observes.

In conclusion, HERMES has extracted the first 5-flavor polarized-quark distributions for the nucleon. Additionally, a second analysis was performed to extract the polarized strange density using a different, simulation-independent technique. The results of the two analyses are fully consistent, as well as being compatible with LO (NLO) QCD fits, within the measured x region.

References

1. F. Stock et al., Nucl. Instrum. Meth. **A343**, 334 (1994).
2. K. Ackerstaff et al., Nucl. Instrum. Meth. **A417**, 230 (1998).
3. N. Akopov et al., Nucl. Instrum. Meth. **A479**, 511 (2002).
4. G. Ingelman, A. Edin, and J. Rathsman, Comput. Phys. Commun. **101**, 108 (1997).
5. T. Sjostrand, Comput. Phys. Commun. **82**, 74 (1994).
6. A. Hillenbrand, Prepared for 11th International Workshop on Deep Inelastic

Scattering (DIS 2003), St. Petersburg, Russia, 23-27 Apr 2003.
7. J. Pumplin et al., JHEP **07**, 012 (2002).
8. E. Leader and D. B. Stamenov, Phys. Rev. **D67**, 037503 (2003).
9. S. F. Pate, Eur. Phys. J. **A24S2**, 67 (2005).

ELECTROWEAK PHYSICS AT HERA

J. LIST
DESY
Notkestr. 85
22603 Hamburg, Germany
E-mail: jenny.list@desy.de

At HERA, electroweak physics can be investigated in deep inelastic scattering. The H1 experiment determined QCD and electroweak paramters simultaneously for the first time in a global fit to the cross sections measured at HERA I. At HERA II, the longitudinal polarisation of the lepton beam gives access to the chiral properties of neutral and charged current interactions at high momentum transfers. Due to the higher luminosity, the number of candidate events for single W production has increased, but still more statistics is needed to draw final conclusions.

1. Combined QCD and EW Fit of HERA I data

During the HERA I phase, which ended in the year 2000, the H1 and ZEUS experiments collected each about 20 pb^{-1} of e^-p and 100 pb^{-1} of e^+p data at center–of–mass energies of 300 and 320 GeV. The resulting measurements of the inclusive neutral and charged current cross sections span a wide range in x and Q^2 and have been used by H1 to extract for the first time in a global fit the parton density functions (pdfs) as well as electroweak parameters[1], which is possible because they affect the cross sections in different phase space regions in different ways.

At low values of $Q^2 << M_Z^2$, the neutral current ep scattering reaction is dominated by photon exchange and thus allows the extraction of the pdfs of the proton. At $Q^2 \geq M_Z^2$, the Z exchange and its interference with the photon exchange become more significant and allow an extraction of the axial and vector couplings of the Z to the up and down quarks. These measurements are shown in figure 1 together with the CDF and LEP results for the charm and bottom quark, which are expected in the Standard Model to be equal to the up and down quark values, respectively. The charged current cross section depends on the propagator mass of the exchanged boson M_{prop} and its coupling G to the quarks. In the Standard Model they

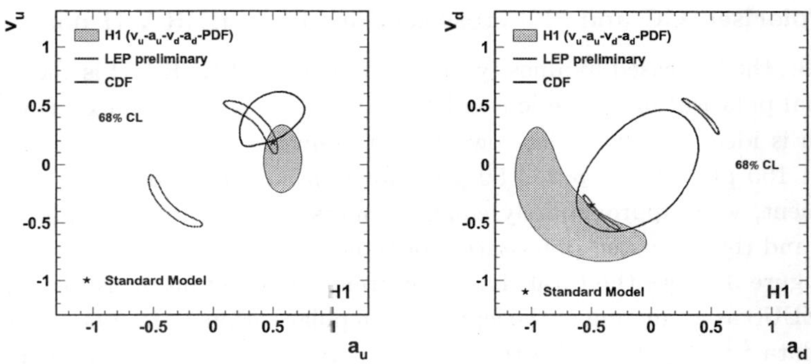

Figure 1. Results of the combined QCD and EW fit for the axial and vector couplings of the light quarks to the Z^0 compared to the Standard Model expectation and the results for c and b quarks from LEP and CDF.

are identical to the mass of the $W^{\pm}$ boson and the Fermi constant G_F. Figure 2 shows the results of a fit in which both parameters are adjusted (shaded ellipse) as well as the propagator mass resulting from a fit where the coupling is fixed to G_F. Both fits agree with the world averages for M_W and G_F.

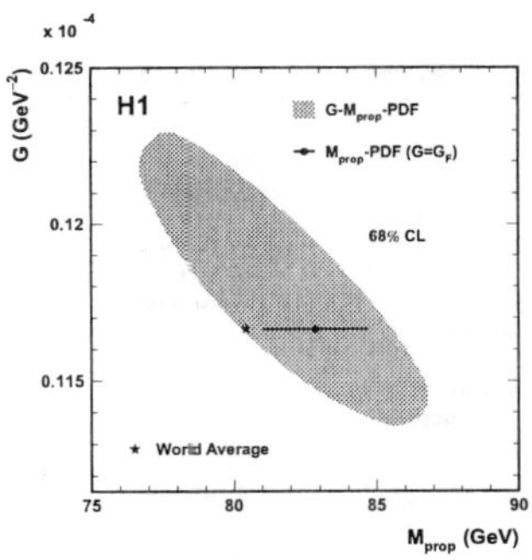

Figure 2. Result of the combined QCD and EW fit for the coupling constant G and the propagator mass M_{prop} compared to the world averages for G_F and M_W.

2. Polarised CC and NC cross sections from HERA II data

Besides the increased luminosity, the main feature of HERA II is the longitudinal polarisation of the lepton beam at the H1 and ZEUS experiments, which is ideal for many electroweak measurements. Currently, samples of about 150 pb^{-1} for e^-p and 50 pb^{-1} for e^+p have been obtained per experiment, with approximately equal amounts of integrated luminosity for left– and right–handed polarisation of typically 30% on average.

Figure 3 shows the total charged current cross section measured by H1 and ZEUS as a function of the lepton beam polarisation P_e for both e^-p and e^+p data [2,3]. Together with the HERA I data at $P_e = 0$, the new HERA II data show a clear linear dependence on the polarisation, which illustrates the chiral structure of the charged current interaction. A common fit to the e^+p data from H1 and ZEUS leads to an extrapolated cross section at $P_e = -1$ of $\sigma_{CC}^{\text{tot}} = -1.0 \pm 1.8_{\text{stat}} \pm 1.1_{\text{sys}}$ pb, which is consistent with the Standard Model expectation of 0.

Figure 4 shows the single differential charged and neutral current cross sections for the left–handed and right–handed e^+p data measured by the ZEUS experiment. For charged current data, the larger cross section for the right–handed sample can clearly be seen for $d\sigma/dQ^2$, $d\sigma/dx$ and $d\sigma/dy$. Effect of the polarisation on the neutral current is visible in the ratio of $d\sigma/dQ^2$ for both helicities and will become more pronounced with increased statistics, especially for high values of Q^2.

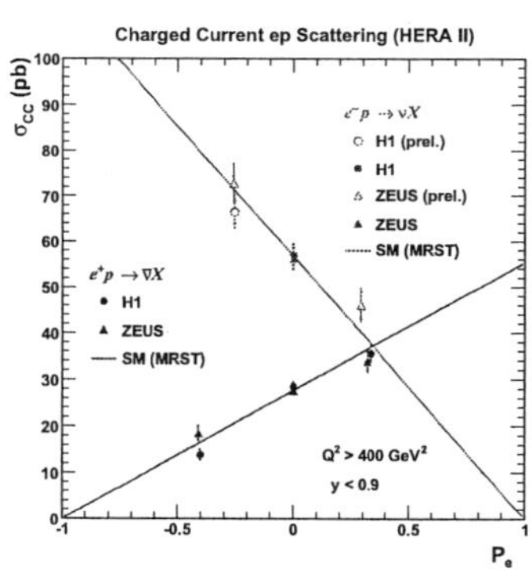

Figure 3. H1 and ZEUS measurements of the charged current cross sections for e^-p and e^+p data as a function of the polarisation P_e of the lepton beam.

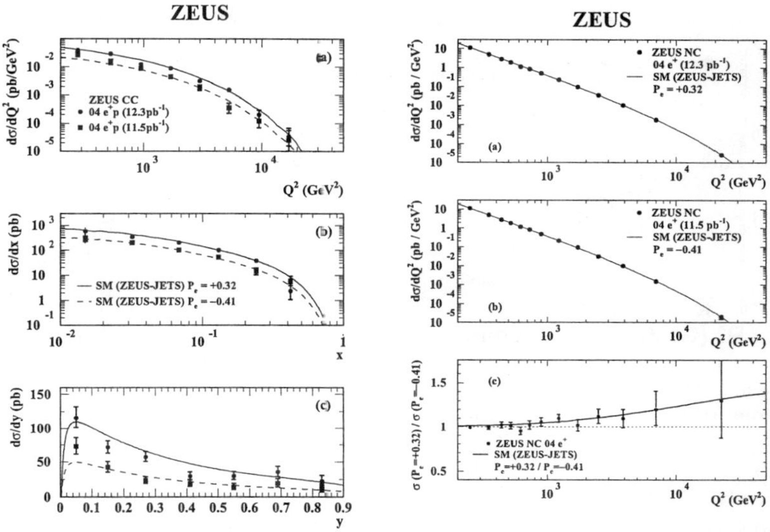

Figure 4. Differential charged (left) and neutral (right) current cross sections for the polarised HERA II datasets measured by the ZEUS experiment.

3. Single $W^{\pm}$ production in HERA I and HERA II data

Real $W^{\pm}$ bosons can be produced at HERA with a cross section of about 1 pb. The H1 experiment analysed all available data with a total integrated luminosity of 279 pb^{-1} for W's decaying into electron or muon plus a corresponding neutrino, whereas the ZEUS experiment has perfomed the analysis in the electron channel on 106 pb^{-1} of $e^{+}p$ data. The signature of such events is an isolated lepton and high missing transverse momentum. The hadronic system often has a low transverse momentum P_T^X and is thus lost in the beampipe, but also spectacular events with a central jet leading to high values of P_T^X are observed. For the electron channel also charged current reactions where a jet is misidentified as electron or neutral current events with a real electron but fake missing E_T contribute to the background expectation.

Figure 5 shows the P_T^X distribution for the $e^{-}p$ (121 pb^{-1}) and $e^{+}p$ (158 pb^{-1}) data as observed by the H1 experiment. The $e^{-}p$ data agree well with the Standard Model expectation: in total 12 events are observed with 15.8±2.2 expected, at $P_T^X > 25$ GeV 2 events are observed compared to an expectation of 4.4±0.7. However, in the $e^{+}p$ data 28 events are observed

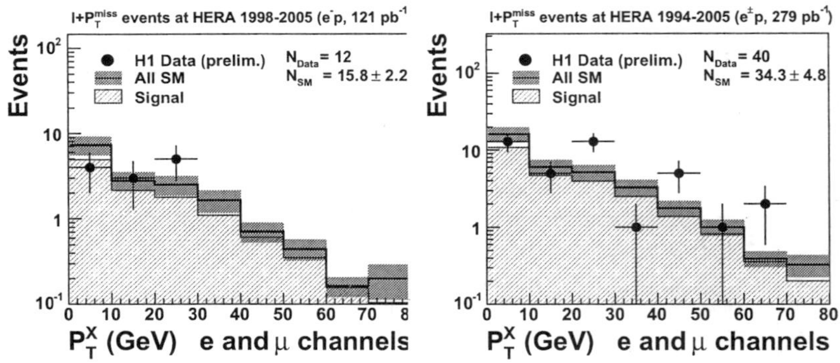

Figure 5. P_T^X distribution observed by the H1 experiment for a) e^-p data and b) e^+p data from HERA I and HERA II running phase combined.

with only 18.5±2.6 expected, at $P_T^X > 25$ GeV 15 events are observed with 4.6±0.8 expected. In 106 pb^{-1} of e^+p data the ZEUS experiment observes 2 events in the electron channel with 3±0.4 expected.

4. Conclusions

Several measurements related to the properties of the electroweak interactions perfomed by the H1 and ZEUS experiments have been presented. Using the full HERA I data set, H1 has performed for the first time a combined fit of QCD and electroweak parameters. The results agree with the expectations from the Standard Model within errors. They will become more precise once HERA II data will be included in this analysis, not only due to the increased statistics at high values of Q^2, but also due to the lepton beam polarisation, which improves the sensitivity to some parameters, for example to the vector couplings of the light quarks. The analysis of the polarised HERA II data has started. Among the first results are the polarisation dependence of the neutral and charged current cross sections and the analysis of events with isolated leptons and missing transverse energy, as they are expected in the Standard Model from the production of real $W^\pm$ bosons. In the e^+p data of HERA I and II combined, H1 observes an excess of these events at high values of $P_T^X > 25$ GeV with 15 events in the data compared to an expectation of 4.6±0.8. The ZEUS experiment does not confirm this excess. More e^+p data are needed to gain further understanding of these events.

References

1. A. Aktas *et al.* [H1 Collaboration], Phys. Lett. B **632** (2006) 35.
2. A. Aktas *et al.* [H1 Collaboration], Phys. Lett. B **634** (2006) 173.
3. S. Chekanov *et al.* [ZEUS Collaboration], arXiv:hep-ex/0602026.

RECENT RESULTS ON RADIATIVE PENGUIN AND LEPTONIC B DECAYS AT BABAR

T. B. MOORE
REPRESENTING THE BABAR COLLABORATION
University of Massachusetts
Amherst, MA 01003-9337, USA
E-mail: tbmoore@slac.stanford.edu

We present recent results on purely leptonic and electroweak penguin decays of the B meson based on data samples of up to 232 million $B\bar{B}$ pairs collected by the BABAR experiment at the PEP-II B-Factory at SLAC. We present measurements of the branching fraction, the direct CP violating asymmetry, and photon energy spectrum of $B \to X_s\gamma$ decays using both a fully inclusive lepton-tagged technique as well as a sum of exclusive final states. Also included are results on the decay rates and asymmetries for the decays $B \to K^{(*)}\ell^+\ell^-$. Finally, we present a search for the fully-leptonic decay $B^+ \to \tau^+\nu_\tau$.

1. $B \to X_s\gamma$

In the Standard Model (SM), the radiative decay $b \to s\gamma$ proceeds via a loop diagram and is sensitive to possible new physics participating in the loop. Next-to-leading order calculations for the branching fraction give $\mathcal{B}(B \to X_s\gamma) = (3.57 \pm 0.30) \times 10^{-4}$ ($E_\gamma > 1.6$ GeV) [1], where E_γ is the photon energy in the B rest frame. The direct CP violating asymmetry is expected to to be smaller than 1% in the SM. Measurement of the E_γ spectrum from $B \to X_s\gamma$ decays allows the determination of the heavy quark effective field theory parameters m_b and μ_π^2 related to the b quark mass and momentum within the B meson, respectively [2]: $\langle E_\gamma \rangle \approx m_b/2$, $\langle (E_\gamma)^2 \rangle - \langle E_\gamma \rangle^2 \approx \mu_\pi^2$. The knowledge of this shape function is a crucial input in the extraction of $|V_{ub}|$ from inclusive semileptonic $B \to X_u\ell\nu$ measurements [3].

1.1. *Inclusive Analysis*

In the fully inclusive analysis [4], the X_s system is not reconstructed. Thus uncertainties in the X_s fragmentation model are avoided but at the cost of higher backgrounds (including a $(4.0\pm1.6)\%$ [5] contamination of $B \to X_d\gamma$ decays). We select events with an isolated photon candidate with $1.5 < E^*_\gamma < 3.5$ GeV where the asterisk indicates the $\Upsilon(4S)$ center-of-momentum frame (CM). Much of the non-$B\bar{B}$ (continuum) background is removed by requiring a high-momentum lepton tag. About 20% of B mesons decay semileptonically to either e or μ and the large mass of the B tends to impart large momentum to its lepton daughter. The reconstructed photon energy spectrum in the CM frame, as well as tables of the first and second moments of this spectrum for various photon energy ranges are presented in Ref. 4 using 88.5 million $B\bar{B}$ pairs.

After correcting for the fact that the cut in photon energy is made on reconstructed E^*_γ in the CM frame rather than on E_γ in the B rest frame we find,

$$\mathcal{B}(B \to X_s\gamma,\ E_\gamma > 1.9\ \text{GeV}) = (3.67 \pm 0.29 \pm 0.34 \pm 0.29) \times 10^{-4}$$

where the uncertainties are statistical, systematic, and model-dependent.

The lepton tag on the non-signal B provides a determination of A_{CP}. The lepton's positive(negative) charge tags the signal side to contain a $b(\bar{b})$ quark. We obtain,

$$A_{CP}(B \to X_{(s+d)}\gamma) = -0.110 \pm 0.115(\text{stat}) \pm 0.017(\text{syst}).$$

1.2. *Sum of Exclusive Modes*

In this analysis [6] we reconstruct the X_s final states in 38 decay modes and their charge conjugates using 88.9 million $B\bar{B}$ events. According to our signal model these modes represent 55% of the total inclusive rate in the region $M(X_s) = 1.1 - 2.8$ GeV$/c^2$. The reconstructed X_s system is combined with a high-energy ($E^*_\gamma > 1.8$ GeV) photon to form a B meson. The identification of $B \to X_s\gamma$ decays makes use of two kinematic variables: the beam-energy substituted mass, $m_{ES} = \sqrt{(\sqrt{s}/2)^2 - p^{*2}_B}$, and the difference between the measured and expected energies of the B candidate, $\Delta E = E^*_B - \sqrt{s}/2$, where E^*_B and p^*_B are the energy and momentum of the B candidate in the CM frame and $\sqrt{s}$ is the total CM energy. Most of the background in this analysis arises from continuum production of a high-energy photon from initial-state radiation or the decays of π^0 and η mesons produced in light-quark jets.

We extract the signal yield in bins of $M(X_s)$ by performing an unbinned maximum likelihood fit to the m_{ES} distribution and correct for the signal efficiency and fraction of missing final states to produce an $M(X_s)$ spectrum. The inclusive branching fraction is obtained from a fit over the full $M(X_s)$ range, corresponding to $E_\gamma > 1.90$ GeV. We obtain,

$$\mathcal{B}(B \to X_s\gamma,\ E_\gamma > 1.9\ \text{GeV}) = (3.27 \pm 0.18^{+0.55}_{-0.40}{}^{+0.04}_{-0.09}) \times 10^{-4}$$

where the errors are statistical, systematic, and model-dependent. The dominant systematic uncertainty is due to the missing X_s fraction.

The relation $E_\gamma = (m_B^2 - m_{X_s}^2)/2m_B$ can be used to convert between the mass of the hadronic system and the photon energy in the B rest frame and produce a photon energy spectrum with an energy resolution of about 5 MeV as shown in Fig. 1. We fit the resulting spectrum to recent QCD predictions in the shape function [7] and kinetic [2] schemes to find the best values of the parameters m_b and μ_π^2. The results of these fits, as well as the measured moments of the energy spectrum are tabulated in Ref. [6]. The agreement between the $b \to s\gamma$ and $b \to X_c\ell\nu$ moments [8] is good and well within the expected theoretical uncertainties.

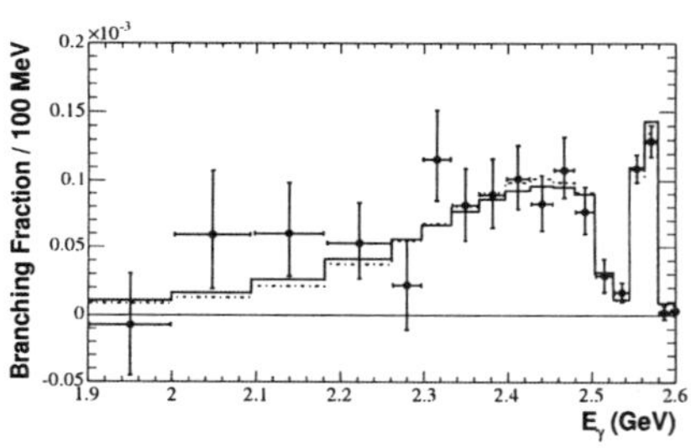

Figure 1. The photon energy spectrum obtained from the sum of exclusive modes. The data points are compared to theoretical predictions (histograms) obtained using the shape function (dashed line) and kinetic (solid line) schemes.

2. $B \to K^{(*)}\ell^+\ell^-$

The decays $B \to K^{(*)}\ell^+\ell^-$, where $\ell^+\ell^-$ are the charged lepton pairs e^+e^- or $\mu^+\mu^-$, result from $b \to s$ flavor-changing neutral currents. In the SM, three amplitudes contribute at lowest order to the $b \to s\ell^+\ell^-$ process: a photon penguin, a Z penguin, and a W^+W^- box diagram. The predicted total branching fraction is $\mathcal{B}(b \to s\ell^+\ell^-) = (4.2 \pm 0.7) \times 10^{-6}$ [9]. The direct CP asymmetries for these decays are expected to be very small in the SM, much less than 1% [10], whereas new physics at the electroweak scale could enhance A_{CP} to values of order one [11].

We select events that include two oppositely charged lepton candidates, a kaon candidate (either $K^{\pm}$ or K^0_S), and, for the $B \to K^*\ell^+\ell^-$ modes, a $\pi^{\pm}$ candidate that, when combined with a kaon candidate, forms a K^* candidate. Signal decays are selected using the m_{ES} and ΔE variables described in section 1.2. The largest backgrounds that peak in m_{ES} and ΔE are B decays to charmonium: $B \to J/\psi K^{(*)}$ and $B \to \psi(2S)K^{(*)}$. We exclude di-lepton pairs consistent with the J/ψ mass or the $\psi(2S)$ mass while accounting for lepton radiation which produces a correlated shift in $m_{\ell^+\ell^-}$ and ΔE.

We perform an unbinned maximum likelihood fit to the distributions of m_{ES}, ΔE and, in $B \to K^*\ell^+\ell^-$ decays, the kaon-pion invariant mass. Using 229 million $B\bar{B}$ events, the combined fits for $B \to K\ell^+\ell^-$ and $B \to K^*\ell^+\ell^-$ yield the branching fraction measurements,

$$\mathcal{B}(B \to K\ell^+\ell^-) = (0.34 \pm 0.07 \pm 0.03) \times 10^{-6}$$
$$\mathcal{B}(B \to K^*\ell^+\ell^-) = (0.78^{+0.19}_{-0.17} \pm 0.12) \times 10^{-6}$$

where the first error is statistical and the second is systematic. The direct CP asymmetry is also extracted from the fit to the modes where the b flavor of signal candidates can be inferred from the charges of the final state $K^{(*)}$ hadrons:

$$A_{CP}(B^+ \to K^+\ell^+\ell^-) = 0.08 \pm 0.22 \pm 0.11$$
$$A_{CP}(B^+ \to K^{*+}\ell^+\ell^-) = -0.03 \pm 0.23 \pm 0.12$$

Fitting the electron and muon channels independently, we determine the ratio of muon to electron branching fractions $R_{K^{(*)}} = \mathcal{B}(B \to K^{(*)}\mu^+\mu^-)/\mathcal{B}(B \to K^{(*)}\mu^+\mu^-)$:

$$R_K = 1.06 \pm 0.48 \pm 0.05, \qquad R_{K^*} = 0.93 \pm 0.46 \pm 0.12$$

In the SM we expect $R_K = 1$ and $R_{K^*} \approx 0.75$ [9].

3. $B^+ \to \tau^+\nu$

In the SM, $B^+ \to \tau^+\nu_\tau$ proceeds via W-boson annihilation with a branching fraction which is sensitive to the product $f_B \cdot |V_{ub}|$ where f_B is the B decay constant. Currently, the $\sim 16\%$ uncertainty on f_B from lattice QCD calculations [12] limits the precision with which $|V_{td}|$ can be extracted from precision measurements of Δm_d. Assuming $f_B = 196 \pm 32$ MeV [12] and $|V_{ub}| = (3.67 \pm 0.47) \times 10^{-3}$ [13], the predicted value is $\mathcal{B}(B^+ \to \tau^+\nu_\tau) = (9.3 \pm 3.6) \times 10^{-5}$. The most stringent published limit is $\mathcal{B}(B^+ \to \tau^+\nu_\tau) <$

4.2×10^{-4} at the 90% confidence level [14]. Physics beyond the SM, such as supersymmetry or two-Higgs-doublet models, could enhance $B(B^+ \to \tau^+\nu_\tau)$ up to the current experimental limit [15].

The $B^+ \to \tau^+\nu_\tau$ decay produces multiple neutrinos in the final state. Thus, the analysis [16] proceeds by first reconstructing the B^- in an $\Upsilon(4S) \to B^+B^-$ event in a semileptonic mode $B^-_{sl} \to D^{*0}\ell^-\bar{\nu}_\ell$ ($\ell = e$ or μ). All remaining charged and neutral particles in the event are then examined for evidence of $B^+ \to \tau^+\nu_\tau$. We identify the τ lepton from the signal decay in the channels: $e^+\nu_e\bar{\nu}_\tau$, $\mu^+\nu_\mu\bar{\nu}_\tau$, $\pi^+\bar{\nu}_\tau$, $\pi^+\pi^0\bar{\nu}_\tau$, and $\pi^+\pi^-\pi^+\bar{\nu}_\tau$.

The sum of the laboratory-frame energies of the neutral electromagnetic calorimeter clusters which are not associated with either the B^-_{sl} or the τ decay is denoted $E_{\rm extra}$. This variable peaks near 0 for signal while for background it takes on larger values. The $E_{\rm extra}$ selection region defines the signal region for each channel.

With 232 million $B\bar{B}$ pairs, we see no significant excess of observed events. Therefore, we set an upper limit on the branching fraction of $\mathcal{B}(B^+ \to \tau^+\nu_\tau) < 2.8\times10^{-4}$ at the 90% CL. A previous *BABAR* search based on a sample of 88.9×10^6 $B\bar{B}$ pairs where the B^- meson is reconstructed in a set of hadronic modes yielded $\mathcal{B}(B^+ \to \tau^+\nu_\tau) < 4.2 \times 10^{-4}$ [14]. These results are statistically independent and may be combined to give $\mathcal{B}(B^+ \to \tau^+\nu_\tau) < 2.6 \times 10^{-4}$ at the 90% CL.

References

1. P. Gambino and M. Misiak, *Nucl. Phys. B* **611**, 338 (2001).
2. D. Benson, I.I. Bigi, and N. Ultrasev, *Nucl. Phys. B* **710**, 371 (2005).
3. I.I. Bigi and N. Ultrasev, *Int. J. Mod. Phys.* **A17**, 4709 (2002).
4. *BABAR* Collaboration, B. Aubert *et al.*, hep-ex 0507001.
5. S. Chen *et al.*, *Phys. Rev. Lett.* **87**, 251807 (2001).
6. B. Aubert *et al.*, *Phys. Rev. D* **72**, 052004 (2005).
7. S.W. Bosch, B.O. Lange, M. Neubert, and G. Paz, *Nucl. Phys. B* **699**, 335 (2004).
8. B. Aubert *et al.*, *Phys. Rev. Lett.* **93**, 011803 (2004).
9. A. Ali *et al.*, *Phys. Rev. D* **66**, 034002 (2002).
10. F. Kruger, L.M. Sehgal, N. Sinha and R. Sinha *Phys. Rev. D* **61**, 114028 (2000).
11. F. Kruger and E. Lunghi, *Phys. Rev. D* **63**, 014013 (2001).
12. A.S. Kronfeld, *Nucl. Phys. Proc. Suppl.* **129**, 46 (2004).
13. Particle Data Group, S. Eidelman *et al.*, *Phys. Lett. B* **592**, 1 (2004).
14. B. Aubert *et al.*, *Phys. Rev. Lett.* **95**, 041804 (2005).
15. W.-S. Hou, *Phys. Rev. D* **48**, 2342 (1993).
16. B. Aubert *et al.*, *Phys. Rev. D* **73**, 057101 (2006).

THE COBRA DOUBLE BETA DECAY EXPERIMENT *

B. MORGAN

Department of Physics,
University of Warwick,
Coventry CV4 7AL, England
E-mail: Ben.Morgan@warwick.ac.uk

The COBRA experiment aims to use a large quantity of CdZnTe semiconductor detectors to search for neutrinoless double beta decay. The current status of the experiment is discussed, and new limits on several double beta modes are presented. Future plans for a large scale experiment are also described.

1. Introduction

Neutrinoless double beta decay ($0\nu\beta\beta$) is a second order weak process involving the decay of two nucleonic neutrons with the emission of two electrons:

$$(A, Z) \rightarrow (A, Z+2) + 2e^- \tag{1}$$

The signature for this decay is a sharp line in the electron sum energy spectrum at the transition energy Q of the decay. However, $0\nu\beta\beta$ decay is only allowed if neutrinos are massive Majorana particles, the half-life $T^{0\nu}_{1/2}$ being related to the Majorana mass of the neutrino, $\langle m_{ee} \rangle$, via

$$(T^{0\nu}_{1/2})^{-1} = G^{0\nu}(Q, Z)|NME|^2 \frac{\langle m_{ee} \rangle^2}{m_e^2}, \tag{2}$$

where $G^{0\nu}(Q, Z)$ is the phase space factor for transition energy Q of nucleus Z, NME is the nuclear matrix element and m_e is the electron mass[1]. Thus observing $0\nu\beta\beta$ decay would confirm neutrinos as Majorana in nature and also provide a measurement of the absolute neutrino mass scale. Experiments are now aiming for sensitivities to $\langle m_{ee} \rangle \sim 50$meV. Such an

*http://cobra.physik.uni-dortmund.de

experiment is however challenging as this mass only ~ 1 decay per 100kg of material per year is expected. A large mass of double beta isotopes and a low background are therefore crucial.

The COBRA experiment plans to use a large array of Cadmium-Zinc-Telluride (CdZnTe) semiconductor detectors to search for $0\nu\beta\beta$ decay[2]. CdZnTe contains 5 of the 35 known isotopes able to undergo double beta decay via $\beta^-\beta^-$ emission, including the high Q value isotopes ^{130}Te($Q =$ 2529keV) and ^{116}Cd($Q = 2805$keV). In addition it contains 4 isotopes that can decay via $\beta^+ EC$ (positron emission plus electron capture) and $ECEC$ (double electron capture) modes, with ^{106}Cd($Q = 2771$keV) also allowed to decay via the $\beta^+\beta^+$ mode. COBRA's use of CdZnTe has many advantages:

- Simple experimental setup: isotopes of interest intrinsic to detector itself.
- Modular, easily scalable design.
- High Q values maximize double beta decay rate whilst background rate decreases with increasing Q.
- Commercially available, well understood semiconductor detector.
- Very low contamination with radioisotopes and provides excellent energy resolution ($O(1\%)$ at 2-3MeV).
- Room temperature operation, no need for complex cryogenic systems.

Using a modular design also provides for further background reduction as high energy gamma rays are likely to hit several crystals, whereas double beta events are confined to a single crystal. Pixelisation of CdZnTe crystal readout has also been demonstrated, offering the possibility of tracking and vertex reconstruction of events.

2. Current Status

Between Summer 2003 and Autumn 2005, four 1cm^3 CdZnTe detectors supplied by eV PRODUCTS were installed at Gran Sasso for R&D on background levels and electronics. These detectors were mounted in a pertinax holder housed in a copper brick that was placed inside a passive shield of $20 \times 20 \times 20$cm cube of electropolished copper plus a further 15cm thick layer of lead. This setup was installed inside a Faraday cage surrounded by 7cm of boron-loaded polyethylene to provide shielding against neutrons.

Contamination levels in the various components were measured in the Gran Sasso HPGe facility. These showed the passivation material painted on the surfaces of the crystals to be the dominant source of background. Comparison of simulations of the setup with data confirmed the limiting background to be the passivation paint. Investigations of alternatives to replace the passivation paint are currently underway.

Table 1. Preliminary 90% C.L.s on half-lives of $0\nu\beta\beta$ decays of Cd, Te and Zn.

Isotope	Decay	$T_{1/2}$ Limit/yr
	$0\nu\beta^-\beta^-$ Decays	
^{70}Zn	to ground state(g.s.)	1.58×10^{17}
^{130}Te	to g.s.	5.67×10^{19}
^{130}Te	to 536keV	4.63×10^{19}
^{128}Te	to g.s.	3.11×10^{19}
^{116}Cd	to g.s.	1.18×10^{19}
^{116}Cd	to 1294keV	4.26×10^{18}
^{116}Cd	to 1757keV	8.71×10^{18}
^{116}Cd	to 2112keV	1.85×10^{19}
^{116}Cd	to 2225keV	6.47×10^{19}
	$0\nu\beta^+\beta^+$ Decays	
^{64}Zn	$0\nu\beta^+EC$ to g.s.	5.07×10^{18}
^{64}Zn	$0\nu ECEC$ to g.s.	9.52×10^{16}
^{120}Te	$0\nu\beta^+EC$ to g.s.	4.89×10^{16}
^{120}Te	$0\nu ECEC$ to g.s.	1.91×10^{15}
^{120}Te	$0\nu ECEC$ to 1171keV	3.84×10^{15}
^{106}Cd	$0\nu\beta^+\beta^+$ to g.s.	3.57×10^{17}
^{106}Cd	$0\nu\beta^+EC$ to g.s.	2.48×10^{19}
^{106}Cd	$0\nu ECEC$ to g.s.	5.34×10^{16}
^{106}Cd	$0\nu\beta^+\beta^+$ to 512keV	7.45×10^{16}
^{106}Cd	$0\nu\beta^+EC$ to 512keV	1.93×10^{18}

A search for decays of all the double beta isotopes with peaks above 600keV was performed using the 3.82kg.days of data collected by the 4 crystals. Preliminary limits on the half-lives of these decays, including modes where the decay is to an excited state of the daughter isotope, are shown in Table. 1. Of these, the limits on the decays of ^{64}Zn and ^{120}Te are world leading at present.

3. The 64 Array

The next step towards a large scale COBRA detector is the 64 crystal array, shown in Fig. 1. This provides 418g of CdZnTe in a scalable, modular design and the larger array will enable the exploration of the use of multi-crystal events for background reduction.

Figure 1. Artist's impression of the the COBRA 64 crystal array, courtesy of D. Dobos.

At present, the crystals are being tested, and the first stage of installation at Gran Sasso has been carried out.

4. Future Plans

As discussed earlier, the target for double beta decay searches is sensitivity to Majorana masses of $\sim$ 50meV. Fig. 2 shows predictions of the half-life sensitivity of a 64000 CdZnTe crystal array (418kg of CdZnTe) with different background rates and energy resolutions. Enrichment to 90% in ^{116}Cd is neccessary, but sensitivity to half-lives of 10^{26}yr($\langle m_{ee}\rangle \sim$ 50meV) is posssible provided the background in the 2.8MeV region can be limited to $< 10^{-3}$keV^{-1}kg^{-1}yr^{-1}.

Background studies of a 64000 crystal array are now underway using

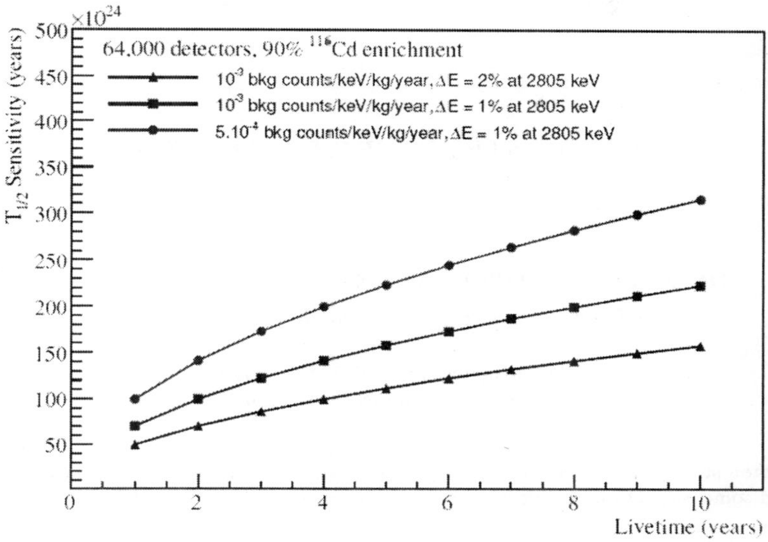

Figure 2. Expected half-life sensitivity for $0\nu\beta\beta$ decay of ^{116}Cd in an array of 64000 1cm^3 crystals with 90% enrichment of ^{116}Cd for different background rate and energy resolution scenarios.

Geant4 based simulations. A simple background model has been developed that describes:

- Radioisotope contamination in crystals.
- Radioisotope contamination of construction materials.
- Cosmic ray activation of CdZnTe prior to installation underground.
- Shielding of external muons and neutrons.

Work is now in progress to improve the accuracy of the model, and determine acceptable contamination levels and shielding configurations.

Experimental and simulation studies of CdZnTe pixellized readouts are also underway. The differing ranges of α, β and γ particles in CdZnTe combined with the tracking of particles with pixels offer unique signatures that could be used to reject backgrounds with $\sim$ 100% efficiency.

References

1. Y. Zdesenko, *Rev. Mod. Phys.* **74**, 663 (2002).
2. K. Zuber, *Phys. Lett.* **B519**, 1 (2001).

PROTON STRUCTURE FROM HERA

K. NAGANO
ON BEHALF OF THE H1 AND ZEUS COLLABORATIONS
High Energy Accelerator Research Organization, KEK,
1-1 Oho,
Tsukuba, Ibaraki 305-0801, Japan
E-mail: knagano@post.kek.jp

Proton structure as measured up until now by *ep* collisions at HERA is reviewed, and some issues to be tackled further are addressed.

1. Introduction

Deep inelastic scattering (DIS) of leptons off nucleons is a straightforward tool to probe the content of the nucleons. We measure DIS cross sections double-differentially in the Bjorken scaling variable, x, and the negative of the four-momentum transfer squared, Q^2, each of which corresponds to the momentum fraction carried by the struck parton, and the scale where the nucleon is probed, respectively. The cross section can be written with the structure functions (SFs) F_2 and F_L which "parameterize" the proton structure, i.e how far from point-like [a] , as

$$\frac{d^2\sigma^{NC}}{dxdQ^2} = \frac{2\pi\alpha^2}{xQ^4}\{y^2({F_2}^{NC} - {F_L}^{NC})/x + 2(1-y){F_2}^{NC}/x\}, \qquad (1)$$

where α is the fine structure constant, and y is the inelasticity [b]. The F_2 contributes to the cross section dominantly while the F_L becomes significant only in high-y region. The perturbative QCD (pQCD) tells that the SFs can be written by means of the parton distribution functions (PDFs). For example at the leading order, F_2 can be expressed as a sum of all quark PDFs multiplied each with charge-squared. In low-x region, the DGLAP evolution formalism ascribes the Q^2-dependence of F_2 ("scaling violation")

[a]The cross section can be written for collisions between two spin-1/2 point-like particles as $\frac{d^2\sigma}{dxdQ^2} = \frac{2\pi\alpha^2}{xQ^4}\{y^2 + 2(1-y)\}$

[b]The contribution from F_3 is neglected in the formula.

and also a non-zero value of the F_L as largely owning to gluon splitting into $q\bar{q}$-pair such that both $\partial F_2/\partial \ln Q^2$ and F_L are almost proportional to $\alpha_s g(x, Q^2)$.

HERA at DESY is a unique facility which collides electrons (or positrons) with protons. The large center-of-mass energy of about 320 GeV allows an extension of the explorable kinematic phase-space by two orders of magnitude both in Q^2 and x. Figure 1 shows the HERA F_2 measure-

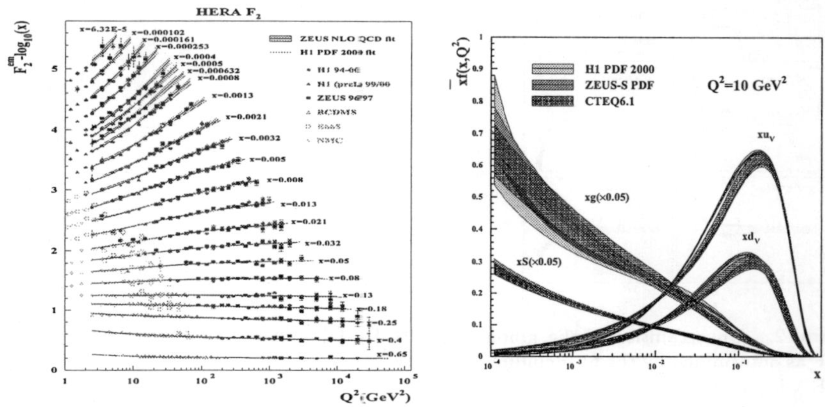

Figure 1. HERA F_2 measurements and the extracted PDFs

ments together with those by fixed-target DIS experiments in the left plot. The rapid rise of the F_2 at low-x was first observed at HERA and the scaling violation was such precisely measured, both of which can be explained by the pQCD as a consequence of large gluon PDF at low-x. The right plot shows the extracted PDFs in which the gluon PDF is determined with typical precision of $5-10\%$.

2. Direct determination of gluon PDF

The gluon PDF is determined mainly from the scaling violation, i.e. indirectly. Therefore, testing universality in other processes that are directly sensitive to gluon PDF is a vital issue to be addressed.

NLO QCD fit including HERA jet data

ZEUS has performed a next-to-leading order QCD analysis on their DIS inclusive cross sections together with their jet production cross sections in

DIS and photoproducion which are directly sensitive to gluon [1]. This is the first fit that utilizes HERA jet data.

Figure 2 shows the precision of the determined gluon PDF compared to that of their previous fit without using jet data, showing a clear improvement at middle-x, $0.01 \lesssim x \lesssim 0.3$.

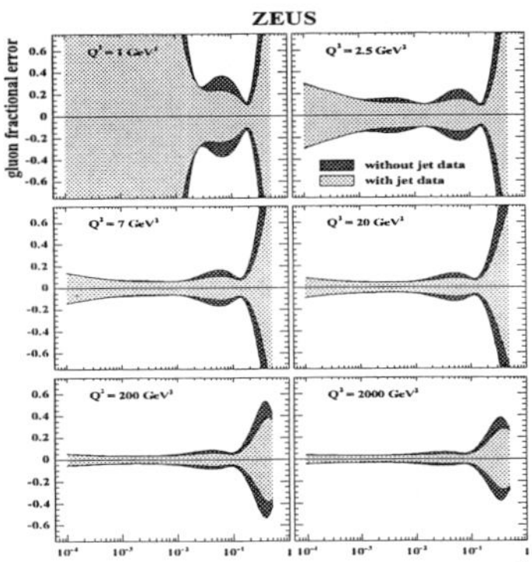

Figure 2. Uncertainties of the gluon PDF as determined by the fit including HERA jet data

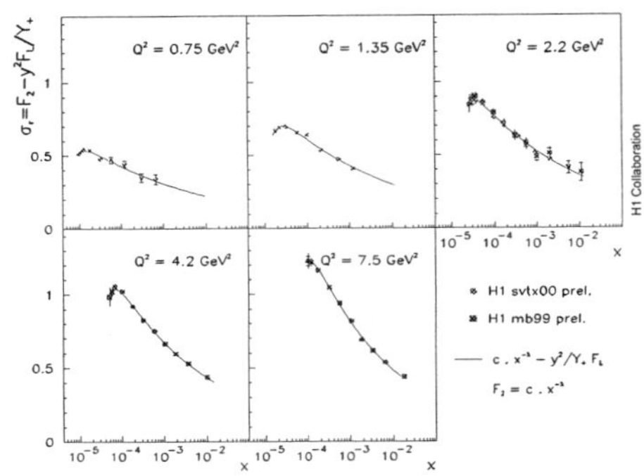

Figure 3. DIS cross sections at very low $Q^2 \approx 1$ GeV^2

$\underline{F_L}$

Figure 3 shows the H1 measurement of DIS cross sections at very low Q^2, around $Q^2 \approx 1$ GeV^2 [2]. A turn-over is seen at very low x, i.e. large y, where F_L is expected to give a significant and negative contribution to the cross section (see eq. (1)). The dotted lines are the F_2 contribution extrapolated from high-x by assuming exponential shape. The difference can be regarded as F_L, which is straightforwardly sensitive to gluon. Notice that this is a model-dependent extraction. A model-independent extraction will be possible if cross sections with different y, i.e. different center-of-mass energy, are available for same (x, Q^2). Such low-energy runs are currently being considered at HERA.

3. Flavor decomposition of quark PDFs

The flavor composition of quark PDFs is not yet fully understood. For example, the d-quark PDF is less constrained than that of the u-quark [c],

[c]Notice that $F_2 \sim (4u + d)$ and moreover $u > d$.

and the s-quark PDF is almost unknown [d].

Charged current DIS

The charged current (CC) DIS, $ep \to \nu X$, provides a powerful discrimination of quark flavors. Due to charge selecting nature, the e^+p (e^-p) CC occurs only with down-type (up-type) quarks and up-type (down-type) anti-quarks. In addition, the HERA data is free from heavy-target correction and higher-twist and target-mass effects. Figure 4 shows the measured

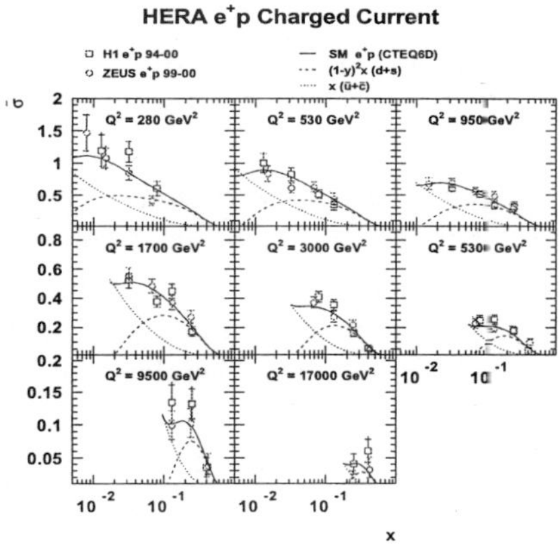

Figure 4. e^+p charged current DIS cross sections

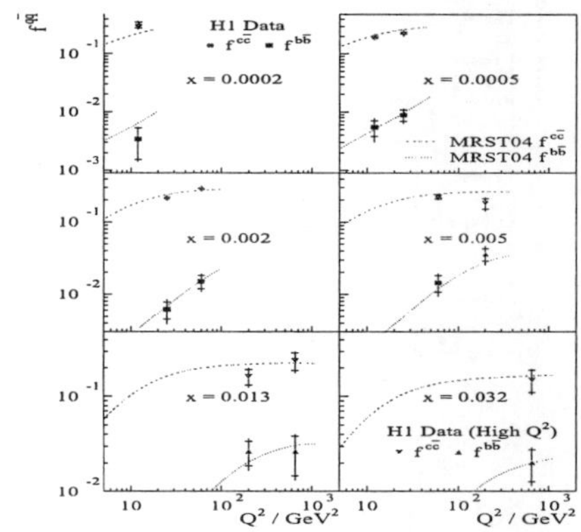

Figure 5. Heavy-quark contribution to F_2

e^+p CC DIS cross sections. It is greatly hoped that the large luminosity at HERA-II will bring precise determination of d-quark PDF through the e^+p CC DIS.

Heavy-flavor PDFs

Figure 5 shows the heavy-flavor contribution to the F_2, both of charm and bottom quarks, as measured by H1 [3]. It was shown that the charm owes about 30% and the bottom owes about 3% of F_2, both of which increase with Q^2. This is the first measurement of $F_2^{b\bar{b}}$. For tagging these flavors in final state, signed impact parameter measured using the Vertex Detector was used to identify displaced vertex. The $F_2^{c\bar{c}}$ measurement using the Vertex Detector gives a consistent result with that measured using the technique of

[d] The strange sea is assumed to be suppressed as $s = \bar{s} = 0.2(\bar{u}+\bar{d})$ so as to be consistent with di-muon data from neutrino-beam fixed-target experiments.

tagging D^* with slow π. There are big prospects on this topic at HERA-II as ZEUS also installed a similar vertex detector.

4. PDFs at large Q^2 and large x

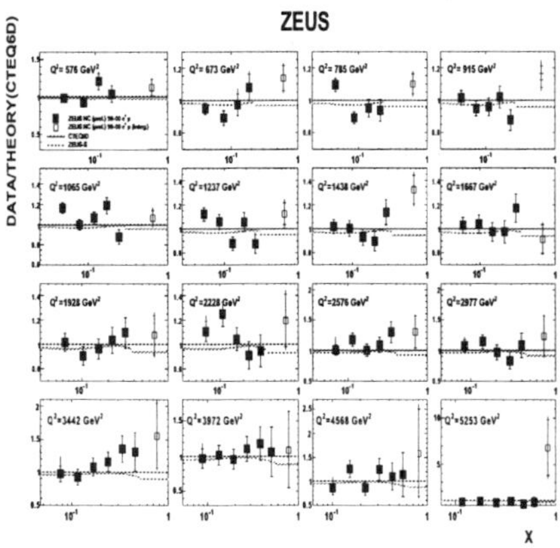

Figure 6. F_2 at large x

It is important to validate the DGLAP evolution until the largest-Q^2 HERA can reach, which will be a vital input to the LHC.

Also high x is of interest. ZEUS developed a new method to explore ultimately large-x region [4]. Events with no jet reconstructed within fiducial volume are assumed to come from x above a certain value x_{Edge}. An integrated cross section corresponding to this region, $x_{Edge} < x < 1$, is measured. As shown in Figure 6, the newly measured cross-sections at the largest x are in general in good agreement with the current PDFs.

5. Summary

HERA has provided most precise inclusive SF measurements, which brought significant improvements to our knowledge on proton structure. For further comprehensive understanding, new analysis with new techniques and large amount of luminosity are on-going.

References

1. ZEUS collab., *Eur. Phys. J* **C42**, 1 (2005),
2. H1 collab., contributed paper #161 for the "32nd International Conference on High Energy Physics (ICHEP04)",
3. H1 collab., *Eur. Phys. J* **C45**, 23 (2006),
4. ZEUS collab., contributed paper #261 for the "32nd International Conference on High Energy Physics (ICHEP04)",

THE LHCB TRACKING SYSTEM AND ITS PERFORMANCE

J. NARDULLI
On Behalf of the LHCb Collaboration

NIKHEF,
Kruislaan 409,
1098 SJ Amsterdam, Netherlands
E-mail: jnardull@nikhef.nl

LHCb is a next-generation forward spectrometer for CP violation measurements, using the Large Hadron Collider at CERN. In order to achieve its goals a high overall track reconstruction performance is needed. The LHCb tracking system comprises three main sub-systems: the vertex locator, the trigger tracker and the downstream tracking stations.

1. Introduction

LHCb aims to study CP violation and rare B-meson decays with high precision, using the Large Hadron Collider (LHC), where all B-meson families are produced in 14 TeV pp collisions. In these events the $b\bar{b}$ pairs are typically produced in the same forward (or backward) direction. The LHCb detector is a single-arm spectrometer with a forward coverage from 10 $mrad$ to 300 $mrad$ in the horizontal plane (i.e., the bending plane of the magnet). The acceptance lies between 10-250 $mrad$ in the vertical plane (non-bending plane). The detector layout in the bending plane is shown in Figure 1.

2. Tracking System

The LHCb Tracking system can be divided into three sub-systems.

- The VErtex LOcator (VELO) covers the area around the interaction point and is made of silicon sensors.
- The Trigger Tracker (TT), also built with silicon sensors, is downstream the interaction point, but in front of the large dipole mag-

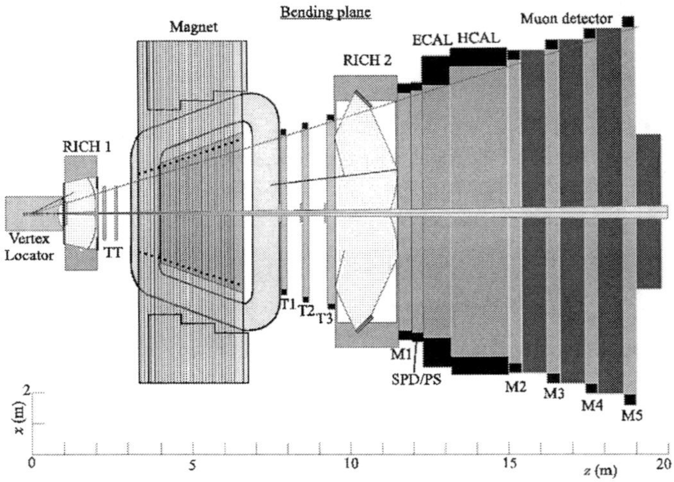

Figure 1. The LHCb setup with the different sub-detectors shown in the horizontal plane — also referred to as the bending plane of the magnet.

net.

- The Tracker Stations behind the magnet are divided in two parts: Inner Tracker (IT) and Outer Tracker (OT). The IT consists of silicon strip detectors, while the OT is made from straw tube drift cells.

2.1. *The Vertex Locator*

The Vertex Locator (VELO)[1,2] is made of 21 stations, placed along and perpendicular to the beam axis. Two different types of sensors are used: one measures the r coordinate and the other measures the ϕ coordinate. The half-disk sensors, shown in Figure 2, are arranged in pairs of r and ϕ sensors and mounted back-to-back.

The r-ϕ geometry is used for having a projection in the r-z plane and for fast 2D-tracking in the trigger. The VELO has two main roles: it has to provide precise coordinate measurements to allow determination of primary and secondary vertices, and it is the main tracking detector before the magnet.

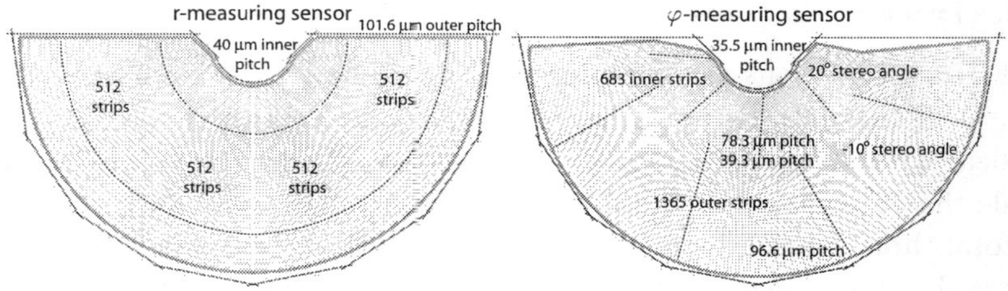

Figure 2. Layout of the VELO r and ϕ measuring sensors.

2.2. *The Trigger Tracker*

The Trigger Tracker (TT) is located downstream the first RICH (Ring Imaging CHerenkov) detector and just in front of the magnet. It consists of two stations separated by a distance of 27 cm. Every station consists of two layers, for a total of four layers, with angles between the y and the z axes in the following configuration: $0^o, +5^o, -5^o, 0^o$.

The Trigger Tracker main function is to provide momentum information to the trigger system; it is therefore mostly used for the so-called *VELO-TT Tracking* and to get a first estimate of the momentum, as it feels the fringe field of the magnet (see Figure 3).

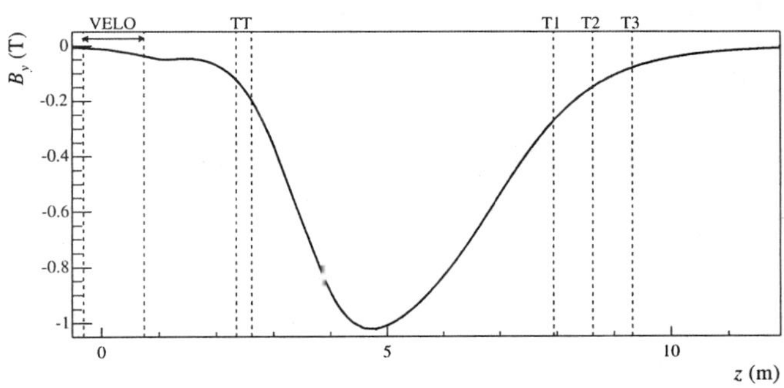

Figure 3. The main component of the magnetic field strength (B_y) along the z axis.

The silicon sensors are 500 μm thick and the strip pitch is 183 μm. The maximum occupancy is approximately 2%.

2.3. *Tracker Stations*

The Tracker Stations located after the magnet cover an area of about 6×5 m^2. Over this area there is a large difference in the particle flux. In order to deal with this, the innermost part is covered with silicon strips (IT), while the outer part is covered with straw tube drift cells (OT). There are in total three Tracker Stations with four layers each. For the four different layers the stereo angles between the y and z axis are $0^o, +5^o, -5^o, 0^o$.

2.3.1. *The Inner Tracker*

The Inner Tracker[3] covers approximately 2% of the area of the Tracker Stations, which corresponds to about 20% of the particle flux. The detectors are divided in boxes with 320 or 410 μm thick silicon. The strip pitch is 198 μm. The maximum occupancy is approximately 2%.

2.3.2. *The Outer Tracker*

In the Tracker Stations, the Outer Tracker (OT)[4] covers the large region outside the acceptance of the Inner Tracker. The Outer Tracker consists of straw tube detectors with a diameter of 5 mm. The gas mixture in use is Ar(70%)/CO_2(30%). The signal collection is within 3 LHC beam crossings (75 ns). Each of the three stations consists of four double layers of straws for a total of approximately 53.000 channels. The average occupancy is approximately 4.5%, while the maximum occupancy is approximately 9%.

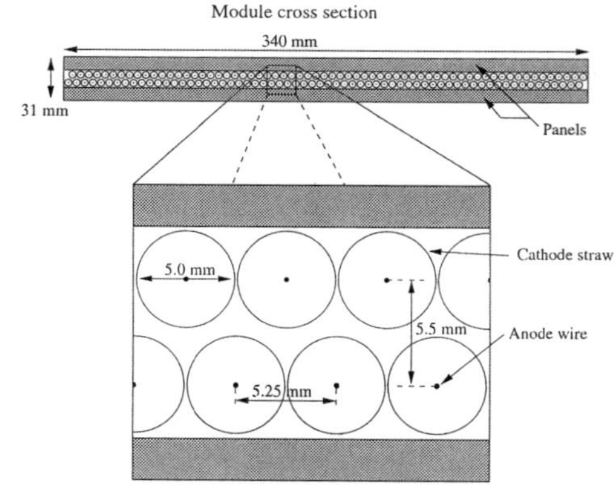

Figure 4. Cross section of an OT module (128 straws). A small region containing a few straws is magnified.

The wires in the 4.8 m modules are split in the middle and the read-out is both at the top and at the bottom. A cross section of an Outer Tracker module is shown in Figure 4. The spatial resolution obtained in beam tests[5] with this gas is approximately 200 μm.

3. Tracking Strategy and Performance

LHCb is a challenging environment for tracking. Typically a particle sees 40% of a radiation length up to RICH2. A robust tracking strategy is therefore needed. Track finding starts by reconstructing tracks in the VELO. These tracks are extrapolated through the detector and matched to hits in the T Stations to form "long tracks". The unused hits are then considered to reconstruct the T seeds in the Tracker Stations. These T seeds are also matched to tracks in the VELO which did not get T hits added at the beginning. Upstream tracks are found with hits in the VELO and in the TT, while tracks which have hits in the TT and in the T stations are called downstream tracks. The various track types are sketched in Figure 5.

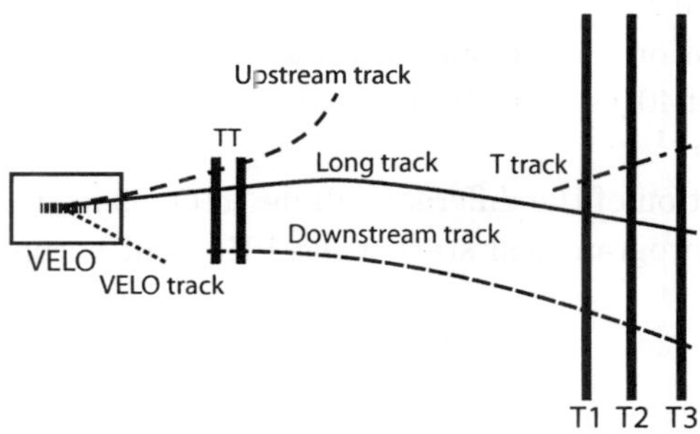

Figure 5. Sketch of the five different track types in the LHCb tracking system.

On average an event contains 72 reconstructed tracks, which can be divided into 26 long, 11 upstream, 4 downstream, 5 T and 26 VELO tracks. In Figure 6 the efficiency for finding long tracks and the ghost rate are shown as a function of the momentum.

The overall track reconstruction performance together with the excellent momentum resolution ($\Delta p/p = 0.38\%$) leads to a typical resolution on the B_s^0-meson mass of 14 MeV/c^2. The momentum resolution together with the impact parameter resolution results in proper time resolution better then $40 fs$ which should allow LHCb to make a 5σ measurement of the B_s^0 mixing parameter Δm_s up to 68 ps^{-1} [6].

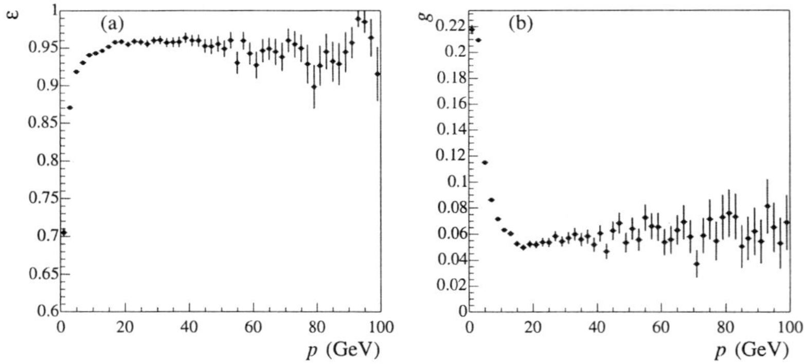

Figure 6. Tracking efficiency (a) and ghost rate (b) for long tracks as a function of momentum, p.

4. Conclusion

The general LHCb tracking performance in terms of efficiency, ghost rate, momentum resolution ($\Delta p/p = 0.38\%$) and vertex resolution leads to a typical resolution on the B_s^0-meson mass of 14 MeV/c^2 and to a proper time resolution of 40 fs. This proper time resolution is sufficient to observe Δm_s up to 68 ps^{-1}.

The construction of the different sub-detectors is in steady progress and the full physics program will start when LHC will become operational in 2007.

References

1. The LHCb Collaboration, P.R.Barbosa *et al.*, *LHCb VELO Technical Design Report*, CERN-LHCC/2001-011, May 2001.
2. The LHCb Collaboration, Nobrega R. *et al.*, *LHCb Reoptimized Detector Design and Performance Technical Design Report*, CERN-LHCC/2003-030, Sept. 2003.
3. The LHCb Collaboration, P.R.Barbosa *et al.*,*LHCb Inner Tracker Technical Design Report*, CERN-LHCC/2002-029, Nov. 2002.
4. The LHCb Collaboration, P.R.Barbosa *et al.*,*Outer Tracker Technical Design Report*, CERN-LHCC/2001-024, Sept. 2001.
5. G.v.Apeldoorn *et al.*,*Beam Tests of Final Modules and Electronics of the LHCb Outer Tracker in 2005*, LHCb Note 2005-076, Oct. 2005.
6. O. Leroy, in these proceedings (2006).

NEW STATES AT BABAR

M. NEGRINI*

Dipartimento di Fisica dell'Università di Ferrara - INFN,
Via Saragat 1,
44100 Ferrara, ITALY
E-mail: negrini@fe.infn.it

The BaBar experiment at the PEP-II B-factory offers excellent opportunities in quarkonium spectroscopy. Investigation of the properties of new states like the $X(3872)$ and $Y(4260)$ are performed, aiming to shed light on their nature. Recent BaBar results in this field are presented.

1. Introduction

The B-factories are copious sources of charmonium states, which can be produced through several mechanisms, i.e. color suppressed B decay ($B \to K + c\bar{c}$), two photons production ($\gamma^*\gamma^* \to c\bar{c}$, where the $c\bar{c}$ has positive C-parity), double charmonium production ($e^+e^- \to c\bar{c} + c\bar{c}$), and in e^+e^- interaction after initial state radiation (ISR) of a photon, lowering the effective center of mass energy of the e^+e^- annihilation ($e^+e^- \to \gamma_{ISR} + c\bar{c}$, where the $c\bar{c}$ has the quantum numbers $J^{PC} = 1^{--}$).

Some states with masses in the region around 4 GeV have been recently discovered by experiments at the B-factories, Belle and BaBar, thanks to the high integrated luminosity (hundreds of inverse femtobarns) available today, reviving the interest in quarkonium spectroscopy.

Several interpretations for these states exist. The mass values suggest that they could be conventional charmonium [1,2,3], but also other interpretations like $D^0\bar{D}^{*0}$ molecule [4,5] or diquark-antidiquark states [6] among many others have been proposed.

More experimental results are needed in order to shed light on the nature of these states. The BaBar experiment [7], at the PEP-II B-factory, is designed to perform precision measurements of CP violation in the B me-

*On behalf of the BaBar Collaboration.

son system but has a much broader physics reach and an extensive program of quarkonium spectroscopy. Recent BaBar results on the investigation of the $X(3872)$ and $Y(4260)$ are reported here.

2. Investigation of the $X(3872)$

The $X(3872)$, have been discovered by the Belle collaboration in the decay $B^{\pm} \to K^{\pm} J/\psi\pi^{+}\pi^{-}$, where a narrow signal in the $J/\psi\pi^{+}\pi^{-}$ invariant mass is observed at 3872 MeV/c^2 [8]. The observation was then confirmed by CDF [9], D0 [10] and BaBar [11].

The width of the $J/\psi\pi^{+}\pi^{-}$ invariant mass peak measured by Belle is compatible with the experimental resolution, therefore Belle determined an upper limit on the $X(3872)$ width $\Gamma_{X(3872)} < 2.3$ MeV/c^2 at 90% confidence level (CL). The favored quantum number assignment is $J^{PC} = 1^{++}$. The mass value and the fact that $B^{\pm} \to K^{\pm} c\bar{c}$ is a typical decay mode of the B meson, suggest the possible charmonium nature of the state. However, its mass is not compatible with existing predictions from potential models, and its likely $J/\psi\rho^0$ decay mode would be forbidden for charmonium states because of isospin symmetry. Moreover the mass value is exactly at the threshold for $D^0\bar{D}^{*0}$ production.

The decays $B^- \to K^- J/\psi\pi^{+}\pi^{-}$ and $B^0 \to K^0_s J/\psi\pi^{+}\pi^{-}$ have been studied by BaBar using a sample of 211 fb^{-1} of data, corresponding to 232 millions of $B\bar{B}$ pairs [12]. We obtain 61 ± 15 and 8.3 ± 4.5 $X(3872)$ decay events in B^- and B^0 decay respectively, with signal significance of 6.1σ and 2.5σ, including systematic uncertainty. Using these events it was possible to determine the branching fractions $BR(B^- \to K^- X(3872), X \to J/\psi\pi^{+}\pi^{-}) = (10.1 \pm 2.5 \pm 1.0) \times 10^{-6}$ and $BR(B^0 \to K^0 X(3872), X \to J/\psi\pi^{+}\pi^{-}) = (5.1 \pm 2.8 \pm 0.7) \times 10^{-6}$. The process $B^0 \to K^0 X(3872)$ is predicted to be suppressed by one order of magnitude in the $D\bar{D}^*$ molecule hypothesis [13]. In the diquark-antidiquark interpretation, two states with quark content $X_u = (cu)(\bar{c}\bar{u})$ and $X_d = (cd)(\bar{c}\bar{d})$ will exist, with a predicted mass difference $\Delta M \sim 7$ MeV/c^2 [6]. These states should be produced differently in $B^{\pm}$ and B^0 decay. ¿From our data we measured $\Delta M = 2.7 \pm 1.3$ MeV/c^2. Although no conclusion can be drawn, the method seems promising for a future reanalysis when more data will become available.

If the $X(3872)$ belongs to an isospin multiplet, as suggested by the possibility of the decay $X(3872) \to J/\psi\rho^0$, there must exist charged partners, that could be observed through $X(3872)^{\pm} \to J/\psi\rho^{\pm}$. A search of charged partners have been performed by BaBar in the $B^- \to \bar{K}^0 J/\psi\pi^-\pi^0$ and

$B^0 \to K^+ J/\psi\pi^-\pi^0$ channels [14]. No charged partner was observed, and we determined the upper limits $BR(B^0 \to K^+X^-, X^- \to J/\psi\pi^-\pi^0) < 5.4 \times 10^{-6}$ and $BR(B^- \to \bar{K}^0X^-, X \to J/\psi\pi^-\pi^0) < 22 \times 10^{-6}$ at 90% confidence level.

3. Inclusive charmonia in B decay

Charmonium states can be produced at B-factories through the two body B decays $B \to Kc\bar{c}$, therefore it is possible to observe charmonium states by looking at the mass recoiling against the K in the B reference frame. In $B\bar{B}$ events, when one of the two Bs is fully reconstructed it is possible to determine the momentum of the recoiling (unreconstructed) B by using the momentum of the reconstructed B and the beam parameters. This technique allows to observe charmonium states independently from their decay mode and to perform absolute measurements of the $BR(B \to Kc\bar{c})$. The disadvantage is the large K background that is present in B decays.

BaBar performed the analysis using 211 fb^{-1} of data [15] and observing the momentum spectrum of charged kaons. The observed momentum spectrum of $K^\pm$ is shown in Figure 1, where it is possible to observe the production of J/ψ, η_c, χ_{c1}, η_c', ψ', and ψ'' in B decays, althought some of them with low statistical significance. No other state has been observed. The results on the branching fraction measurements are reported in Table 1.

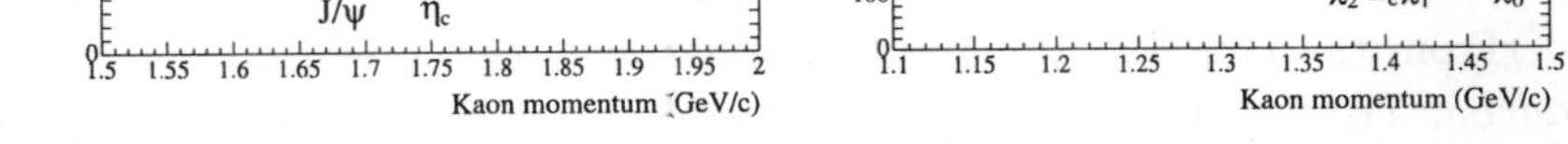

Figure 1. $K^\pm$ momentum distribution in the recoiling $B^\pm$ reference system. Arrows indicate the momentum corresponding to the masses of the known charmonia states. In the low mass region (a) two peaks corresponding to the η_c and J/ψ are clearly visible; in the high mass region (b) χ_{c1}, η_c', ψ', and ψ'' peaks can be observed.

¿From the upper limit on $BR(B^\pm \to K^\pm X(3872))$ and the average of Belle and BaBar results $BR(B^\pm \to K^\pm X(3872)) \cdot BR(X(3872) \to$

Table 1. Measured yields and branching fractions $BR(B^{\pm} \to K^{\pm} c\bar{c})$.

Particle	Yield	BR (10^{-4})	Particle	Yield	BR (10^{-4})
η_c	273 ± 43	$8.4 \pm 1.3 \pm 0.8$	η_c'	98 ± 52	$3.4 \pm 1.8 \pm 0.3$
J/ψ	259 ± 41	$8.1 \pm 1.3 \pm 0.7$	ψ'	139 ± 44	$4.9 \pm 1.6 \pm 0.4$
χ_0	9 ± 21	< 1.8	ψ''	99 ± 69	$3.5 \pm 2.5 \pm 0.3$
χ_1	227 ± 40	$8.0 \pm 1.4 \pm 0.7$	$X(3872)$	15 ± 39	< 3.2
χ_2	0 ± 36	< 2.0			

$J/\psi\pi^+\pi^-) = (13.3 \pm 2.5) \cdot 10^{-6}$ it is possible to extract a lower limit $BR(X(3872) \to J/\psi\pi^+\pi^-) > 4.2\%$ at 90% CL.

A search of charged partners was has been done using the same technique on neutral B decays. No charged partner have been observed therefore we determined the upper limit $BR(B^0 \to K^{\pm}X(3872)^{\mp}) < 5 \times 10^{-4}$ at 90% CL for the production of $X(3872)$ charged partners in B decays.

4. The $Y(4260)$

The reaction $e^+e^- \to \gamma_{ISR} J/\psi\pi^+\pi^-$ has been studied using 233 fb^{-1} of data, and the corresponding $J/\psi\pi^+\pi^-$ invariant mass distribution for the selected events is shown in Figure 2. A broad enhancement at 4260 MeV/c^2, called $Y(4260)$, is clearly observed [16]. An unbinned fit with a Breit-Wigner signal function and a second order polynomial background yields 125 ± 23 $Y(4260) \to J/\psi\pi^+\pi^-$ events, with a mass $M_Y = 4259 \pm 8(\text{stat})^{+2}_{-6}(\text{syst})$ MeV/c^2 and a width $\Gamma_Y = 88 \pm 23(\text{stat})^{+6}_{-4}(\text{syst})$ MeV.

A search for $Y(4260)$ in B decays has been performed by studying the decay $B^{\pm} \to K^{\pm}J/\psi\pi^+\pi^-$ using 211 fb^{-1} of data. Fixing M_Y and Γ_Y to the values measured in the ISR production process, an excess of 128 ± 42 events compatible with $B^{\pm} \to K^{\pm}Y(4260)$, $Y(4260) \to J/\psi\pi^+\pi^-$ is observed [12].

¿From the fact that the $Y(4260)$ is not observed in the total $e^+e^- \to$ *hadrons* cross section, it is possible to infer that $\Gamma(Y(4260) \to e^+e^-)$ is much smaller with respect to the corresponding widths of all other known $J^{PC} = 1^{--}$ charmonium resonances, while $\Gamma(Y(4260) \to J/\psi\pi^+\pi^-)$ is much larger [17]. The observation of other decay modes may help in understanding the nature of this state. BaBar searched for $Y(4260) \to p\bar{p}$ signal in ISR $e^+e^- \to \gamma_{ISR}p\bar{p}$ events. No signal has been found, so it is possible to determine the upper limit $BR(Y(4260) \to p\bar{p})/BR(Y(4260) \to J/\psi\pi^+\pi^-) < 13\%$ at 90% CL [18].

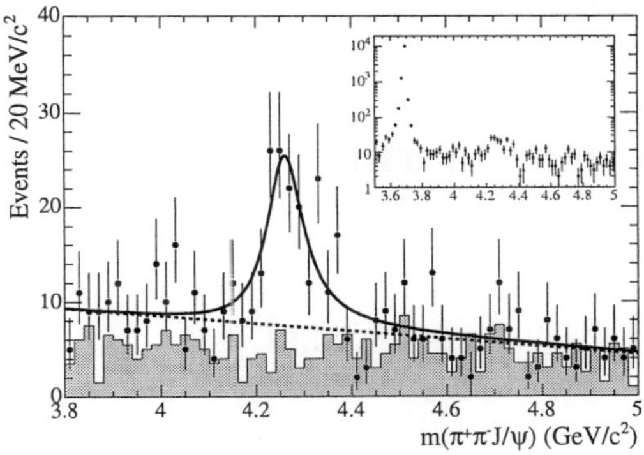

Figure 2. Invariant mass distribution for initial state radiation $J/\psi\pi^+\pi^-$ events in the region 3.8-5 GeV/c^2. The inset shows a larger region, where the ψ' peak is clearly visible. The solid line is the fit of the distribution with a relativistic Breit-Wigner and a second order polynomial background (dashed line). The histogram represents the background estimated from the J/ψ sideband regions.

References

1. T. Barnes and S. Godfrey, Phys. Rev. D **69**, 054008 (2004).
2. E. J. Eichten, K. Lane and C. Quigg, Phys. Rev. D **69**, 094019 (2004).
3. M. Suzuki, Phys. Rev. D **72**, 114013 (2005).
4. E. S. Swanson, Phys. Lett. B **588**, 189 (2004).
5. N. A. Tornqvist, Phys. Lett. B **590**, 209 (2004).
6. L. Maiani, F. Piccinini, A. D. Polosa and V. Riquer, Phys. Rev. D **71**, 014028 (2005).
7. B. Aubert *et al.* [BABAR Collaboration], Nucl. Instrum. Meth. A **479**, 1 (2002).
8. S. K. Choi *et al.* [Belle Collaboration], Phys. Rev. Lett. **91**, 262001 (2003).
9. D. Acosta *et al.* [CDF II Collaboration], Phys. Rev. Lett. **93**, 072001 (2004).
10. V. M. Abazov *et al.* [D0 Collaboration], Phys. Rev. Lett. **93**, 162002 (2004).
11. B. Aubert *et al.* [BABAR Collaboration], Phys. Rev. D **71**, 071103 (2005).
12. B. Aubert *et al.* [BABAR Collaboration], Phys. Rev. D **73**, 011101 (2006).
13. E. Braaten and M. Kusunoki, Phys. Rev. D **71**, 074005 (2005).
14. B. Aubert *et al.* [BaBar Collaboration], Phys. Rev. D **71**, 031501 (2005).
15. B. Aubert *et al.* [BABAR Collaboration], Phys. Rev. Lett. **96**, 052002 (2006).
16. B. Aubert *et al.* [BABAR Collaboration], Phys. Rev. Lett. **95**, 142001 (2005).
17. F. E. Close and P. R. Page, Phys. Lett. B **628**, 215 (2005).
18. B. Aubert *et al.* [BABAR Collaboration], Phys. Rev. D **73**, 012005 (2006).

MEASUREMENT OF THE CKM ANGLE γ AT *BABAR*: STATUS AND PROSPECTS.

N. NERI

(representing the BABAR Collaboration)
INFN - Sezione di Pisa
Largo B. Pontecorvo,3
56127, Pisa, Italy
E-mail: nicola.neri@infn.it

We present the results for the measurement of γ in the *BABAR* experiment which exploit the interference between the V_{ub} and V_{cb} mediated amplitudes in the $B \to DK$ and $B \to D\pi$ systems. Experimental details along with future prospects are discussed.

1. Introduction

The primary goal of the *BABAR* experiment is the study of the *CP* violation in the B meson system via the precise measurement of the Cabibbo-Kobayashi-Maskawa (CKM) matrix elements. The measurement of the CKM matrix elements is one of the most stringent tests of the Standard Model (SM) and for this reason it also represents a window into New Physics. The precise measurement of the angle $\gamma \equiv \arg(-V_{ud}V_{ub}^*/V_{cd}V_{cb}^*)$, of the unitarity triangle [1] is a crucial test of the *CP* violation picture in the SM.

2. An Overview of Methods for the γ Measurement

There are several decay channels that can be used to extract information on γ, and each has distinct merits and drawbacks. The most sensitive methods are based on studies of the decay $B^{\pm} \to DK^{\pm}$, [a] which can be reached through two different quark-level processes (see Fig. 1). The interference between the quark-level processes $b \to c\bar{u}s$ and $b \to u\bar{c}s$

[a] We will refer in the following to the combination of D^0 and $\overline{D}^0$ with the symbol D unless explicitely stated.

(respectively $B^- \to D^0 K^-$ and $B^- \to \bar{D}^0 K^-$), introduces a relative phase γ in the decay amplitude. Here, we will discuss the following measurements: $B^\pm \to D^{(*)} K^{(*)\pm}$ with D^0_{CP} decay modes [b] (GLW method), $B^\pm \to D^{(*)} K^\pm$ with D^0 doubly Cabibbo suppressed (ADS method), and $B^\pm \to D^{(*)} K^{(*)\pm}$ with $D^0 \to$ 3-body ("Dalitz method"). The sensitivity of the different methods to γ depends on the magnitude of the ratio $r_B = |A(b \to u\bar{c}s)|/|A(b \to c\bar{u}s)| \simeq 0.1$ of the $b \to u\bar{c}s$ amplitude with respect to the $b \to c\bar{u}s$ one.

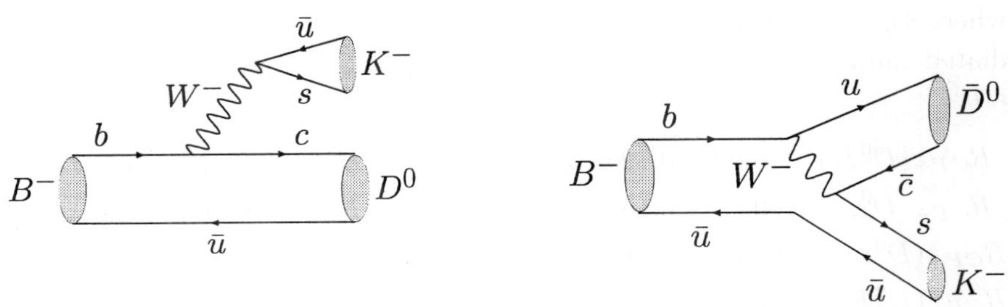

Figure 1. Diagrams contributing to $B^\pm \to DK^\pm$ and related modes. The diagram (a) proceeds via V_{cb} transition, while the diagram (b) proceeds via V_{ub} transition and is color suppressed.

2.1. *The GLW method: $B^\pm \to D^{(*)} K^{(*)\pm}$ with D^0_{CP} Decays*

This method considers the decays $B^\pm \to D^{(*)} K^{(*)\pm}$ where the D decays to a *CP* eigenstate [2] (D^0_{CP}). It takes advantage of the interference into the final common D^0_{CP} eigenstates. The main advantage of this method is that γ can be extracted in a theoretically-clean manner if one reconstructs D^0_{CP}-even and D^0_{CP}-odd decays, however, an 8-fold ambiguity on the value of γ is not resolved.

[b] The $D^0_{CP_+}$ modes are $D^0 \to \pi\pi, KK$ while the $D^0_{CP_-}$ modes are $D^0 \to K^0_S\pi^0, K^0_S\omega, K^0_S\phi$.

The CP observables are:

$$R_{CP\pm} \equiv \frac{\mathcal{B}(B^- \to D^0_{CP\pm}K^-) + \mathcal{B}(B^+ \to D^0_{CP\pm}K^+)}{\mathcal{B}(B^- \to D^0K^-) + \mathcal{B}(B^+ \to \overline{D}^0K^+)} \equiv 1 + {r_B}^2 \pm 2r_B \cos\delta_B \cos\gamma \tag{1}$$

$$A_{CP\pm} \equiv \frac{\mathcal{B}(B^- \to D^0_{CP\pm}K^-) - \mathcal{B}(B^+ \to D^0_{CP\pm}K^+)}{\mathcal{B}(B^- \to D^0_{CP\pm}K^-) + \mathcal{B}(B^+ \to D^0_{CP\pm}K^+)} \equiv \frac{\pm 2\, r_B \sin\delta_B \sin\gamma}{R_{CP\pm}} \tag{2}$$

where δ_B is the strong phase difference between the V_{ub} and the V_{cb} mediated amplitudes. The *BABAR* results [3], using 232 million $B\overline{B}$ pairs [c] are:

$$\begin{aligned}
R_{CP+}(D^0K) &\equiv 0.90 \pm 0.12 \pm 0.04 & A_{CP+}(D^0K) &\equiv 0.35 \pm 0.13 \pm 0.04\\
R_{CP-}(D^0K) &\equiv 0.86 \pm 0.10 \pm 0.05 & A_{CP-}(D^0K) &\equiv -0.06 \pm 0.13 \pm 0.04\\
R_{CP+}(D^0K^*) &\equiv 1.96 \pm 0.40 \pm 0.11 & A_{CP+}(D^0K^*) &\equiv -0.08 \pm 0.19 \pm 0.08\\
R_{CP-}(D^0K^*) &\equiv 0.65 \pm 0.26 \pm 0.08 & A_{CP-}(D^0K^*) &\equiv -0.26 \pm 0.40 \pm 0.12\\
R_{CP+}(D^{*0}K) &\equiv 1.06 \pm 0.26 \pm 0.10 & A_{CP+}(D^{*0}K) &\equiv -0.10 \pm 0.23 \pm 0.04.
\end{aligned}$$

The precision of this measurement alone does not significantly constrain the value of γ, but it can be combined with the other measurements described below.

2.2. *The ADS method: $B^\pm \to D^{(*)}K^{(*)\pm}$ with Double-Cabibbo-Suppressed D Decays)*

The CP asymmetry is potentially larger in modes f such that $D^0 \to f$ is doubly Cabibbo suppressed while $\bar{D^0} \to f$ is Cabibbo allowed [4]. We will consider $f = K^+\pi^-$ in the following. As a result, the two interfering amplitudes become comparable. Explicitly, the direct CP asymmetry is given by

$$A_{ADS} \equiv \frac{\mathcal{B}(B^- \to [K^+\pi^-]_D K^-) - \mathcal{B}(B^+ \to [K^-\pi^+]_D K^+)}{\mathcal{B}(B^- \to [K^-\pi^+]_D K^-) + \mathcal{B}(B^+ \to [K^+\pi^-]_D K^+)} \equiv \frac{2\, r_B r_D \sin\gamma \sin(\delta_B + \delta_D)}{{r_B}^2 + {r_D}^2 + 2r_B r_D \cos\gamma \cos(\delta_B + \delta_D)} \tag{3}$$

[c] For the $D^{*0}K$ mode, the analysis is performed on 123 million $B\overline{B}$ pairs.

where δ_D is the strong phase in the D^0 decay and r_D is the magnitude of the ratio of amplitudes between the Cabibbo-suppressed and Cabibbo-allowed D^0 decays. With a data sample of 232 million $B\overline{B}$ pairs there is no evidence of signal in these modes. These measurements [5] set an upper limit on the r_B values: $r_B(D^0K) < 0.23$ (90% CL), $r_B{}^2(D^{*0}K) < (0.16)^2$ (90% CL), $r_B(D^0K^*) = 0.28^{+0.06}_{-0.10}$.

2.3. *The Dalitz method: $B^\pm \to D^{(*)}K^{(*)\pm}$ Decays with a Dalitz Analysis of $D^0 \to K^0_S\pi^+\pi^-$*

The primary advantage of this method [6] is that it involves the entire resonant structure of the three-body decay, with interference between doubly Cabibbo-suppressed, Cabibbo-allowed and *CP*-eigenstate amplitudes providing the sensitivity to γ. It is the most sensitive method for the extraction of the γ value and it suffers from only a two-fold ambiguity. The extraction of the angle γ is performed through a fit to the Dalitz distribution of the D^0. The $D^0 \to K^0_S\pi^+\pi^-$ amplitude is parametrized as the sum of 13 Breit-Wigner functions representing two-body D^0 decays, plus a constant term representing the non-resonant amplitude. The parameters of the model are extracted from a fit to a highly pure $D^0 \to K^0_S\pi^+\pi^-$sample selected from D^*-tagged events. A frequentist analysis of the *CP* measured quantities with the same data sample of 232 million $B\overline{B}$ pairs yields [7] $\gamma = (67 \pm 28 \pm 13 \pm 11)^\circ$. The first error is statistical, the second systematic and the third is due to the parametrization of the $D^0 \to K^0_S\pi^+\pi^-$decay amplitude.

3. Alternative Paths to γ

Alternative methods have been explored to better constrain the angle γ. So far, there is no evidence of new methods with better sensitivity with the available statistics.

3.1. *$B^0 \to D^{(*)0}K^{(*)0}$, $B^0 \to D^0K^-\pi^+$, $B^0 \to D_S^{(*)+}a^-_{0,2}$ Decay Modes*

In principle it is possible to measure $\sin(2\beta + \gamma)$ with a time-dependent analysis which exploits the interference of decays with and without mixing [8]. In $B^0 \to D^{(*)0}K^{(*)0}$ decays, we expect a larger r_B value since the V_{ub}- and V_{cb}-mediated amplitudes are both color suppressed. Using 226 million $B\overline{B}$ pairs, we have measured the branching ratio of the following decay modes [9]: $\mathcal{B}(B \to D^0\overline{K}^0) = (5.3 \pm 0.7 \pm 0.3) \cdot 10^{-5}$, $\mathcal{B}(B \to D^{*0}\overline{K}^0) = (3.6 \pm 1.2 \pm 0.3) \cdot 10^{-5}$, $\mathcal{B}(\overline{B}^0 \to D^0\overline{K}^{*0}) = (4.0 \pm 0.7 \pm 0.3) \cdot 10^{-5}$, $\mathcal{B}(\overline{B}^0 \to \overline{D}^0\overline{K}^{*0}) < 1.1 \cdot 10^{-5}$ (90 % CL). The latter measurements set an upper limit on the value of $r_B < 0.4$ (90% CL) for the $\overline{B}^0 \to D^0\overline{K}^{*0}$ decay mode, smaller than the naive theoretical expectations.

Larger *CP* violation effects are possible also in the $B^0 \to D^0K^-\pi^+$ decay modes. In this case, color-allowed diagrams are present in the V_{ub}-mediated transition. We have measured the branching ratio of the following decay modes: $\mathcal{B}(B^0 \to \overline{D}^0K^+\pi^-) = (88 \pm 15 \pm 9) \cdot 10^{-6}$, $\mathcal{B}(B^0 \to D^0K^+\pi^-) < 19 \cdot 10^{-6}$ (90 % CL). There is no evidence of the V_{ub}-mediated process and we constrain the *CP* asymmetry to be smaller than expected. In principle, it is possible to have large *CP* asymmetry even in the decay modes $B^0 \to D^{(*)+}a^-_{0,2}$. We have performed a search for decays related by SU(3), but found no evidence for signal. We set the following upper limits: $\mathcal{B}(B^0 \to D^+_S a^-_0) < 1.9 \cdot 10^{-5}$, $\mathcal{B}(B^0 \to D^+_S a^-_2) < 19 \cdot 10^{-5}$, $\mathcal{B}(B^0 \to D^{*+}_S a^-_0) < 3.6 \cdot 10^{-5}$, $\mathcal{B}(B^0 \to D^{*+}_S a^-_2) < 20 \cdot 10^{-5}$ (90 % CL). These measurements reduce the theoretical expectations for the *CP* asymmetry in $B^0 \to D^{(*)+}a^-_{0,2}$ decay modes and therefore the interest for the γ measurement with these modes.

4. Conclusions

We have presented the status of the γ related measurement in *BABAR* for the GLW, ADS and Dalitz methods. A Bayesian combination of these measurements [14] yields: $\gamma = (73 \pm 29)^\circ \cup (-107 \pm 29)^\circ$. The precision of these measurements is very sensitive to the value of r_B, which is not well determined. Assuming an average r_B=0.1 for the $B^\pm \to D^{(*)}K^{(*)\pm}$ modes, it is expected to measure the angle γ with an error of $< 10^\circ$ with an integrated data sample of 1 ab^{-1}. New methods and decay modes have been explored, but contibute relatively little to the sensitivity to γ.

Acknowledgments

I wish to thank SLAC for the kind hospitality and Istituto Nazionale di Fisica Nucleare (INFN) for the support.

References

1. N. Cabibbo, *Phys. ReV. Lett.* **10**, 531 (1963); M. Kobayashi and T. Maskawa, *Prog. Theor. Phys.* **49**, 652 (1973).

2. M. Gronau and D.London, *Phys. Lett.* **B253**, 483 (1991); M. Gronau and D. Wyler, *Phys. Lett.* **B265**, 172 (1991).
3. B. Aubert *et al.*, *Phys. Rev. Lett.* **73**, 051105 (2006); B. Aubert *et al.*, *Phys. Rev. Lett.* **72**, 071103 (2005); B. Aubert *et al.*, *Phys. Rev. D* **71**, 031102 (2005)
4. D. Atwood, I. Dunietz and A. Soni, *Phys. Rev. Lett.* **78**, 3257 (1997).
5. B. Aubert *et al.*, *Phys. Rev. D* **72**, 032004 (2005); B. Aubert *et al.*, *Phys. Rev. D* **72**, 071104 (2005)
6. A. Giri, Yu. Grossman, A. Scffer and J. Zupan, *Phys. Rev. D* **68**, 054018 (2003).
7. B. Aubert *et al.*, *Phys. Rev. Lett.* **95**, 121802 (2005); B. Aubert *et al.*,arXiv:hep-ex/0507101.
8. B. Kayser, D. London, *Phys. Rev. D* **61**, 116013 (2001).
9. B. Aubert *et al.*,arXiv:hep-ex/0604016.
10. R. Aleksan, T. Petersen, *Phys. Rev. D* **67**, 096002 (2003); M. Gronau, *Phys. Lett.* **B557**, 198 (2003).
11. B. Aubert *et al.*, *Phys. Rev. Lett.* **96**, 011803 (2006);
12. M. Diehl, G. Hiller, *Phys. Lett.* **B517**, 125 (2001); M. Diehl, G. Hiller, *JHEP* **0106**, 067 (2001); C.S. Kim, J.P. Lee, S. Oh, *Phys. Rev. D* **67**, 014011 (2003).
13. B. Aubert *et al.*,arXiv:hep-ex/0512031.
14. M. Ciuchini et al., *JHEP 0107* **013** (2001) (hep-ph/0012308). for the updates see the web page : http:\\www.utfit.org

SEARCH FOR NEW PHENOMENA AT THE DØ EXPERIMENT

C. NOEDING

University of Freiburg,
Institute of Physics,
Hermann-Herder-Str. 3,
79104 Freiburg, Germany
E-mail: noeding@fnal.gov

Searches for New Physics beyond the Standard Model have been performed with the DØ detector at the Tevatron $p\bar{p}$ collider (Fermi National Accelerator Laboratory) at $\sqrt{s} = 1.96$ TeV. Using a Run II data set of ~ 350 pb^{-1}, various analyses have covered a multitude of signatures, reaching sensitivity in parameter regions well beyond existing limits. No significant deviation from the Standard Model has been observed while searching for supersymmetry, leptoquarks and neutral long-lived particles.

1. Introduction

The Standard Model describes successfully all phenomena observed in collision experiments at current energy scales. However there are various indications that it is only an effective low-energy theory of a more fundamental theory: the hierarchy between the electroweak scale and the Planck scale, the evolution of the coupling constants and the pending integration of gravity.

In the following, a brief overview of recent results from DØ in Run II is given.[1] All reported cross section limits are derived at 95% C.L. A detailed description of the DØ detector can be found elsewhere.[2]

2. Search for Associated Chargino/Neutralino Production

Due to the small background from Standard Model processes, the associated production of chargino/neutralino with subsequent decay into three charged leptons is a promising channel for SUSY discovery at the Tevatron. Assuming R-parity, the lightest neutralino is stable and escapes detection, leading to large $\not\!\!E_T$. Using 320 pb^{-1}, four different selections are defined:

$ee\ell$, $\mu\mu\ell$, $e\mu\ell$ and like-sign $\mu\mu$ To maximize the signal yield, the third lepton is identified as an isolated track. The optimization of the selection is based on mSUGRA signals with low slepton masses and low chargino and neutralino masses at $\tan\beta = 3$. Combining all four analyses, the expected Standard Model background sums up to $2.93 \pm 0.54(\text{stat}) \pm 0.64(\text{sys})$ events, while 3 events remain in data. An upper limit on $\sigma \times \text{BR}(3\ell)$ is extracted and compared with three different benchmark scenarios. The resulting chargino mass limit (see Figure 1 (left)) extends the LEP limits of 103 GeV for SUSY scenarios with enhanced leptonic branching ratios (3l-max).

3. Search for Squarks and Gluinos in Jets and $\not\!\!E_T$ Topology

The most copiously produced supersymmetric particles in hadron collisions should be – if sufficiently light – squarks ($\tilde{q}$) and gluinos ($\tilde{g}$). Assuming R-parity, the particles are produced in pairs and the event topology consists of multijet events with significant $\not\!\!E_T$ stemming from the LSP. The search uses 310 pb^{-1} of data, and the selection criteria are optimized in three benchmark scenarios using the scalar sum of all jets (H_T) and $\not\!\!E_T$ to minimize the expected upper limit on the cross section. In the mass region $m_{\tilde{g}} > m_{\tilde{q}}$, di-jet events are selected with $H_T > 250$ GeV and $\not\!\!E_T > 175$ GeV. Three jet events with $H_T > 325$ GeV and $\not\!\!E_T > 100$ GeV cover the mass region $m_{\tilde{g}} \sim m_{\tilde{q}}$, while four jet events with $H_T > 250$ GeV and $\not\!\!E_T > 75$ GeV cover $m_{\tilde{g}} < m_{\tilde{q}}$. Limits are set for an mSUGRA model with $\tan\beta = 3$, $A_0 = 0$ and $\mu < 0$ and are shown in Figure 1 (right).

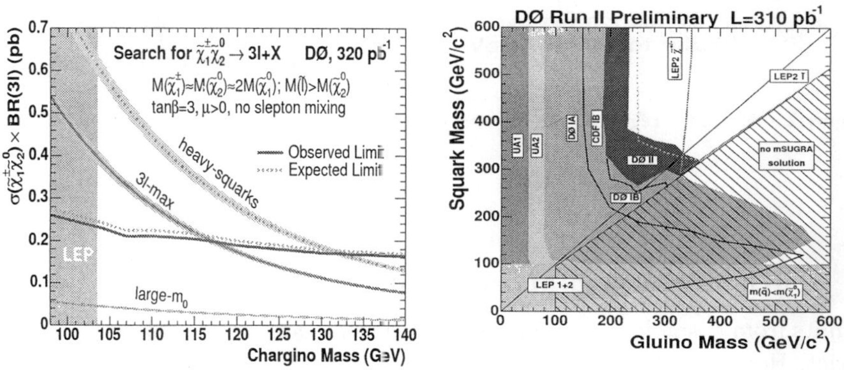

Figure 1. *Left:* Resulting limit on $\sigma \times \text{BR}(3\ell)$ as a function of the chargino mass from the search for associated production of chargino/neutralino. *Right:* Excluded region in the $(m_{\tilde{g}}, m_{\tilde{q}})$-plane from the search for squarks and gluinos.

4. Search for Direct Production of Scalar Top and Bottom Quarks

Large Higgs Yukawa couplings to the third quark generation induce a strong mixing between the supersymmetric partners of the two chirality states of the top and bottom quark, resulting in two different physical states: $\tilde{t}_1$ ($\tilde{b}_1$) and $\tilde{t}_2$ ($\tilde{b}_2$). Therefore the lightest scalar top quark ($\tilde{t}_1$) might be the lightest squark. At large values of $\tan\beta$, a light $\tilde{b}_1$ is also expected.

A combination of two analyses which search for the scalar top in the $e^{\pm}\mu^{\mp}b\bar{b}+\not{E}_T$ ($e\mu$ selection) and $\mu^+\mu^-b\bar{b}+\not{E}_T$ ($\mu\mu$ selection) final states from the $\tilde{t}_1$ pair production is performed using 339 pb^{-1} ($\mu\mu$), resp. 350 pb^{-1} ($e\mu$). This result supersedes the result shown at the conference. The excluded region in the $(m_{\tilde{t}}, m_{\tilde{\nu}})$-plane is shown in Figure 2 (left). Using 310 pb^{-1}, the sbottom pair production with subsequent decays $\tilde{b} \to b\tilde{\chi}_1^0$ is examined. The resulting exclusion regions in the $(m_{\tilde{b}}, m_{\tilde{\chi}^0})$ are shown in Figure 2 (right).

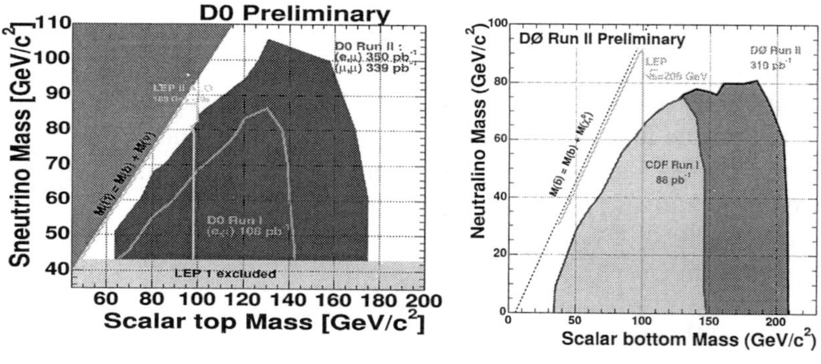

Figure 2. Excluded region in the $(m_{\tilde{t}}, m_{\tilde{\nu}})$-plane *(left)* and the $(m_{\tilde{b}}, m_{\tilde{\chi}^0})$-plane *(right)*.

5. Search for Charged Massive Stable Particles (CMSP)

Limits from cosmology on new particles that are absolutely stable are quite strict. However, these restrictions do not apply to particles that decay outside the detector region. The detector signal of such a CMSP resembles a muon signal. Using timing information from the muon system, the relatively slow moving CMSPs can be distinguished from real muons. DØ has performed a search for the pair production of CMSPs using 390 pb^{-1}. In models with anomaly-mediated supersymmetry breaking, the lifetime of the lightest chargino can be long enough to escape the detector if the mass difference between the lightest chargino and the lightest neutralino is small. In the present analysis, mass limits for a higgsino-like chargino and a gaugino-like chargino can be set at 145 GeV and 175 GeV.

6. Search for Neutral Long-Lived Particles (NLLP)

DØ has searched for the pair production of NLLPs using 383 pb^{-1}. The NLLP is assumed to have a mass as low as several GeV and a decay length in the order of a few centimeters, before decaying to two muons and a neutrino. As a theoretical model, the MSSM with R-parity violating decays of neutralinos is used, where the RPV couplings are expected to be small resulting in long lifetimes. Studies of the decay $K_s^0 \to \pi^+\pi^-$ are performed to demonstrate the ability to reconstruct long-lived particles up to a radius of 20 cm with reasonable efficiency. The DØ result is compared to the NuTeV result (Figure 3), and it improves on the NuTeV limit by several orders of magnitude at long lifetimes and adds coverage at lower lifetimes.

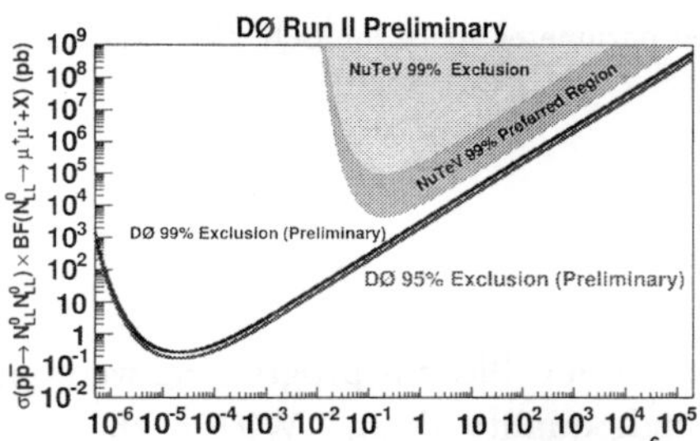

Figure 3. Limit on $\sigma \times$ BR for pair-production of NLLPs as a function of lifetime. The NuTeV exclusion has been converted to a $p\bar{p}$ cross section at $\sqrt{s} = 1.96$ TeV.

7. Search for First and Second Generation Leptoquarks

At the Tevatron, leptoquarks would be produced in pairs through $q\bar{q}$ annihilation and gluon fusion. A search for first generation leptoquarks in the $ejej$ and the $ej\nu j$ channels based on 252 pb^{-1} shows no excess of data over Standard Model background. The limits on the cross sections are translated into lower mass limits on LQ_1 as a function of the branching ratio into eq (see Figure 4 (left)). Using data corresponding to an integrated luminosity of 294 pb^{-1}, DØ has searched for second generation leptoquarks. Assuming a branching ratio of $BR(LQ_2 \to \mu q) = 1$, the final state consists of two muons and two quarks. Since no excess in data is found, lower limits on $m(LQ_2)$ are determined as a function of the branching ratio into μq. The results of a complementary Run I analysis in the $\mu j \nu j$ channel are combined with the current results, and the exclusion is shown in (Figure 4 (right)). Both LQ_1 and LQ_2 searches result in the most stringent limits on leptoquarks from direct measurements for large branching ratios.

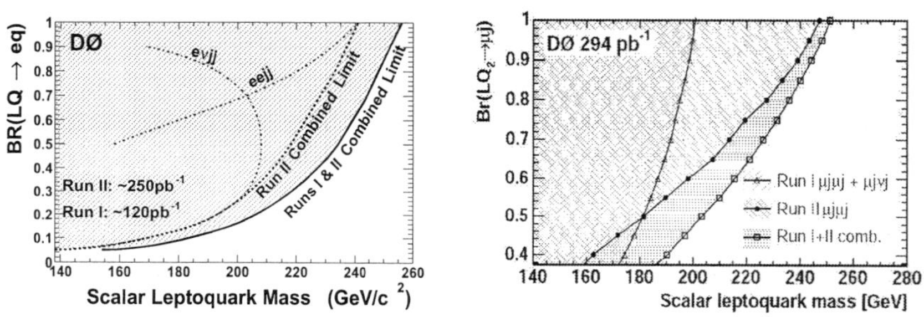

Figure 4. Excluded parameter space for LQ_1 *(left)* and LQ_2 *(right)* searches.

8. Summary

The Run II of the Tevatron collider is progressing well, and DØ is collecting a large amount of high quality data. Within the analyzed data set of 350 pb^{-1}, no deviation from the Standard Model expectation has been found. A number of restrictive upper limits on new physics beyond the Standard Model has been derived, which are the most constraining to date. Currently 1 fb^{-1} of data is being analyzed, which should lead to substantial improvements of the current limits if no deviation is found.

References

1. `http://www-d0.fnal.gov/Run2Physics/WWW/results/np.htm`
2. DØ Collaboration, V. Abazov *et al.*, physics/0507191, submitted to Nucl. Instr. Methods in Phys. Res. A (2005).

CHARM PHYSICS AT *BaBar*

A. OYANGUREN
BaBar Collaboration

LAL-Orsay, France
E-mail: oyangur@lal.in2p3.fr

With the large amount of data recorded by the *BABAR* experiment at the $\Upsilon(4S)$ energy, and in particular, using charm pairs coming from the continuum, a diversity of charm physics analyses can be addressed. In this talk, recent results on the Dalitz plot analysis of the $D^0 \to K_s K^+ K^-$ decay, and on charm baryon properties are presented.

1. Introduction

Thanks to the high luminosity reached by the PEP-II machine, who has delivered an integrated luminosity of 340 fb^{-1}, and to the high $e^+e^- \to c\bar{c}$ cross section at the $\Upsilon(4S)$ energy, the *BaBar* experiment is, in addition to a B-factory, an excellent laboratory to study the production and decay properties of charm hadrons. With more than 850 million charm decays, measurements of the D-mixing, Dalitz plots, leptonic and semileptonic charm decays, rare decays, charm baryons and new D_{sJ} states are ongoing at present. In the following, a small content of this vast program is described. It concerns results on the Dalitz plot analysis of the $D^0 \to K_s K^+ K^-$ decay channel, the measurement of the Λ_c mass and the study of the Ω_c^0 baryon properties.

2. Dalitz plot of the $D^0 \to K_s K^+ K^-$ decay

The light-quark scalar sector in the Standard Model is not well understood. Controversy exists, for instance, on the existence of the $\kappa(800)$ and $\sigma(500)$ resonances [1]. This is due to the fact that scalar states are difficult to isolate since they are broad, have similar masses, and are usually near or below the decay threshold. Theory predictions of the light scalar mesons involve $q\bar{q}$ nonets, glueballs and $4q$ states near the $K\bar{K}$ threshold [2]. A detailed study of the $K\bar{K}$ system has been performed at the *BABAR* experiment through Dalitz plot and partial wave analyses of the $D^0 \to K_s K^+ K^-$ decay channel [3]. It aims to clarify the scalar sector composition. $D^0 \to K_s K^+ K^-$ (with the K_s into $\pi^+\pi^-$) and $D^0 \to K_s \pi^+\pi^-$ decay channels have been reconstructed using 91 fb^{-1} of *BABAR* data. D^{*+} mesons decaying into $D^0\pi^+$ are used to determine the D flavour[a] and, using the $m_{D^*} - m_D$ mass difference, to reduce the background. The Dalitz plot for the $D^0 \to K_s K^+ K^-$ decay channel is shown in Fig. 1. From the selected events, and using the efficiencies obtained from MC, the ratio between the decay partial width of the two channels has been measured: $\Gamma(D^0 \to K_s K^+ K^-)/\Gamma(D^0 \to K_s \pi^+\pi^-)$=$(15.8 \pm 0.1 \pm 0.5)\%$.

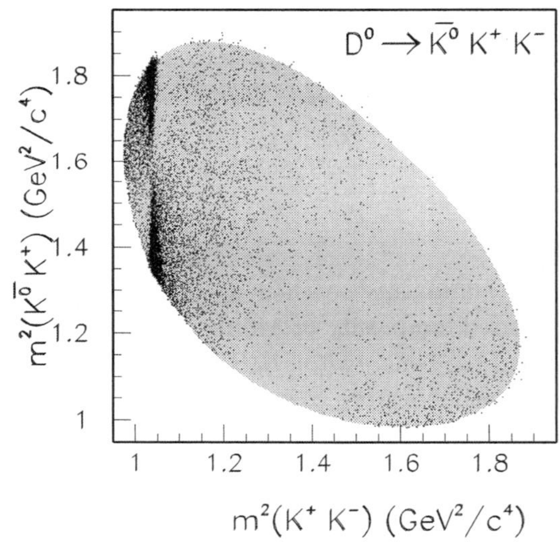

Figure 1. Dalitz plot of the $D^0 \to K_s K^+ K^-$ decay channel.

A partial wave analysis of the $D^0 \to K_s K^+ K^-$ decay channel has been performed to separate the contribution of the scalar ($a_0(980)/f_0(980)$) and vector ($\phi(1020)$) components. Events have been reweighted, using the helicity angle, θ_K, by the efficiency corrected spherical harmonics. Assuming a relativistic P-wave Breit-Wigner, the mass and width of the $\phi(1020)$ have been fitted. Considering, in the K^+K^- mass, only the scalar contribution of the $a_0(980)^0$, it has been described as a coupled Breit-Wigner to the $\bar{K}K$

[a] neglecting the D^0 double-Cabibbo-suppressed decays

and $\eta\pi$ channels since its mass is close to the $\bar{K}K$ threshold. The coupling to the $\bar{K}K$ channel has been obtained: $g_{\bar{K}K} = 473 \pm 29 \pm 40\,\mathrm{MeV}^{1/2}$. This value is larger than previous measurements [4]. A strong S-P interference has also been observed in the $\phi(1020)$ mass region. Performing an unbinned likelihood fit of the Dalitz plot, the relative amplitudes and phases for each component have been obtained. The main contributions have been found to be from the $D^0 \to K^0 a_0(980)$ (66%), $D^0 \to K^0\phi(1020)$ (46%) and $D^0 \to K^- a_0(980)^+$ (13%). The $f_0(980)^0$ contribution has been found to be small. The remaining contribution is compatible with the tail of a broad resonance (for instance the $f_0(1400)$).

3. Charm baryons

With the large data sample registered by the *BABAR* detector, and thanks to the good vertexing, tracking reconstruction and particle identification, very precise measurements of charm baryon characteristics, production and decay properties can be performed. In the following results on the Λ_c and Ω_c baryons are presented.

3.1. *Measurement of the Λ_c baryon mass*

The mass of the Λ_c baryon has been determined using 232 fb^{-1} of *BABAR* data [5]. This measurement is performed by exclusively reconstructing the $\Lambda_c^+ \to \Lambda K_s^0 K^+$ and $\Lambda_c^+ \to \Sigma K_s^0 K^+$ decay channels, which have the advantage of smaller Q value[b], and thus smaller uncertainty in the reconstructed mass, as compared to other channels. Fig. 2 shows the reconstructed mass for the $\Lambda_c^+ \to \Lambda K_s^0 K^+$ decay channel. In addition, Λ_c^+ decays into $pK^-\pi^+$ and pK_s^0 are used as control samples to evaluate systematic uncertainties. The main source of systematics has been found to come from energy-loss of particles in the tracking system. Combining the results of the fitted reconstructed mass for the $\Lambda_c^+ \to \Lambda K_s^0 K^+$ and $\Lambda_c^+ \to \Sigma K_s^0 K^+$ decay channels, the most precise value has been determined: $m_{\Lambda_c^+} = 2286.46 \pm 0.14\,\mathrm{MeV/c}^2$.

3.2. *Production and decay of the Ω_c baryon*

Production and decay characteristics of the Ω_c baryon have been studied with 230 fb^{-1} of *BABAR* data [6]. This baryon has the particular property

---

[b] $Q = m(\Lambda_c) - \sum_i m(x_i)$ with summation over the decay products.

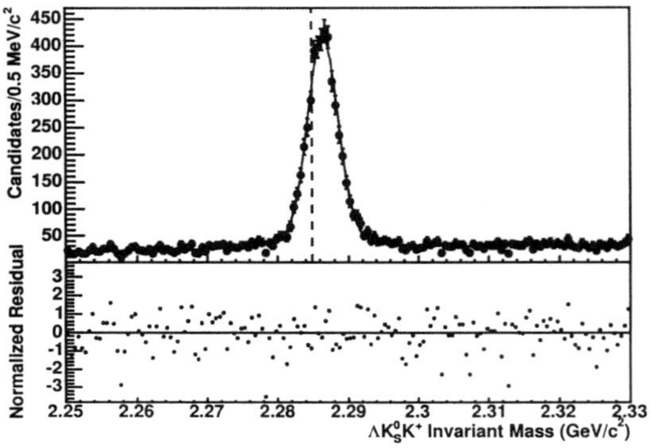

Figure 2. Mass distribution of the $\Lambda_c^+ \to \Lambda K_s^0 K^+$ decay channel. A fit using two common-mean Gaussians for the signal and a linear function for the background is superimposed. The dashed line indicates the present PDG value, which is significantly lower.

of a short lifetime, which is expected to come from a constructive Pauli interference between the produced and the two spectator strange quarks. The Ω_c has been reconstructed in the $\Omega^-\pi^+$, $\Omega^-\pi^+\pi^-\pi^+$ and $\Xi^-K^-\pi^+\pi^+$ decay channels. A clear signal is observed for the first channel in Fig. 3 (left).

Branching ratios relative to the $\Omega^-\pi^+$ channel have been found to be $0.31 \pm 0.15 \pm 0.04$ for the $\Xi^-K^-\pi^+\pi^+$ decay, and less than 0.30 at 90%CL for the $\Omega^-\pi^+\pi^-\pi^+$ decay channel. From on-peak and off-peak data, signal $\Omega_c \to \Omega^-\pi^+$ candidates have been fitted in different bins of momentum to study the Ω_c production mechanism. As it is shown in Fig. 3 (right), a peak at high $p^*(\Omega_c)$ momentum is compatible with a production rate of 10^{-3} from continuum events. The signal at low momentum shows a clear evidence for Ω_c production from $B\bar{B}$ events.

4. Summary

With the large statistics recorded by the BABAR experiment and its good performance, precise measurements of charm physics can be achieved. Examples of such analyses are the first significant Dalitz plot and partial wave analyses of the $D^0 \to K_sK^+K^-$ decay, and accurate measurements of charm baryon production and decay properties.

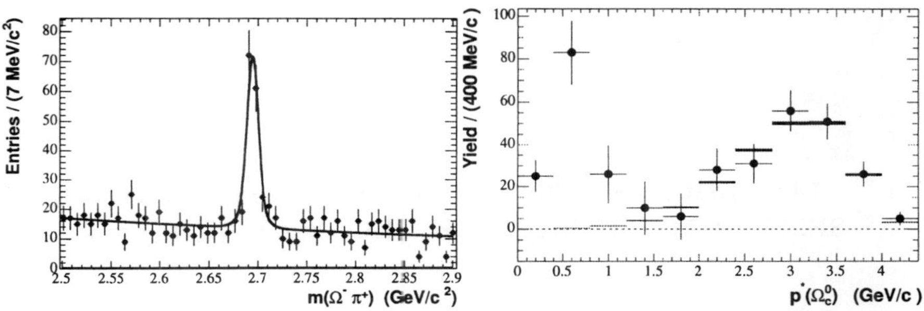

Figure 3. Left: Ω_c mass distribution for the $\Omega^-\pi^+$ decay channel. Right: Signal yield as function of the Ω_c momentum in the CMS (not corrected by efficiency). Data at high momentum are compatible with events coming from the continuum, as solid lines indicate. The signal at low momentum shows a clear evidence of Ω_c production from $B\bar{B}$ events.

Acknowledgments

We are grateful for the excellent luminosity and machine conditions provided by our PEP-II colleagues, and for the substantial dedicated effort from the computing organizations that support *BABAR* . The collaborating institutions wish to thank SLAC for its support and kind hospitality. This work is supported by DOE and NSF (USA), NSERC (Canada), IHEP (China), CEA and CNRS-IN2P3 (France), BMBF and DFG (Germany), INFN (Italy), FOM (The Netherlands), NFR (Norway), MIST (Russia), and PPARC (United Kingdom). Individuals have received support from the A. P. Sloan Foundation, Research Corporation, and Alexander von Humboldt Foundation.

References

1. E. M. Aitala *et al.* [E791 Collaboration], Phys. Rev. Lett. **89** (2002) 121801 [arXiv:hep-ex/0204018].
2. F. E. Close and N. A. Tornqvist, J. Phys. G **28** (2002) R249 [arXiv:hep-ph/0204205].
3. B. Aubert *et al.* [BABAR Collaboration], Phys. Rev. D **72** (2005) 052008 [arXiv:hep-ex/0507026].
4. A. Abele *et al*, Phys. Rev. **D57** (1998) 3860.
5. B. Aubert *et al.* [BABAR Collaboration], Phys. Rev. D **72** (2005) 052006 [arXiv:hep-ex/0507009].
6. B. Aubert *et al.* [BABAR Collaboration], arXiv:hep-ex/0507011.

MASS-DEPENDENT $\mathcal{O}(\alpha_S^2)$ CORRECTIONS TO SEMILEPTONIC B-QUARK DECAY AT $M(\ell\bar{\nu}_\ell) = 0$

ALEXEY PAK

Department of Physics, University of Alberta, Edmonton AB T6G 2J1, Canada

Two-gluon radiative corrections to the decay rate $b \to c\ell\bar{\nu}$ in the kinematical limit $m(\ell\bar{\nu}) = 0$ have been calculated as an expansion around $\frac{m_c}{m_b} = 0$. In combination with known expansion around $\frac{m_c}{m_b} = 1$, these results describe the corrections in the whole range of final quark mass with accuracy much better than 1%.

1. Introduction

Presently the precision of theoretical description of b-quark decays into c-quark is limited by the complexity of perturbative calculations [1]. Together with precision measurements by B-factories, multiloop results may lead to more accurate determination of fundamental parameters of the Standard Model, such as $|V_{cb}|$. However, the same hierarchy of b and c-quark masses which simplifies non-perturbative contributions, makes multiloop calculations very challenging. In this paper we demonstrate some techniques of dealing with non-zero mass of c-quark on the example of decay $b \to c\ell\bar{\nu}_\ell$ rate in the kinematic limit of $m(W^*) = m(\ell\bar{\nu}_\ell) = 0$ and discuss the prospects for complete $\Gamma(b \to c\ell\bar{\nu}_\ell)$ calculation.

Differential rate $d\Gamma(Q \to qW^* \to q\ell\bar{\nu}_\ell)/dm^2(W^*)$ at $m^2(W^*) = 0$ by only a constant factor differs from the decay rate $Q \to qW^*$ with $m(W^*) = 0$. Two-gluon corrections to the rate of decay $Q \to qW^*$ have been studied in various kinematic limits [2], [3], [4]. Here we present the calculation of $\mathcal{O}(\alpha_s^2)$ corrections to this process as an expansion in terms of small parameter $\frac{m_c}{m_b} \ll 1$ and combine the results with expansion [3] in terms of $1 - \frac{m_c}{m_b} \ll 1$.

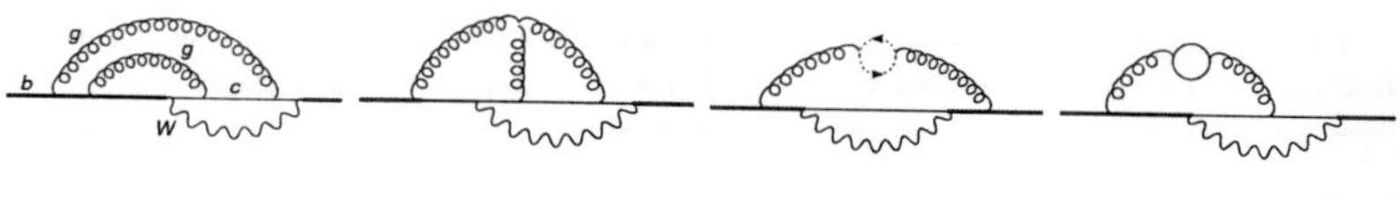

Figure 1. Examples of diagrams contributing to $\mathcal{O}(\alpha_s^2)$ corrections to b-decay

2. $\mathcal{O}(\alpha_s^2)$ corrections to $b \to cW^*$ decay width.

The rate of $b \to cW^*$ decay expanded in the strong coupling constant α_s and mass ratio $\rho = \frac{m_c}{m_b}$ may be parameterized as follows:

$$\Gamma(b \to cW^*) = \Gamma_0 \left[X_0 + C_F \frac{\alpha_s}{\pi} X_1 + C_F \left(\frac{\alpha_s}{\pi}\right)^2 X_2 + \mathcal{O}(\alpha_s^3) \right], \tag{1}$$

$$\Gamma_0 = \frac{G_F |V_{cb}|^2 m_b^3}{8\sqrt{2}\pi}, \quad X_0 = 1 - 3\rho^2 + 3\rho^4 - \rho^6,$$

$$X_1 = \frac{5}{4} - \frac{\pi^2}{3} + \left[-\frac{11}{4} + \frac{\pi^2}{3} - 9\ln\rho \right]\rho^2 + \left[\frac{1}{4} + \frac{\pi^2}{3} + 6\ln\rho \right]\rho^4$$
$$+ \left[\frac{5}{36} - \frac{\pi^2}{3} - \frac{5}{3}\ln\rho \right]\rho^6 + \left[\frac{65}{72} - \frac{5}{6}\ln\rho \right]\rho^8 + \mathcal{O}(\rho^{10}),$$

$$\text{and} \quad X_2 = T_R N_L X_L + T_R N_H X_H + T_R N_C X_C + C_F X_A + C_A X_{NA}.$$

Here N_L represents a number of massless quarks (3 in QCD), and $N_H(N_C)$ label the contributions of $b(c)$-quarks. Top quark contribution is suppressed by $\left(\frac{m_b}{m_t}\right)^2$ and we neglect it. In $SU(3)$, the color factors take values $T_R = \frac{1}{2}$, $C_F = \frac{4}{3}$, and $C_A = 3$.

2.1. *Calculation*

Taking advantage of the well-known optical theorem, for X_L, X_H, X_C, X_A, and X_{NA} of Eq. (1) we need to evaluate 39 propagator-type three-loop diagrams, such as on Fig. 1, and 19 one- and two-loop renormalization contributions. To treat UV and IR divergencies, we use dimensional regularization with dimension $D = 4 - 2\epsilon$.

Contributing diagrams depend on two scales: m_b and m_c. To reduce the problem to single-scale integrals, we apply the so-called "asymptotic operation" [5]. Table 1 presents an example of asymptotic operation applied to a two-loop topology. In each region loop momenta are either "hard" ($|k| \gg m_c$) or "soft" ($|k| \sim m_c$), and propagators may be Taylor expanded,

Table 1. Example of expansion of a double-scale topology. Dotted lines represent massless, thin solid lines – soft scale, and thick solid lines – hard scale propagators.

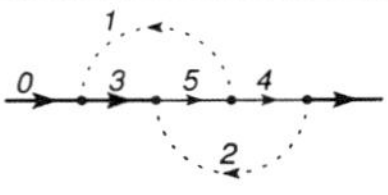	$I(a_1,a_2,a_3,a_4,a_5) = \int \frac{[d^D k_1][d^D k_2]}{[1]^{a_1}[2]^{a_2}[3]^{a_3}[4]^{a_4}[5]^{a_5}}$, $[1]=(k_1-k_2)^2$, $[2]=(k_2-p)^2$, $[3]=(p+k_1-k_2)^2+m_b^2$, $[4]=k_2^2+m_c^2$, $[5]=k_1^2+m_c^2$, $(p^2=-m_b^2)$
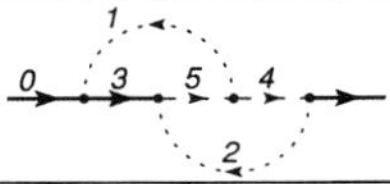	Region 1: $\mathbf{\lvert k_1\rvert, \lvert k_2\rvert \gg m_c}$ $[4]\to k_1^2$, $[5]\to k_2^2$ (two-loop single-scale topology)
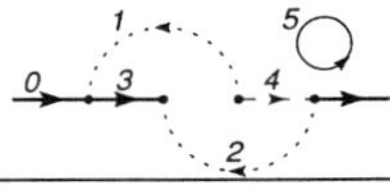	Region 2: $\mathbf{\lvert k_2\rvert \gg m_c, \lvert k_1\rvert \sim m_c}$ $[1]\to k_2^2$, $[3]\to (p-k_2)^2+m_b^2$, $[5]\to k_1^2$ (vacuum bubble × one-loop topology)
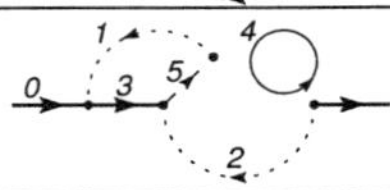	Region 3: $\mathbf{\lvert k_1\rvert \gg m_c, \lvert k_2\rvert \sim m_c}$ $[1]\to k_1^2$, $[2]\to p^2$, $[3]\to (p+k_1)^2+m_b^2$, $[5]\to k_1^2$ (purely real topology)
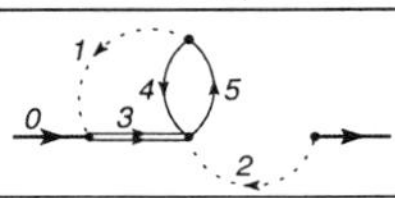	Region 4: $\mathbf{\lvert k_1\rvert, \lvert k_2\rvert \sim m_c}$ $[2]\to p^2$, $[3]\to 2p(k_1-k_2)+i0$ (purely real eikonal topology)

for example:

$$|k_1| \sim m_c,\ |k_2| \gg m_c,\ \frac{1}{(p+k_1-k_2)^2+m_b^2} = \sum_{n=0}^{\infty} \frac{(2k_1k_2-2pk_1-k_1^2)^n}{(k_2^2-2pk_2)^{n+1}} \tag{2}$$

This algorithm generally produces a large number of integrals with various powers of denominator factors. Each of them may be reduced to a combination of a small set of "master integrals" using the so-called integration-by-parts identities. For an arbitrary integrand $I(a_1,...)$ with propagator powers $a_1,...$, integral of D-dimensional divergence should vanish, i.e. $\frac{d}{dp_i^\mu}\left(p_j^\mu I\right) = 0$. Such relations for loop momenta p_i and loop/external momenta p_j may be expressed as combinaions of integrals with different propagator powers: $\sum c(a_1,...,\epsilon) I(a_1 \pm 1,...) = 0$. In some cases, these relations may be used to construct general formulas, reducing the integrals to simpler topologies or to master integrals. For this calculation, reduction of one-, two-, and three-loop topologies has been performed with FORM [6].

To calculate higher terms of m_c expansion, which contain the highest powers of denominator factors, we employed the algorithm [7] of automated reduction done by specialized computer algebra system based on GiNaC [8] library. To obtain terms up to $\left(\frac{m_c}{m_b}\right)^9$, reduction took about 10 days on four-Opteron system.

3. Results

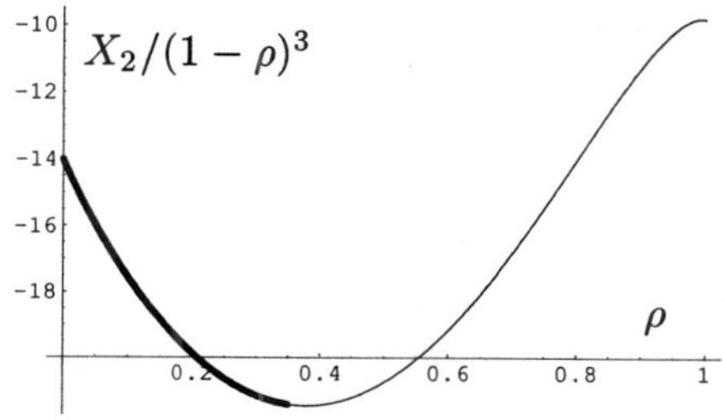

Figure 2. Correction X_2 of Eq. (1) as combination of expansions around $\rho = 0$ (thick line) and $\rho = 1$ (thin line).

Our results for the contributions to Eq. (1) are obtained as series in $\rho = \frac{m_c}{m_b}$. We have obtained terms up to ρ^9 of the expansions and present here contributions to X_2 through terms of order ρ^3:

$$X_L = -\frac{4}{9} + \frac{23\pi^2}{108} + \zeta_3 + \left[\frac{28}{9} - \frac{101\pi^2}{108} - \zeta_3 + \frac{13}{2}\ln\rho - 3\ln^2\rho\right]\rho^2, \quad (3)$$

$$X_H = \frac{12991}{1296} - \frac{53\pi^2}{54} - \frac{\zeta_3}{3} + \left[-\frac{135467}{6480} + \frac{113\pi^2}{54} + \frac{\zeta_3}{3}\right]\rho^2, \quad (4)$$

$$X_C = -\frac{4}{9} + \frac{23\pi^2}{108} + \zeta_3 - \frac{3\pi^2}{4}\rho + \left[\frac{65}{18} + \frac{133\pi^2}{108} - \zeta_3 + \frac{13}{2}\ln\rho - 3\ln^2\rho\right]\rho^2 - \frac{25\pi^2}{18}\rho^3, \quad (5)$$

$$X_A = 5 - \frac{119\pi^2}{48} - \frac{53}{8}\zeta_3 - \frac{11\pi^4}{720} + \frac{19\pi^2}{4}\ln 2 + \left[-\frac{315}{8} + \frac{497\pi^2}{48} + \frac{151}{8}\zeta_3 + \frac{37\pi^4}{360} - \frac{57\pi^2}{4}\ln 2 + \left(\pi^2 - \frac{75}{8}\right)\ln\rho - \frac{27}{2}\ln^2\rho\right]\rho^2 - \frac{4\pi^2}{3}\rho^3, \quad (6)$$

$$X_{NA} = \frac{521}{576} + \frac{505\pi^2}{864} + \frac{9}{16}\zeta_3 + \frac{11\pi^4}{1440} - \frac{19\pi^2}{8}\ln 2 + \left[-\frac{2315}{576} - \frac{2119\pi^2}{864} - \frac{107}{16}\zeta_3 - \frac{\pi^4}{144} + \frac{57\pi^2}{8}\ln 2 - \frac{185}{8}\ln\rho + \frac{33}{4}\ln^2\rho\right]\rho^2 + \frac{2\pi^2}{3}\rho^3. \quad (7)$$

Plot on Fig. 2 presents X_2 calculated to $\mathcal{O}(\rho^{10})$ around $\rho = 0$, and expansion from [3] calculated through $\mathcal{O}((1-\rho)^{21})$.

For convenience we present here the numerical fit result for X_2, providing accuracy better than 0.01 for $0 \leq \rho \leq 1$:

$$X_2 \approx (1-\rho)^3 \left[-14.03 - 42.51\rho + 82.63\rho^2 - 109.3\rho^3 + 151.2\rho^4 + 58.08\rho^5 - 422.3\rho^6 + 447.3\rho^7 - 160.8\rho^8\right]. \quad (8)$$

4. Conclusion

With the approach demonstrated in this paper, and with increased availability of computational resources it becomes a feasible task to calculate the complete decay rate $\Gamma(b \to c\ell\bar{\nu})$ at $\mathcal{O}(\alpha_s^2)$. Recent developments in master integral evaluation techniques, such as Mellin-Barnes transformation [9], and new high-performance algebra systems, such as ParFORM [10] and GiNaC [8], give hope that soon many new high-precision results will appear.

Acknowledgements: Author would like to thank Maciej Slusarczyk and Ian Blokland for many helpful discussions.

References

1. A. Manohar and M. Wise, *Heavy quark physics*, (Cambridge University, 2000).
2. A. Czarnecki and K. Melnikov, *Phys. Rev. Lett.* **88**, 131801 (2002), hep-ph/0112264.
3. A. Czarnecki and K. Melnikov, *Phys. Rev.* **D56**, 7216 (1997), hep-ph/9706227.
4. I. Blokland, A. Czarnecki, M. Slusarczyk, and F. Tkachov, *Phys. Rev.* **D71**, 054004 (2005), hep-ph/0503039.
5. F. Tkachov, *Sov. J. Part. Nucl.* **25**, 649 (1994).
6. J.A.M.Vermaseren, New features of FORM, (2000), math-ph/0010025.
7. S. Laporta, *Int. J. Mod. Phys.* **A15**, 5087 (2000), hep-ph/0102033.
8. J. Vollinga, GiNaC: Symbolic computation with C++, (2005), hep-ph/0510057.
9. M. Czakon, Automatized analytic continuation of Mellin-Barnes integrals, (2005), hep-ph/0511200.
10. Tentyukov, M. *et al*, ParFORM: Parallel Version of the Symbolic Manipulation Program FORM, (2004), cs.sc/0407066

TOP PHYSICS AT CDF

ENRIQUE PALENCIA*†

Instituto de Física de Cantabria (CSIC-Universidad de Cantabria)
Edificio Juan Jordá, Avenida de los Castros, s/n
E-39005 Santander
Cantabria, Spain
E-mail: palencia@fnal.gov

The top quark is the most massive fundamental particle observed so far, and the study of its properties is interesting for several reasons ranging from its possible special role in electroweak symmetry breaking to its sensitivity to physics beyond the Standard Model (SM). This article will focus on the latest top physics results from CDF based on 320-750 pb^{-1} of $p\bar{p}$ collision data at $\sqrt{s} = 1.96$ TeV. The $t\bar{t}$ cross section and the top mass have been measured in different decay channels and using different methods. We have also searched for massive $t\bar{t}$ resonances.

1. Introduction: top at CDF

The CDF detector, upgraded [1] for Run II of the Tevatron, has recorded $\approx$1.2 fb^{-1} of $p\bar{p}$ collision data at $\sqrt{s} = 1.96$ TeV. At the Tevatron, top is produced predominantly in $t\bar{t}$ pairs through the strong interaction by quark-antiquark annihilation (85%) or by gluon gluon fusion (15%).

In the SM, top decays almost 100% of the times to Wb. Therefore, the final state of a $t\bar{t}$ event is given by the decay mode of the W bosons. Events where both Ws decay to e or μ are called "dilepton" events. This mode is relatively clean, with a S/N of about 1.5 to 3.5, but it has a low branching ratio ($\sim$5%) due to the small leptonic branching fraction of W. Events in which one W decays to e or μ and the other decays to quarks are called "lepton+jets" events. This decay channel has a higher branching ratio ($\sim$30%), but it has worse S/N (0.3 to 3). Finally, events in which both Ws decay to quarks are called "all hadronic" events. This channel has the largest branching ratio ($\sim$44%) but is also the one with more backgrounds.

*Speaker
†On behalf of the CDF collaboration

2. Top Cross-Section

The measurement of the $t\bar{t}$ production cross section in the different decay modes and with different methods is a great tool to look for new physics.

In the lepton+jets decay mode several cross section measurements have been performed. The basic event selection requires one identified lepton, large missing transverse energy and 3 or more energetic jets.

Counting analyses require at least one of the jets to be tagged as a b-jet in order to further reduce the otherwise dominant W+jets and QCD backgrounds.

The most precise determination of the $t\bar{t}$ cross section at CDF uses ≈695 pb^{-1} of lepton+jets events and requires at least one jet to be b-tagged by a secondary vertex [2] algorithm, which looks at tracks associated to the jet. The resulting cross section is

$$\sigma_{t\bar{t}} = 8.2 \pm 0.6(\text{stat}) \pm 1.0(\text{sys})\ \text{pb}\ (\text{m}_\text{t} = 175\ \text{GeV}).$$

There is another measurement that uses ≈318 pb^{-1} and applies the same event selection but uses the Jet Probability [3] tagging algorithm in order to identify b-jets. The obtained cross section in this case is

$$\sigma_{t\bar{t}} = 8.9 \pm 1.0(\text{stat}) \pm 1.1(\text{sys})\ \text{pb}\ (\text{m}_\text{t} = 178\ \text{GeV}).$$

In both cases, 6% luminosity uncertainty is included in the systematic uncertainty. Figure 1 left (right) shows the distribution of the b-tagged events, $t\bar{t}$ signal and the predicted background rates as a function of jet multiplicity for the Secondary Vertex (Jet Probability) analysis.

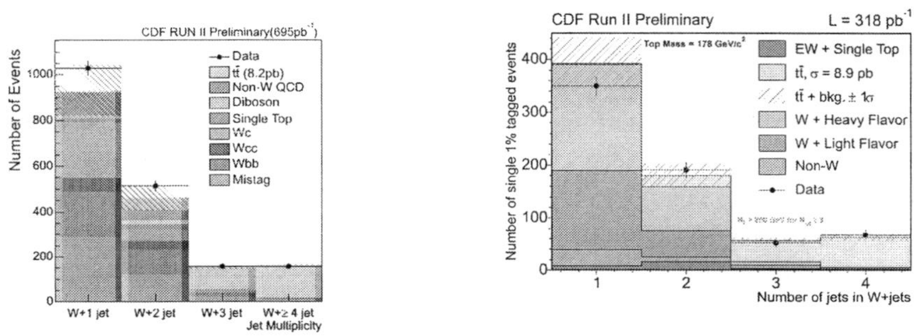

Figure 1. Left (right): b-tagged lepton+jets events, $t\bar{t}$ signal and expected background as a function of jet multiplicity for the Secondary Vertex (Jet Probability) analysis.

Other analyses in the lepton+jets channel do not use b-tagging. Instead, they make use of kinematic information in the $t\bar{t}$ candidate events. A neural net (NN), which includes information from seven different kinematic distributions, is built to separate top signal from the background. The performance of the NN is shown in Fig. 2 (left). The NN analysis uses $\approx$760 pb^{-1} and observes 325 (pretag) $t\bar{t}$ events from the fit. The measured cross section is

$$\sigma_{t\bar{t}} = 6.0 \pm 0.6(\text{stat}) \pm 0.9(\text{sys})\ \text{pb}\ (\text{m}_\text{t} = 175\ \text{GeV}).$$

In the dilepton channel, the basic event selection requires two identified leptons (e or μ), large missing transverse energy and 2 or more energetic jets. Figure 2 (right) shows the distribution of candidate events, $t\bar{t}$ signal and the predicted background rates as a function of jet multiplicity for this analysis. The resulting cross section is

$$\sigma_{t\bar{t}} = 8.3 \pm 1.5(\text{stat}) \pm 1.0(\text{sys}) \pm 0.5(\text{lum})\ \text{pb}\ (\text{m}_\text{t} = 175\ \text{GeV}).$$

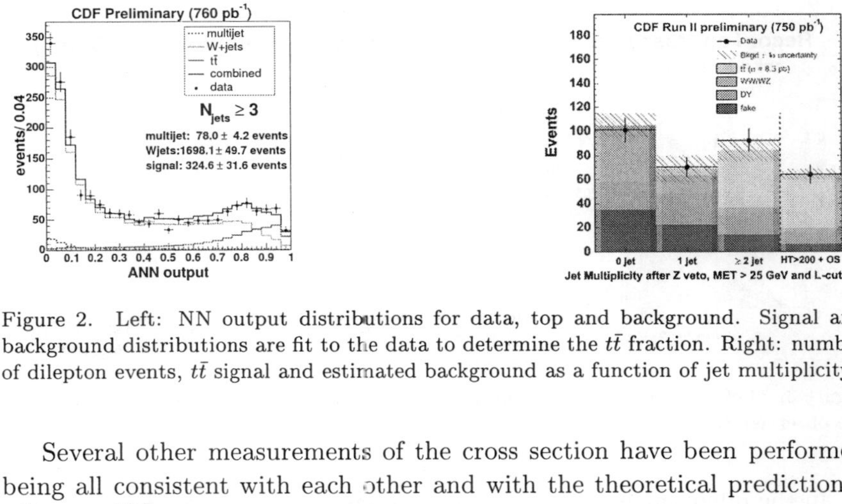

Figure 2. Left: NN output distributions for data, top and background. Signal and background distributions are fit to the data to determine the $t\bar{t}$ fraction. Right: number of dilepton events, $t\bar{t}$ signal and estimated background as a function of jet multiplicity.

Several other measurements of the cross section have been performed being all consistent with each other and with the theoretical prediction [4] $\sigma_{t\bar{t}}^{\text{th}} = 6.7^{+0.7}_{-0.9}$ pb for $m_t = 175$ GeV.

3. Top Mass

The top mass is a fundamental parameter of the SM since it is a dominant parameter in higher order radiative corrections to other SM observables. In particular, an accurate determination of the top mass, combined with precision electroweak measurements, helps to constrain the mass of the elusive SM Higgs.

Top mass measurements are difficult for several reasons: there are many jet-parton possible assignments; in the dilepton channel, the undetected neutrinos cause the event kinematics to be under-constrained; and the jet energies must be known with high accuracy. The dominant systematic uncertainty for all top mass measurements is the jet energy determination.

In the lepton+jets channel, the top mass has been measured using "template" analyses, in which a value for the top mass is reconstructed for each event, with ≈680 pb^{-1} of data. The resulting m_t distribution is then compared to Monte Carlo m_t templates simulated at various top masses (shown in Fig. 3 left). Since reconstructed top mass is sensitive to the Jet Energy Scale, JES, m_t is determined by a simultaneous fit of the templates to the observed distribution, as a function of m_t and Δ_{JES} using $W \to jj$ decays. Figure 3 (right) shows the result of the fit. The obtained top mass is

$$m_t = 173.4 \pm 2.5(\text{stat} + \text{JES}) \pm 1.3(\text{sys}) \text{ GeV}.$$

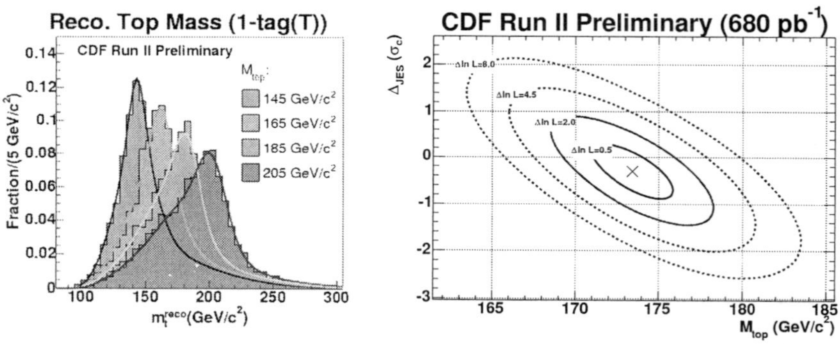

Figure 3. Left: Monte Carlo m_t templates. Right: result of the fit of the templates to the observed distribution, as a function of m_t and Δ_{JES} using $W \to jj$ decays.

In the dilepton channel, the top mass has been measured using a "matrix element" technique with ≈750 pb^{-1} of data. The reconstructed top mass for each event is obtained convoluting matrix elements with resolution functions. The top mass is determined to be

$$m_t = 164.5 \pm 4.5(\text{stat}) \pm 3.1(\text{sys}) \text{ GeV}.$$

Other measurement that does not need full event reconstruction has been developed. This analysis, with ≈695 pb^{-1}, makes use of the correlation between the top mass and the decay length of the b hadrons from the top decay. Although is a statistically limited method, it has the advantage that does not depend on the JES. The measured top mass is

$$m_t = 183.9^{+15.7}_{-13.9}(\text{stat}) \pm 5.6(\text{sys}) \text{ GeV}.$$

4. Search for $t\bar{t}$ Resonances

If we move beyond the SM, $t\bar{t}$ can also be produced via an unknown heavy resonant state, or through some other process such as Topcolor-Assisted Technicolor [5]. CDF has performed a search for a heavy resonance decaying into $t\bar{t}$ in the lepton+jets channel using $\approx$680 pb^{-1} of data. No evidence has been observed. Upper limits have been set on the production cross section at the 95% CL. For one leptophobic topcolor production mechanism we exclude masses up to 725 GeV.

5. Conclusions

Top physics program at CDF is very rich. No evidence of non-SM top quark has been found so far. The $t\bar{t}$ production cross section has been measured in different top decay channels and using different techniques. The top mass has also been measured using different samples and techniques achieving a total uncertainty of $\sim$2 GeV. We expect to have more precise measurements by this summer, with 1 fb^{-1} of data.

Acknowledgments

The results shown here represent the work of many people. I thank my CDF colleagues for their efforts to carry out these challenging physics analyses. I thank the conference organizers for a very nice week of physics. Finally, I thank the colleagues of my research institution, IFCA, for all their help.

References

1. The CDF Collaboration, *The CDF II Detector Technical Design Report*, FERMILAB-Pub-96/390-E.
2. D. Acosta *et al.* (CDF Collaboration), Phys. Rev. D **71**, 052003 (2005).
3. A. Affolder *et al.* (CDF Collaboration), Phys. Rev. D **64**, 032002 (2001), Erratum-ibid:, Phys. Rev. D **67**, 119901 (2003).
   D. Buskulic *et al.* (ALEPH Collaboration), Phys. Lett. B **313**, 535 (1993).
4. M. Cacciari *et al.* JHEP **0404** (2004) 068, hep-ph/0303085.
5. C. T. Hill, Phys. Lett. **B266**:419-424, 1991; Phys.Lett.**B345**:483-489, 1995. hep-ph/9411426. B. A. Dobrescu, C. T. Hill, Phys. Rev. Lett. **81**:2634-2637, 1998. e-Print Archive: hep-ph/9712319.

DI-ELECTRON WIDTHS OF THE $\Upsilon(1S)$, $\Upsilon(2S)$, AND $\Upsilon(3S)$ FROM THE CLEO-III DETECTOR *

JIM PIVARSKI

Cornell University,
Ithaca, NY 14853, USA
E-mail: mccann@lepp.cornell.edu

We determine the di-electron widths of the $\Upsilon(1S)$, $\Upsilon(2S)$, and $\Upsilon(3S)$ resonances with better than 2% precision by integrating the cross-section of $e^+e^- \to \Upsilon$ over the e^+e^- center-of-mass energy. Using e^+e^- scans of the Υ resonances at the Cornell Electron Storage Ring and measuring Υ production with the CLEO detector, we find di-electron widths of 1.252 ± 0.004 ($\sigma_{\rm stat}$) $\pm$ 0.019 ($\sigma_{\rm syst}$) keV, 0.581 $\pm$ 0.004 $\pm$ 0.009 keV, and 0.413 $\pm$ 0.004 $\pm$ 0.006 keV for the $\Upsilon(1S)$, $\Upsilon(2S)$, and $\Upsilon(3S)$, respectively.

1. Introduction and Motivation

The Υ meson is a bottom quark and an anti-bottom quark, Strongly bound in a vector state labeled by excitation number ($1S$, $2S$, and $3S$), and its di-electron width, Γ_{ee}, is the decay rate of the Υ into e^+e^-. This width is a basic parameter of each Υ state, as it characterizes the spatial size of the meson. Non-perturbative QCD is required to calculate Γ_{ee}, so experimental measurements of Γ_{ee} test non-perturbative techniques.

Recent advances in Lattice QCD, in which quark and gluon fields are represented on a lattice in a computer, have made few-percent calculations of many quantities[1]. Di-electron width calculations of the $\Upsilon(1S)$, $\Upsilon(2S)$, and $\Upsilon(3S)$ put extreme demands on the continuum limit, as Γ_{ee} is proportional to the quark wavefunction at the origin. To test these calculations, we have measured Γ_{ee} for the first three Υ resonances with 1–2% precision.

Agreement between Lattice calculations and our measured Γ_{ee} would lend credence to calculations of similar quantities. A quantity of particular interest is the B meson decay constant, f_B, which obfuscates measurements

*This work is supported by the A.P. Sloan Foundation, the National Science Foundation, and the U.S. Department of Energy.

of the CKM element V_{td} because it is poorly known. If an f_B calculation may be trusted within a few percent of itself, our present knowledge of V_{td} would shrink from 20% to a few percent.

2. Measurement Method

We determine the $\Upsilon \to e^+e^-$ decay rate from the $e^+e^- \to \Upsilon$ production cross-section:

$$\Gamma_{ee} = \frac{M_\Upsilon{}^2}{6\pi^2} \int \sigma(e^+e^- \to \Upsilon)\, dE_{\rm CM} \tag{1}$$

where M_Υ is the Υ mass and $E_{\rm CM} = \sqrt{s}$ is the center-of-mass energy of colliding e^+e^- beams. We integrate by sampling the cross-section at several $E_{\rm CM}$ values near the Υ resonance peak and fitting to a parameterized function whose integral is known. Our fit function is a Breit-Wigner resonance peak convoluted by a Gaussian to simulate beam energy spread, also convoluted by an initial state radiation distribution[2]. Additional terms account for backgrounds, including interference between the resonance peak and the continuum backgrounds for $q\bar{q}$ and $\tau^+\tau^-$ final states.

To measure the Υ production cross-section, we count hadronic events and divide by integrated luminosity, where we define a hadronic event to be any beam-beam collision that results in a final state other than e^+e^-, $\mu^+\mu^-$, and $\tau^+\tau^-$. A well-measured fraction[3] of Υ mesons decay into e^+e^-, $\mu^+\mu^-$, and $\tau^+\tau^-$ (about 7%), so we use this fraction to correct our cross-section measurement.

2.1. *Backgrounds*

Depending on $E_{\rm CM}$, 30% to 100% of our hadron counts are not Υ events. These backgrounds have a flat dependence on $E_{\rm CM}$, whereas the signal is highly peaked (see Figure 1), so signal and background are distinguished by the fit function. The majority of the backgrounds are continuum processes which depend on $E_{\rm CM}$ as $1/s$, and 8% of the continuum (at 9 GeV) are two-photon fusion events ($e^+e^- \to e^+e^-X$) which have a $\log s$ dependence. Radiation to lower-energy resonances contributes at the $\sim \frac{1}{2}$% level.

Cosmic rays, collisions between beam particles and gas (beam-gas), and collisions between beam particles and the wall of the beampipe (beam-wall) are not proportional to integrated luminosity, so they must be subtracted from the hadron counts. We determine the number of cosmic rays at each $E_{\rm CM}$ by normalizing to a control sample acquired with no colliding beams, and similarly estimate beam-gas and beam-wall with single-beam samples.

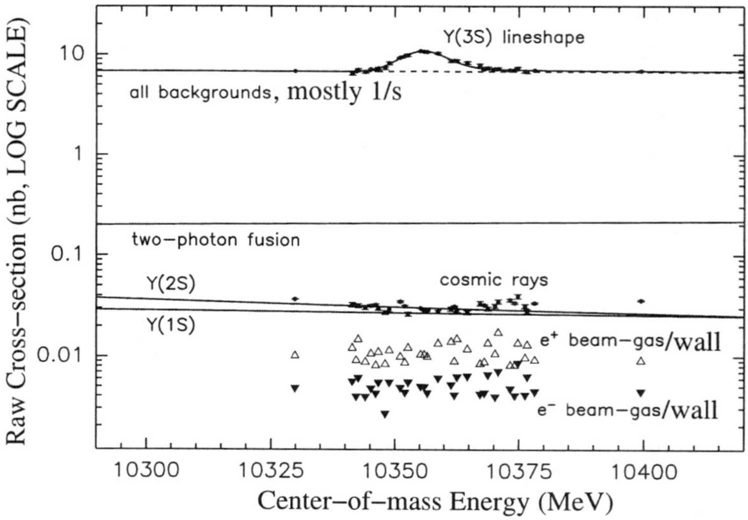

Figure 1. The $\Upsilon(3S)$ resonance peak, presented in log scale to highlight backgrounds. "Raw cross-section" is hadronic event yield per nb^{-1}, which includes backgrounds, but not efficiency corrections or corrections for e^+e^-, $\mu^+\mu^-$, and $\tau^+\tau^-$ final states.

2.2. *Efficiency*

We determine the efficiency of our triggers and event selection for $\Upsilon(1S)$ events without assuming a decay model for the Υ meson. From our 1.3 fb^{-1} $\Upsilon(2S)$ sample, we select $\Upsilon(2S) \to \pi^+\pi^-\Upsilon(1S)$ events by requiring the $\pi^+\pi^-$ to recoil against an object with the $\Upsilon(1S)$ mass. This way, we trigger and select events independently of way the $\Upsilon(1S)$ decays, leaving the set of $\Upsilon(1S)$ decays unbiased. Applying our triggers and event selection to the $\Upsilon(1S)$ decays reveals a hadronic efficiency of $(97.8 \pm 0.5)\%$. We extrapolate this efficiency from the $\Upsilon(1S)$ to the $\Upsilon(2S)$ and $\Upsilon(3S)$ using Monte Carlo and measurements of $\Upsilon(nS) \to X\Upsilon(mS) \to X\mu^+\mu^-$ branching fractions. This procedure is discussed in greater detail elsewhere[4 5].

2.3. *Integrated Luminosity*

To measure integrated luminosity, we count Bhabha events ($e^+e^- \to e^+e^-$) and divide by the efficiency-weighted Bhabha cross-section, determined from Monte Carlo simulations[6 7]. We evaluate the systematic uncertainty in the efficiency-weighted cross-section (1.3%) by comparing Bhabha-derived integrated luminosities with values determined from $e^+e^- \to \mu^+\mu^-$ and $e^+e^- \to \gamma\gamma$, following the method used in CLEO-II[8].

2.4. *Beam Energy Uncertainty*

The potential effect of fluctuations or drifts in the calibration of the beam energy measurement was minimized by splitting the data-taking into short, 10-hour scans, and by alternating measurements above and below the resonance peak. The point of maximum slope ($d\sigma/dE_{\rm CM}$) was measured twice to bound fluctuations in the calibration through their effect on cross-section between the two measurements, leading to a 0.2% uncertainty in Γ_{ee}.

3. Fit Results

The $\Upsilon(1S)$, $\Upsilon(2S)$, and $\Upsilon(3S)$ fits are presented in Figure 2, which yield the following values for Γ_{ee}. They are consistent with and more precise than the current world averages[9].

$\Gamma_{ee}(1S)$	$1.354 \pm 0.004 \pm 0.020$ keV
$\Gamma_{ee}(2S)$	$0.619 \pm 0.004 \pm 0.010$ keV
$\Gamma_{ee}(3S)$	$0.446 \pm 0.004 \pm 0.007$ keV

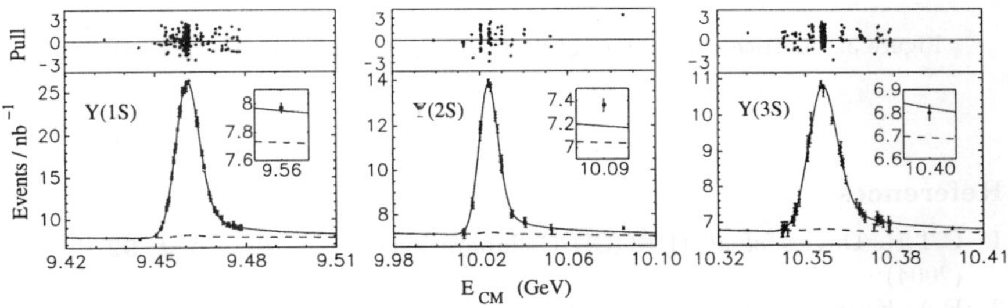

Figure 2. The Υ lineshape fits used to determine Γ_{ee}. Points with errorbars are data, the solid line is the fit, the dashed line represents backgrounds, and the pull (fit residual divided by uncertainty) of every point is presented above the plots. Measurements 100 GeV, 60 GeV and 45 GeV above the three resonance maxima are shown in insets.

Uncertainty in the hadronic efficiency (0.5%) and in the overall luminosity scale (1.3%) dominate the systematic uncertainty, but they are also shared by all three resonances and therefore cancel in ratios of Γ_{ee}. The systematic uncertainty in $\Gamma_{ee}(mS)/\Gamma_{ee}(nS)$ is 0.9%–1.0%.

4. Conclusions

Though Lattice QCD calculations of Γ_{ee} have not been corrected for lattice renormalization yet[10], this effect largely cancels in the ratio of $\Gamma_{ee}(2S)$ to

$\Gamma_{ee}(1S)$. The Lattice calculation compares favorably with our measurement (see Figure 3), though its uncertainty is 10% due to the steep dependence on lattice spacing size. We eagerly await the full calculations.

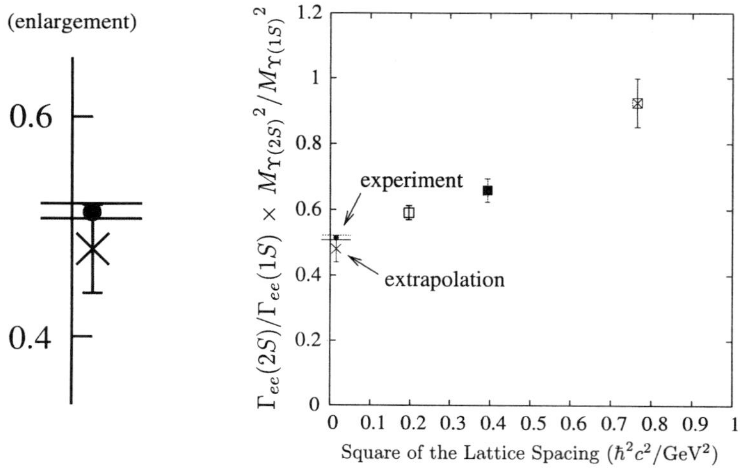

Figure 3. Lattice calculations of $\Gamma_{ee}(2S)/\Gamma_{ee}(1S)$ with our result overlaid.

References

1. C.T.H. Davies *et al.* (HPQCD Collaboration), *Phys. Rev. Lett.* **92**, 022001 (2004).
2. E.A. Kuraev and V.S. Fadin, *Sov. J. Nucl. Phys.* **41**, 466 (1985) [Yad. Fiz. **41**, 733 (1985)].
3. G.S. Adams *et al.* (CLEO Collaboration), *Phys. Rev. Lett.* **94**, 012001 (2005).
4. J.L. Rosner *et al.*, *Phys. Rev. Lett.* **96**, 092003 (2006).
5. J. Pivarski, Cornell University, Ph.D. thesis, `hep-ex/0604026` (2006).
6. C.M. Carloni Calame *et al.*, *Nucl. Phys. Proc. Suppl. B* **131**, 48 (2004).
7. R. Brun *et al.*, Geant 3.21, CERN Program Library Long Writeup W5013 (1993), unpublished.
8. G.D. Crawford *et al.* (CLEO Collaboration), *Nucl. Instrum. Methods Phys. Res., Sect A* **345**, 429 (1994).
9. S. Eidelman *et al.* (Particle Data Group), Phys. Lett. B **592**, 1 (2004).
10. A. Gray, I. Allison, C. T. H. Davies, E. Gulez, G. P. Lepage, J. Shigemitsu and M. Wingate, *Phys. Rev. D* **72**, 094507 (2005).

ELECTRON RECONSTRUCTION AND CALIBRATION WITH SINGLE Z AND W PRODUCTION IN CMS AT THE LHC

C. ROVELLI

Università di Milano Bicocca, Dipartimento di Fisica G.Occhialini
Piazza delle Scienze 3
20126 Milano, Italy
E-mail: chiara.rovelli@mib.infn.it

The CMS experiment at the LHC is building an electromagnetic calorimeter with high performance. Preserving high reconstruction efficiency and best four momentum measurements for electrons is a necessity for optimal discovery prospects in the $ZZ^{(*)}$ and $WW^{(*)}$ Higgs boson decay channels. This is challenging in view of the material budget in front of ECAL and of the presence of a strong magnetic field. A new reconstruction strategy for electrons in CMS is described. The usage of electrons from single Z and W production for the ECAL calibration strategy is also discussed.

1. Introduction

The CMS electromagnetic calorimeter (ECAL) [1] will play a central role in the physics CMS plans to do. One of the primary goals of CMS is the search for the Higgs boson in the large range of allowed masses. The golden channel for the search of a light (m_H <150 GeV/c^2) boson is the decay into two photons, while for larger masses the production of $WW^{(*)}/ZZ^{(*)}$ which decay leptonically becomes important.
ECAL is made of lead tungstate scintillating crystals organized in a barrel and two endcaps. The small radiation length (X_0 = 0.89 cm) and Molière radius (R_M = 2.2 cm) allow the construction of a compact and highly granular calorimeter. Detailed reconstruction algorithms have been developed to best exploit the characteristics of ECAL.

2. Electron reconstruction

The reconstruction of electrons in CMS is made challenging by the large amount of tracker material which is distributed in front of ECAL and by the solenoidal 4T magnetic field. Electrons traversing the tracker radiate bremsstrahlung photons and this results in a spread in ϕ of the energy reaching ECAL. The bremsstrahlung emission also introduces non-Gaussian contributions to the event fluctuations of tracking and calorimetry measurements. Both the energy clustering and the track reconstruction are designed to take such emissions into account at best.

Energy clustering. The first step in the electron reconstruction is the clustering of the energy which is deposited in the ECAL from the electron and the bremsstrahlung photons. The bremsstrahlung emission causes a spread in the azimuthal direction ϕ only of the energy reaching ECAL, which is therefore recovered collecting clusters along a ϕ road. Two clustering algorithms are used to build the so called super-clusters (clusters of clusters): 'Hybrid' for the barrel and 'Island' for the endcaps [2].
The impact point of the electron on ECAL is computed as the energy weighted mean position of the clusters in the super-cluster. Since the energy density of the shower falls exponentially, each cluster position is computed from the crystals with weights which depend on the logarithm of the fraction of the cluster energy within the crystal.
The clustering algorithms were developed for high p_T electrons to be applied in the High Level Trigger [3]; tuning the parameters allows their best exploitation for lower energy electrons also [4].

Track seeding and reconstruction. A track seed is created when two hits are found in the innermost part of the tracker (pixel detector) which are compatible with the energy weighted position of the super-cluster propagated backward through the magnetic field. If the hits are found the super-cluster is classified as belonging to an electron, else to a photon.
Starting from the seed a trajectory is created. In the offline analysis a dedicated algorithm based on the Gaussian Sum Filter [5] can be used to better deal with the non-Gaussian fluctuations induced by the bremsstrahlung emission. This allows efficient collection of the electron hits while keeping a precise determination of the momentum at vertex. Since the track is followed up to the end, a meaningful estimate of the track parameters at ECAL is possible. This gives the possibility of improving the matching between the tracker and the calorimeter and of estimating the amount of bremsstrahlung radiated by the track using the tracker information only.

Electron classification. Different track-supercluster patterns are observed as a function of the bremsstrahlung emission. Since they imply different measurements errors and electron identification performance, electron classes are defined according to the different topologies [4]. The classification separates 'golden' electrons, having a good matching between the track and the super-cluster, from 'showering' electrons, which emit a large fraction of their energy by bremsstrahlung photons; intermediate cases are also identified. Figure 1 shows the reconstructed energy over the generated

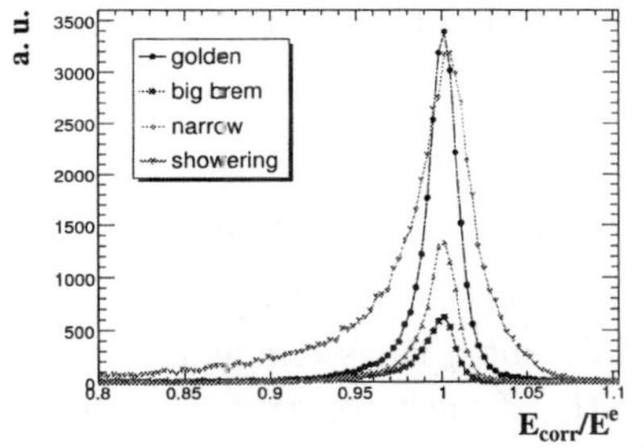

Figure 1. Distribution of the reconstructed electron energy normalized to the generated one for the different electron classes in the ECAL barrel.

one in the ECAL barrel; 'showering' electrons are responsible for most of the tail which is measured by ECAL.

The calorimeter and the tracker measurements are affected by bremsstrahlung in a complementary way, therefore the most appropriate measurement can be used depending on the electron class and the energy range. With such combination the energy resolution for golden electrons is compatible with results obtained using an electron test beam.

3. The ECAL calibration

The ECAL physics reach relies on its very good energy resolution. The crystals intercalibration is the dominant contribution to the constant term of the energy resolution, which has to be kept at the 0.5% level or better. To achieve this task a complex calibration strategy has been developed.

The inital set of intercalibration coefficients will be provided from laboratory measurements of the crystal light yield (with a precision of about 4%) and with cosmic rays (~ 2-3% with one week of data-taking [6]). A few supermodules will be also pre-calibrated with electron beams (~0.5% [7]).

The ultimate limit on the precision will be fixed by the 'in-situ' calibration with physics events. At the beginning of the data taking a fast intercalibration will be performed exploiting the ϕ-symmetry of energy deposition at fixed η [8]. Electrons from $Z \to e^+e^-$ will be exploited to intercalibrate different η rings one against the others and to set the absolute energy scale, then the intercalibration of the different crystals will be performed with $W \to e\nu$ events. The possibility to exploit processes like $\pi^0 \to \gamma\gamma$ and $\eta \to \gamma\gamma$ for the single channel intercalibration is currently under study.
A laser monitoring system will be used to correct for short term variations in the crystal transparency due to radiation damage and the following recovery.

3.1. *Calibration with $Z \to e^+e^-$ events*

Thanks to the large cross section at LHC, $Z \to e^+e^-$ events can be used to intercalibrate the different η rings one against the others. An iterative method which exploits the Z mass constraint has been tested in the ECAL barrel [9]. Stable results have been obtained for different initial miscalibrations. Selecting only events with two 'golden' electrons (to avoid bias due to the limited tracker material knowledge) a 0.6% precision can be obtained with an integrated luminosity of 2 fb^{-1} (figure 2, left); no systematic effect is observed as a function of the pseudorapidity. $Z \to e^+e^-$ events can be also used to set the absolute energy scale. The method has been tested with a variation of the miscalibration scale; a statistical precision of 0.05% with an integrated luminosity of 2 fb^{-1} can be reached.

3.2. *Calibration with $W \to e\nu$ events*

The intercalibration of the different crystals can be obtained with $W \to e\nu$ events using the tracker momentum measurement as a reference [10]. Stable results have been obtained starting with an artificial 4% miscalibration. The calibration precision does not depend on ϕ, but a strong dependence on the pseudorapidity was found. The target 0.5% calibration precision can be achieved with an integrated luminosity of 5 fb^{-1} in the $|\eta| < 1$ region (fig.2, right), while in the endcaps the precision varies between 1% and 2%, following the tracker material budget. Since the calibration with W requires a fully aligned tracker, the time needed to reach such a precision can be longer that what estimated here. As in the Z channel, only low radiating electrons have been selected with a set of appositely tuned cuts.

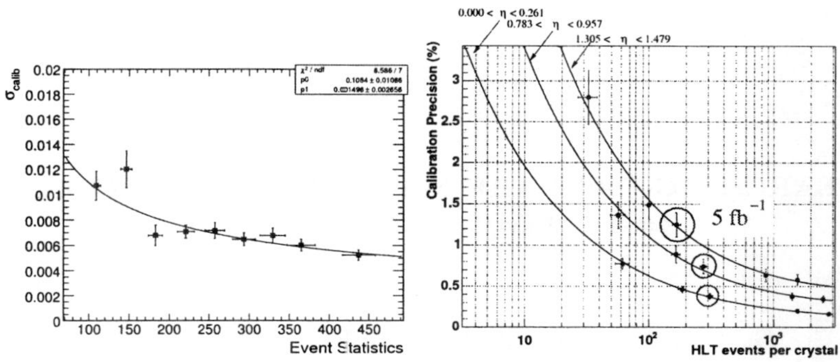

Figure 2. Left) η-rings inter-calibration precision with $Z \to e^+e^-$ events as a function of the event statistics in the ECAL barrel. The point corresponding to 2 fb^{-1} is the one before the last. Right) Inter-calibration precision with $W \to e\nu$ events versus HLT events per crystal for different η regions in the ECAL barrel.

4. Conclusions

An excellent energy resolution is required for the physics of the CMS electromagnetic calorimeter. This is challenging in view of the large amount of tracker material in front of the ECAL and of the strong magnetic field. Robust electron reconstruction algorithms have been developed to recover the energy radiated by bremsstrahlung and to guarantee excellent energy resolution even in the HLT. Such algorithms are largely employed in the calibration procedure, which is a major task for ECAL. Realistic intercalibration studies have been performed, showing that fast and stable results can be expected for the 'in-situ' calibration with Z and W.

References

1. CMS Collaboration, CERN/LHCC 97-33
2. E. Meschi et al., CMS Note 2001/034
3. CMS Collaboration, Vol I, CERN/LHCC 2000-038
4. S. Baffioni et al, CMS Note 2006/040
5. W. Adam et al, CMS Note 2005/001
6. M. Bonesini et al, CMS Note 2005/023
7. The CMS Electromagnetic Calorimeter Group, *The European Physical Journal C* Volume 44, 1-10 (2006)
8. D. Futyan et al, CMS Note 2002/031; D. Futyan et al, CMS Note 2004/007;
9. P. Meridiani et al, CMS Note 2006/039
10. L. Agostino et al, CMS Note 2006/021

SEARCHING FOR DARK MATTER ANNIHILATION IN Z=-1 COSMIC RAYS WITH AMS

G. RYBKA, G.P. CAROSI, P. FISHER, S. XIAO, F. ZHOU
LNS, Physics Department, Massachusetts Institute of Technology
On behalf of the AMS Collaboration

The AMS-01 magnetic spectrometer that flew on the Space Shuttle for ten days and the AMS-02 experiment that will be on the International Space Station for three years offer unique looks at cosmic rays. While the majority of cosmic rays come from astrophysical sources, a small component may be from the annihilation of weakly interacting dark matter particles. We outline our current progress in a search for such a component in the negatively charged particles seen in the AMS-01 detector. This includes simulation of the production of cosmic rays from dark matter annihilation, their propagation through the galaxy, and the signal they would leave in the AMS-01 detector.

1. Introduction

Since Fritz Zwicky proposed it to explain his observations of cluster motion in 1933, the nature of dark matter has been a mystery[1]. Normal baryonic matter both in gaseous and compact forms has been all but ruled out by experiments, leaving a new type of particle as the most attractive candidate.

Some theories, supersymmetry for example, predict that dark matter particles are weakly interacting massive particles (WIMPS) which may annihilate with one another producing an assortment of standard model particles. If this is the case, then there is a chance to detect signs of dark matter by looking for their annihilation products in the cosmic ray spectrum.

Using the AMS-01 cosmic ray experiment, we are attempting to detect the signature of WIMP annihilation. The large background proton flux and and limits on particle identification at high energies in the AMS-01 detector limit us to looking at the sum of fluxes of particles with charge -1, namely, electrons and antiprotons.

2. The AMS-01 Detector

The AMS-01 experiment[2] was a precision spectrometer designed to measure cosmic rays of energies from a few hundred MeV to 300 GeV. It was flown June 2-12 in 1998 on Space Shuttle flight STS-91. AMS-01 was equipped with a 6 layer silicon tracker in a 1.5T magnetic field, a time of flight system and an aerogel Cerenkov counter. The shuttle's flight allowed for roughly 3 days of exposure time at various positions above the Earth.

3. Procedure

The exact process of dark matter annihilation is model dependent. In order to avoid becoming dependent on the predictions of any one model, we have chosen to frame our search in terms of a search for WIMPs annihilating into pairs of heavy standard model particles, such as W^+W^-, ZZ, and $b\bar{b}$, with a center of mass energy equal to the combined mass of two dark matter particles. Once this is achieved, any particular model can be examined by using the predicted branching ratios to these heavy particles and the dark matter candidate mass and comparing these with our results. Currently, our analysis has focused on W^+W^- production as reported in G.P. Carosi's thesis [3].

Fig. 1 shows the electron and antiproton energy spectra from the decay of a W^+W^- pair with a center of mass energy of 400 GeV (two 200 GeV WIMPS), as calculated by the Monte Carlo program Pythia[4].

Charged particles move through our galaxy under the influence of galactic magnetic fields. This makes particle propagation from source to detection a diffusive process. To model how this process will affect the energy spectrum, we assume dark matter is distributed in a standard isothermal sphere[5] through the galaxy and feed this and the annihilation spectra into the program GALPROP[6]. GALPROP takes a number of diffusion model parameters and calculates how particles will propagate through the galaxy.

The dark matter signal is superimposed on the cosmic ray background, which is believed to be from Supernovae remnants. For all particle species it has the property of being a power law in our accessible region. At lower energies, the spectrum drops rapidly from a power-law on account of the solar wind shielding us from low energy particles. The expected signal and background at Earth is shown in Fig. 2 and is characteristically a small deviation from the background power law at energies slightly below the WIMP mass.

Once near Earth, lower energy particles will be turned away by the

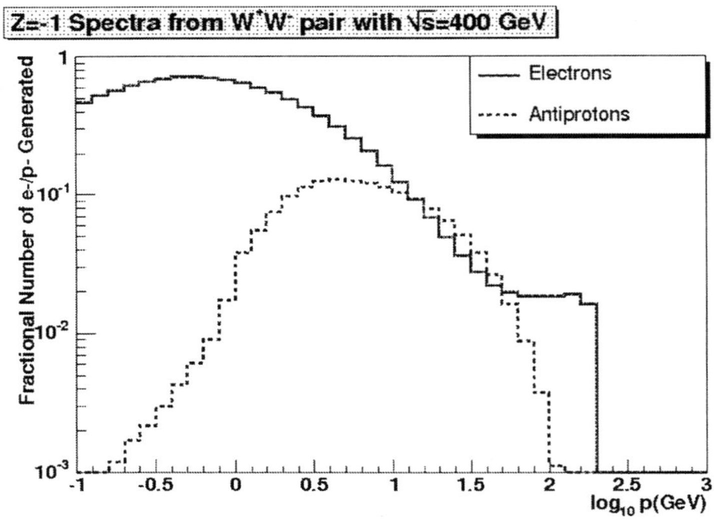

Figure 1. Electron and antiproton spectra from Pythia simulation of W^+W^- pair decay. Horizontal axis is in units of log momentum (GeV), vertical axis is normalized to one pair decay.

Earth's magnetic field, and higher energy particles are so rare as to limit statistical analysis from the AMS-01 dataset, leaving an accessible region of roughly 10-200 GeV. Additionally, wrong-sign charge particle identification must be accounted for. An example fit of the power law background and dark matter signal to AMS-01 data is shown in Fig. 3.

4. Results and Future Work

Current fits of signal+background to data are consistent with no W^+W^- pair production from dark matter annihilation. We have used this to set an upper bound on W^+W^- production (Fig. 4). The limit is given in terms number of pairs produced per second per cm^3 in the vicinity of Earth, assuming dark matter is distributed in a cored isothermal halo with core radius of 2.8 kpc.

Once we have limits on ZZ and $b\bar{b}$ production, we will take branching ratios and mass predictions of supersymmetric theories and relate limits on dark matter density to supersymmetric parameters. The limits we obtain may be limited by uncertainties in the way dark matter distributes itself through the galaxy, and the exact parameters of cosmic ray diffusion. These will become better known with the AMS-02 experiment.

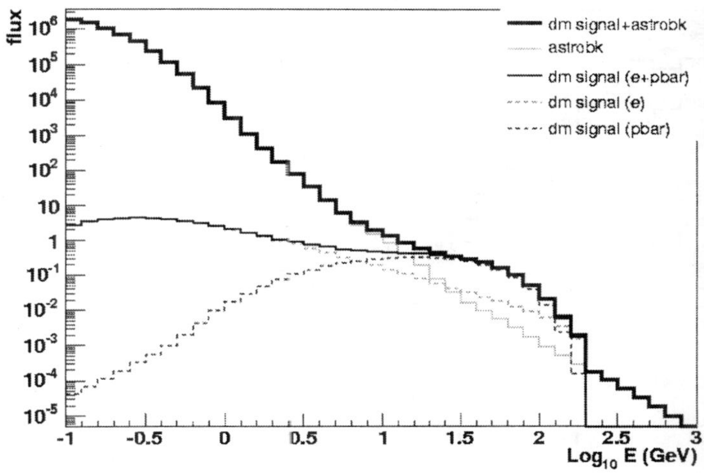

Figure 2. Predicted signal and astrophysical background at Earth (solar modulation effects not included) for a 200 GeV WIMP, given in arbitrary units of flux vs. log energy (GeV).

The AMS-02 experiment is planned to be a follow-up to the AMS-01 experiment. It will be run on the International Space Station for 3 years, and will boast a superconducting magnet, a transition radiation detector and energy calorimeter. AMS-02 will refine cosmic ray propagation models and open up the high-energy positron spectrum to be examined for dark matter annihilation signals.[7].

Acknowledgments

The authors would like to thank the organizers of the Lake Louise Winter Institute.

References

1. F. Zwicky, *Helvetica Physica Acta* (1993).
2. AMS Collaboration, *Physics Rep.* **366(6)** 331 (2002)
3. G.P. Carosi, PhD dissertation, MIT, Physics Department, December 2005.
4. Sjostrand et al., *Computer Phys. Commun.* **135** 238 (2001)
5. J.F. Navarro, C.S. Frenk, and S.D.M. White, *Astroph. J.* **490** 493 (1997)
6. I.V. Moskalenko and A.W. Strong. *Astroph. J.* **493**,694-707 (1998).
7. AMS Collaboration, *Proceedings of the 29th ICRC* 3-11 (2005)

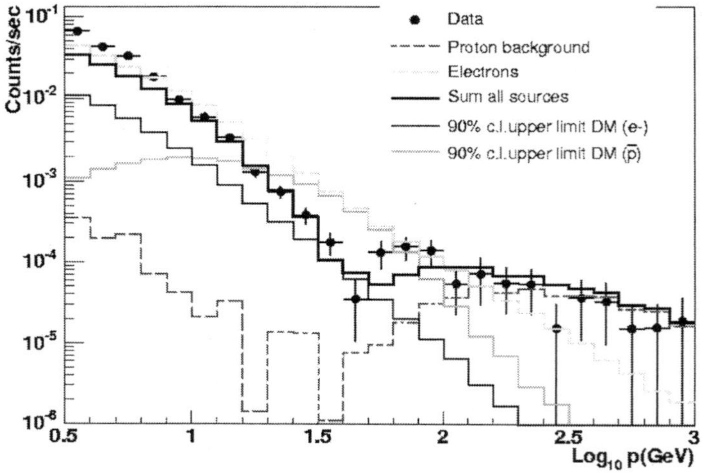

Figure 3. Fit of power law background, misidentified protons, and signal from a 160 GeV WIMP to AMS01 data, given in counts in detector s^{-1}.

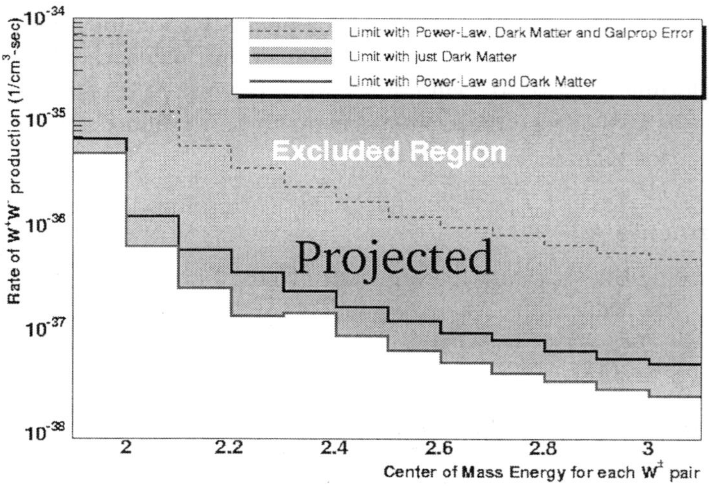

Figure 4. Projected limits on W^+W^- production from WIMP annihilation, as a function of the log combined mass of two WIMPs (GeV). Lines represent limits obtained from only fitting dark matter signal, fitting both signal and background, and after including uncertainties from propagation model.

ATLAS CONSTRUCTION: A STATUS REPORT

ANNA SFYRLA
ON BEHALF OF THE ATLAS COLLABORATION
University of Geneva, Switzerland
E-mail: anna.sfyrla@cern.ch

ATLAS is a general purpose p-p collider detector being constructed for the CERN Large Hadron Collider (LHC). It is located in one of the two high luminosity bunch crossing points (peak luminosity of $10^{34}cm^{-2}s^{-1}$) of the LHC. It consists of 3 main sections. Located close to the beam axis, the tracking system employs pixel detectors, silicon microstrip modules and transition radiation straws, all within a 2 Tesla superconducting solenoid. The tracker is surrounded by the electromagnetic and hadronic calorimeters. In the outer part of the detector, 8 superconducting coils define an open toroidal magnetic field for muon detection. The construction status of the ATLAS detector towards being ready for the first collisions in 2007 will be presented, with particular emphasis on the construction and projected performance of the tracking system.

1. Introduction

The Large Hadron Collider (LHC) is a proton-proton (p-p) collider under construction at CERN. Bunches of protons intersect at 4 points, with a separation of 25ns and center of mass energy of 14TeV. ATLAS, a general purpose collider detector, is placed in one of the two high luminosity intersections (peak luminosity of $10^{34}cm^{-2}s^{-1}$).

The detector concept has been developed in consideration of a broad spectrum of detailed physics studies[1]. The basic design criteria include:

- Multilayered tracking for heavy flavour tagging and high transverse momentum measurements at high luminosity;
- Precision electromagnetic calorimetry for electron and photon identification and measurements, and full-coverage hadronic calorimetry for accurate jet and missing transverse energy measurements;
- Muon spectrometry for good muon identification and high precision momentum measurement.

These considerations lead to the concept of the detector that is pictured in Figure 1. Its design provides large acceptance in pseudorapidity and full azimuthal angle (ϕ) coverage.

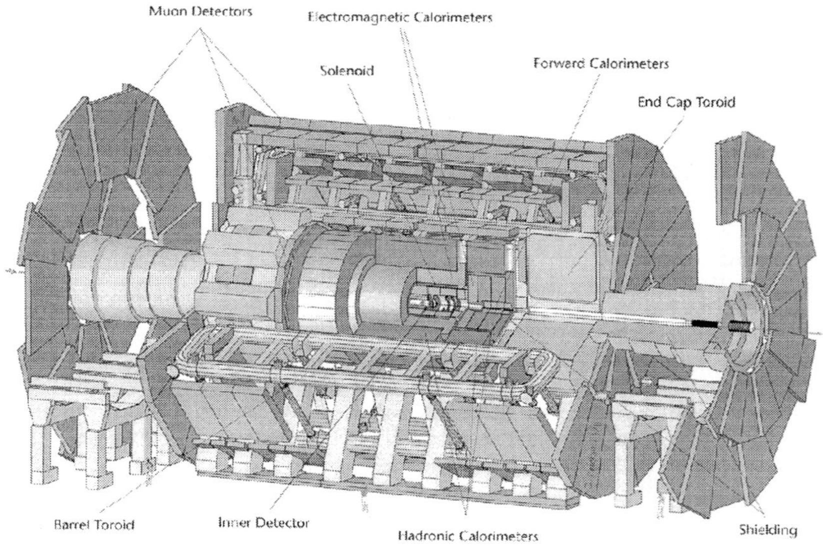

Figure 1. The ATLAS Detector

2. The ATLAS detector description and contruction status

2.1. *Tracking System*

The tracking of ATLAS is provided by a multilayered inner detector (ID) placed within a 2 Tesla superconducting solenoid. The design of the ID had to provide a good time and space resolution with the minimum possible material within the tracking volume.

2.1.1. *Pixel Detector*

The pixel detector is closest to the beam pipe. It consists of 3 barrels and 2 endcaps of 3 disks each, providing 3 tracking points. The proximity to the beam pipe and therefore large amount of radiation makes this detector one of the most challenging parts of ATLAS.

Good progress has been made on the pixel sensors, the electronics and the services. The schedule of the detector construction has been affected by technical problems, with the most important being corrosion leaks in the barrel cooling tubes, necessitating the reconstruction of the pixel staves. The time scale for installation of the detector for the LHC start-up is marginal, but strong attempts are being to achieve this.

2.1.2. *SemiConductor Tracker (SCT)*

The pixel detector is surrounded by the SCT, a detector that consists of more than 4000 double layered silicon strip detectors and therefore more than 6 million readout channels. The strip detectors are arranged in 4 coaxial barrels and 2 endcaps of 9 disks each, providing 4 space points per track.

The silicon barrel and forward module production has taken place at 10 institutions. The SCT barrel modules were assembled onto the barrel structures at Oxford university. Individual barrel tests took place at Oxford and CERN, where the barrels were shipped after assembly. The 4 barrels and the thermal enclosure were fully integrated and tested during summer 2005. At this stage 99.7% of more than 3 million channels were fully functional. The SCT endcap modules were assembled onto the disks in Liverpool and NIKHEF and both endcaps have been delivered to CERN.

2.1.3. *Transition Radiation Tracker (TRT)*

The TRT is the outer tracking layer. It employs straw tubes arranged in a barrel and two endcaps, providing continuous tracking (36 tracking points).

The barrel modules were built and tested in the US and transfered to CERN. Their integration with the FE electronics and the gas and cooling services has been achieved, and acceptance tests that have been successfully completed. Cosmic ray tests started taking place during the summer 2005. The TRT wheels were built and tested in Russia, acceptance tests have already been permormed at CERN and their integration is in progress.

The installation of the SCT and the TRT barrels took place after they were characterised functional. Performance tests of both detectors, individually and as a whole, are now in progress. Once those tests are accomplished, emphasis will be given to cosmic ray runs. The integrated SCT and TRT barrel will be lowered to the ATLAS pit around June/July 2006[2].

2.2. *Calorimetry*

Good electron and photon identification are mainly provided by a lead/liquid argon (LAr) sampling electromagnetic calorimeter with accordion shaped absorber plates in both barrel and endcap regions. The hadronic calorimeter, needed for accurate jet and missing transverse energy measurements, uses LAr sampling in the inner forward part and is surrounded by iron/scintillator sampling (Tiles) in the barrel and outer forward regions.

The calorimeter barrel construction was completed in November 2005, and the barrel is placed in its final position at the center of ATLAS. Cosmic rays runs have successfully been taking place since summer 2005. The installation of the endcaps is in progress and should be completed as scheduled[3].

2.3. *Muon Spectrometer*

The shape of the ATLAS detector is defined by the superconducting air core toroids that provide the 4 Tesla magnetic field for the muon spectrometer. It consists of a large barrel toroid (8 separate coils) and two endcaps (8 coils in a common cryostat).

The mechanical installation of the barrel toroid was completed in October 2005. The electrical and cryogenic connections are now being made and the first full tests are expected in Spring 2006. The endcap toroids are still under contruction and will be completed and installed by summer/fall 2006.

The muon spectrometer itself uses four different types of chambers: the Monitored Drift Tube (MDT) chambers, the Cathode Strip Chambers (CSC), the Resistive Plate Chambers (RPC) and the Thin Gap Chambers (TGC). The MDTs are used for precision tracking mesurements, covering most of the acceptance. They are located in both the barrel and the endcap regions and are supplemented by the CSCs in the very forward region. The RPCs (in the barrel region) and TGCs (in the forward region), arranged in stations located between the tracking chambers, provide the triggering.

The muon barrel construction is very well advanced, and its installation is expected to be complete by August 2006. Cosmic ray runs have already taken place, verifying the good performance of the detectors. Progress in the endcap construction has been delayed by a lack of resources and this problem is now being addressed.

3. Conclusions

During 2005, considerable progress has been achieved in the ATLAS detector construction. The installation is well advanced but the schedule remains very tight. The global commissioning of ATLAS is planned for the period April to June 2007 and this would be followed by combined cosmic ray tests until August 2007. Despite the delays (ex. muon chambers installation) and the still existing critical areas (ex. pixel detector), large collective efforts are taking place to minimize the overall impact of critical issues such that the detector is ready for first collisions in 2007.

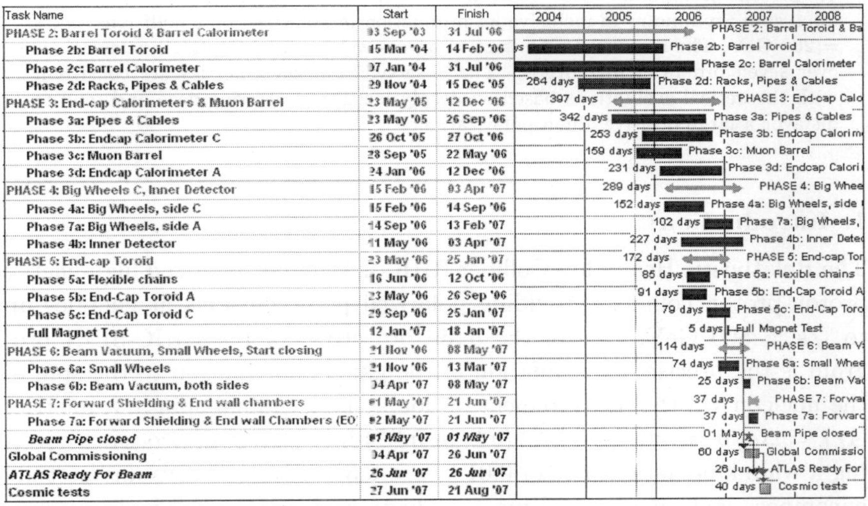

Figure 2. Overall summary installation schedule, version 7.09

References

1. ATLAS Detector and Physics Performance Technical Design Report LHCC 99-14/15.
2. Private communication, H. Pernegger.
3. Private communication, S. Stapnes.

DOUBLE LONGITUDINAL SPIN ASYMMETRY IN INCLUSIVE JET PRODUCTION IN POLARIZED P+P COLLISIONS AT $\sqrt{S} = 200$ GEV

F. SIMON (FOR THE STAR COLLABORATION)

Massachusetts Institute of Technology
77 Massachusetts Avenue
Cambridge, MA 02139, USA
E-mail: fsimon@mit.edu

We present preliminary measurements of the cross section and the double longitudinal spin asymmetry in inclusive jet production in polarized p+p collisions at $\sqrt{s}$ = 200 GeV. The measured cross section agrees well with NLO pQCD calculations over seven orders of magnitude. The observed spin asymmetries are consistent with theoretical evaluations based on deeply inelastic scattering data and tend to disfavor a large positive gluon polarization.

1. Introduction

The Relativistic Heavy Ion Collider RHIC is the first polarized high-energy proton-proton collider, providing polarized p+p collisions at energies up to $\sqrt{s}$ = 500 GeV. One of the main objectives of the RHIC-SPIN program is the precise determination of the gluon contribution to the spin of the nucleon via the measurement of double longitudinal spin asymmetries $A_{LL} = \Delta\sigma/\bar{\sigma} = (\sigma^{++} - \sigma^{+-})/(\sigma^{++} + \sigma^{+-})$ in a variety of processes covering a wide kinematic range of $0.01 < x < 0.3$[1]. The processes under study encompass inclusive jet and pion production, di-jets and di-hadrons, production of heavy quark pairs and photon-jet coincidences. One of the measurements with only modest requirements on the total integrated luminosity is the measurement of the spin asymmetries of inclusive jet production.

With its large acceptance tracking and electromagnetic calorimetry the STAR detector[2] is uniquely capable of full jet reconstruction in p+p collisions at RHIC. The dataset discussed here was recorded during commissioning runs in 2003 and 2004, aimed at developing luminosity and polarization. The recorded luminosity in those two runs is ~0.5 pb^{-1}, with average polarizations of 30% – 40%, allowing an exploratory study of spin asymmetries in inclusive jet production. A first measurement of the inclusive jet cross section is obtained using the ~0.2 pb^{-1} recorded in 2004.

2. Jet Analysis and Inclusive Cross Section

The STAR detector subsystems of principal interest for this analysis are the plastic scintillator beam-beam Counters (BBC), the time projection chamber (TPC) and the barrel electromagnetic callorimeter (BEMC). The BBCs cover $3.3 < |\eta| < 5.0$ in pseudorapidity and are used to trigger on non-singly diffractive (NSD) inelastic reactions. This minimum bias (MB) trigger accepts ~87% of the NSD cross section, corresponding to 26.1 ± 2.0 mb[3]. The TPC provides charged particle tracking over $|\eta| < 1.2$ in full azimuth. The BEMC, partially commissioned in 2004 (2400 out of 4800 towers, full azimuthal coverage) covered $0 < \eta < 1$. In addition to the MB trigger, a high tower (HT) trigger that requires one calorimeter tower above an E_T threshold corresponding to ~2.5 (~3.0) GeV at $\eta = 0$ (1) on top of the MB condition was used. This trigger detects energetic π^0, γ and electrons. In order to enrich the sample of highly energetic processes, the MB trigger was highly prescaled during data taking.

Events were accepted for analysis if the primary vertex was within 60 cm of the center of the detector along the beam axis in order to ensure a uniform tracking efficiency. For HT triggered events an $E_T > 3.5$ GeV in the trigger tower was required to ensure uniform trigger efficiency over the full BEMC acceptance. Jet finding was performed with the midpoint-cone algorithm[4], using all charged tracks originating from within 3 cm of the primary vertex and all calorimeter towers. To reject background events, a sizable contribution from charged tracks to the total jet energy was required. To avoid edge effects at the extreme ends of the detector acceptance, the jet axis was required to be within $0.2 < \eta < 0.8$.

In order to extract the inclusive jet cross section a correction for detector efficiency and resolution is required. The bin-by-bin correction factors were obtained by studying PYTHIA (v6.205)[5] events passed through a GEANT model, a response simulator of the STAR detector and the full data reconstruction frame work. The jet p_T resolution was determined to be 25%, which motivated the choice of the binning. A separate determination of the correction factors was done for MB and HT events. While the MB correction factor is close to unity, the HT correction factor is dominated by the trigger efficiency and varies over two orders of magnitude with p_T. The correction factors for both triggers are shown in figure 1a.

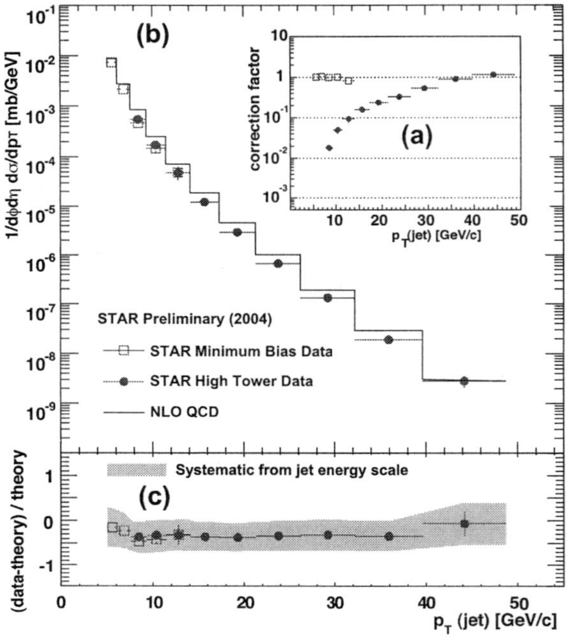

Figure 1. (a) The correction factor for MB and HT data obtained from PYTHIA simulations. Statistical errors are shown. (b) Preliminary inclusive jet cross section compared to NLO pQCD calculation. Statistical uncertainties only. (c) Ratio comparison of data vs. theory. The shaded band represents the dominant systematic uncertainty from the jet energy scale, and an 8% overall normalization uncertainty is not shown. See text for details.

Figure 1b shows the corrected cross section. In the three overlapping bins, good agreement is seen between MB and HT triggered events. The data are compared to NLO pQCD calculations incorporating CTEQ06M PDFs with $\mu_F = \mu_R = p_T$[6]. Figure 1c shows the ratio of data - theory divided by theory. The shaded band indicates the dominant systematic uncertainty (50% change in yield) introduced by a 10% uncertainty in the jet energy scale. For $p_T > 10$ GeV there is a systematic offset between data and theory, but there is good agreement within the large systematic uncertainties. It is also apparent that the spectral shape of the data agrees very well with the theory predictions. Improved calibrations and future measurements of di-jets and γ-jet coincidences are expected to reduce the uncertainties of the jet energy scale.

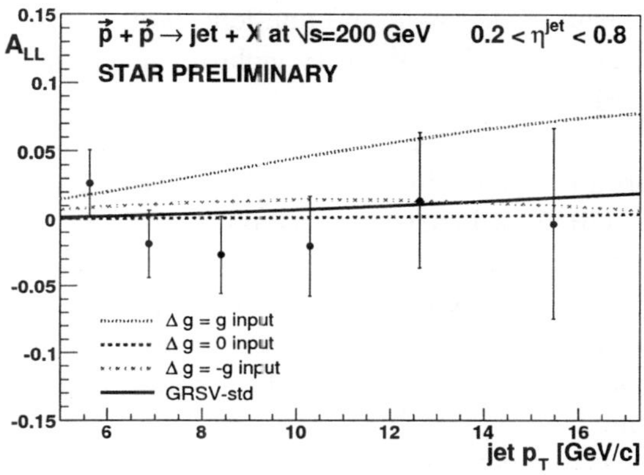

Figure 2. Double longitudinal spin asymmetry A_{LL} for inclusive jet production in polarized p+p collisions at 200 GeV as a function of jet p_t. Errors are statistical only, systematic uncertainties (without polarization) are ∼0.01. The curves show predicted asymmetries for different gluon polarizations, see text for details.

3. Spin Asymmetry

The cross section for inclusive jet production depends on the spin configuration of the two colliding protons. In leading order, $gg \to gg$, $gq \to gq$ and $qq \to qq$ subprocesses contribute to the cross section. The relative contribution of these subprocesses to the spin asymmetry in the cross section is p_T dependent. The dominant contributions in the p_T region covered by STAR are from gg and qg scattering, with gg subprocesses dominating at low p_T and qg at higher p_T[6].

The double longitudinal spin asymmetry is defined as

$$A_{LL}^{jet} = \frac{1}{P_1 P_2} \frac{N^{++} - RN^{+-}}{N^{++} + RN^{+-}} \tag{1}$$

where $N^{++(+-)}$ are the inclusive jet yields for equal (opposite) spin orientations of the colliding protons, R $= L^{++}/L^{+-}$ is the ratio of luminosities for the different spin configurations and $P_{1(2)}$ are the polarizations of the two proton beams. The relative luminosities and the orientation of the polarization vector in the STAR interaction region is measured with the BBCs. The beam polarization is obtained with the RHIC polarimeters[7].

Figure 2 shows preliminary results for the double longitudinal spin asymmetry A_{LL} in inclusive jet production for the datasets taken in 2003 and 2004. Systematic uncertainties originating from the relative luminosity, trigger bias, beam background and non-longitudinal spin contributions have been investigated. Studies of parity violating single spin asymmetries and randomized spin patterns show no evidence for bunch to bunch or fill to fill systematics. The curves in figure 2 show theoretical evaluations of A_{LL}^{jet} in inclusive jet production for different sets of polarized gluon distributions functions[6,8]. The GRSV-std curve is based on the best fit to DIS data, the other curves use a vanishing gluon polarization $\Delta g(x, Q_0^2) = 0$ and maximally positive and negative gluon polarization $\Delta g(x, Q_0^2) = \pm g(x, Q_0^2)$ at an input scale of $Q_0^2 = 0.6$ GeV^2/c^2. The data are consistent with three of these evaluations and tend to disfavor the scenario with a large positive gluon polarization.

So far, the measurements are limited by statistical precision. The data taken in 2005 are in the final stages of analysis and will provide a significant reduction of statistical errors due to higher polarization and an order of magnitude increase in sampled luminosity. Further improvement is expected with increasing integrated luminosity and improved polarization in future runs.

4. Conclusion

We have presented preliminary measurements of the cross section and the double longitudinal spin asymmetry in inclusive jet production in polarized p+p collisions at $\sqrt{s} = 200$ GeV. The cross section is in reasonable agreement with NLO pQCD calculations over seven orders of magnitude, motivating the application of these calculations to interpret the measured spin asymmetries. The preliminary double-longitudinal spin asymmetry is consistent with an evaluation based on a fit to DIS results and disfavors large positive values for the gluon polarization.

References

1. G. Bunce *et al.*, *Annu. Rev. Nucl. Part. Sci.* **50**, 525 (2000).
2. K.H. Ackerman *et al.*, *Nucl. Instrum. Methods* **A499** 624 (2003).
3. J. Adams *et al.*, *Phys. Rev. Lett.* **91**, 172302 (2003).
4. G.C. Blazey *et al.*, arXiv:hep-ex/0005012.
5. T. Sjostrand *et al.*, *Comput. Phys. Commun.* **135**, 238 (2001).
6. B. Jäger, M. Stratmann and W. Vogelsang, *Phys. Rev.* **D70**, 034010 (2004).
7. O. Jinnouchi *et al.*, *AIP Conf. Proc.* **675**, 817 (2003).
8. M. Glück *et al.*, *Phys. Rev.* **D63**, 094005 (2001).

NEW PHENOMENA SEARCHES AT CDF

ARON SOHA*
University of California, Department of Physics
One Shields Avenue, Davis, California 95616, USA
E-mail: soha@fnal.gov

We report on recent results from the Collider Detector at Fermilab (CDF) experiment, which is accumulating data from proton-antiproton collisions with $\sqrt{s} = 1.96\,\text{TeV}$ at Run II of the Fermilab Tevatron. The new phenomena being explored include Higgs, Supersymmetry, and large extra dimensions. We also present the latest results of broad searches for heavy objects, including Z' bosons and heavy quarks.

1. Standard Model Higgs in $h^0 \to WW^{(*)}$

At the Tevatron, the predicted dominant channel for creating Higgs bosons is single neutral h^0 production. For masses of $m_{\text{Higgs}} > 135\,\text{GeV}/c^2$, the largest decay branching ratio is $h^0 \to WW$ (at lower masses, $h^0 \to b\bar{b}$ dominates). In the analysis reported here, one or both of W bosons can be off shell, and both are reconstructed in either of the leptonic decays $W \to e\nu$ or $W \to \mu\nu$. Prior to a series of event selection requirements, the sample is dominated by Drell-Yan events. After the selection requirements, the azimuthal angle between the two final state leptons is used to separate $h^0 \to WW$ signal from the dominant remaining background, which is Standard Model (SM) non-resonant WW production.

This analysis uses $360\,\text{pb}^{-1}$ of CDF Run II [1] data to set a 95% C.L. limit on the cross section times branching ratio, as shown in Figure 1(left). This new result extends the explored mass range to $110 - 200\,\text{GeV}/c^2$, and increases the acceptance beyond that of previous searches. Of the SM Higgs channels being studied at the Tevatron, the $h^0 \to WW$ sensitivity is the closest to the corresponding SM prediction, and is about a factor of two from the prediction of a model with a 4th generation of fermions [2].

*On behalf of the CDF collaboration.

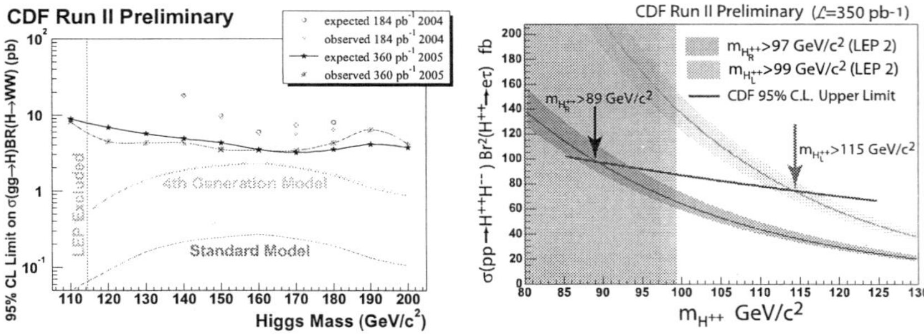

Figure 1. Left: Limit (at 95% C.L.) on the cross section times branching ratio for $gg \to h^0 \to WW$, including expectations and observations for a new result using $360\,\text{pb}^{-1}$ of CDF Run II data. Right: Doubly charged Higgs search results, showing the cross section times branching ratio squared limit, and mass limits for left- and right-handed models.

2. Search for $H^{++}H^{--}$

A search has been conducted for pair produced doubly charged Higgs bosons, which, in certain models [3], may be as light as $100\,\text{GeV}/c^2$ and decay primarily to leptons. The production is through $q\bar{q} \to Z^0/\gamma^* \to H^{++}H^{--}$. The lepton-flavor violating decay is theoretically unconstrained, allowing for the powerful experimental signature of simultaneous $H^{++} \to \tau^+e^+$ and $H^{--} \to \tau^-e^-$ decays. A requirement of three or four isolated leptons reduces the hadronic (QCD) component of the backgrounds. An additional requirement of $\not{E}_T + \Sigma p_T(\text{leptons}) > 190\,\text{GeV}$ and a veto against candidate Z^0 bosons reduce the remaining electroweak backgrounds.

The analysis uses $350\,\text{pb}^{-1}$ of CDF Run II data where, with an expectation of 0.25 events from backgrounds, zero events are observed. This gives a 95% C.L. limit of $\sigma(p\bar{p} \to H^{++}H^{--}) \times Br^2(H^{++} \to e\tau) < 73.5$ fb, and mass limits of $m_{H_L^{++}} > 115\,\text{GeV}/c^2$ and $m_{H_R^{++}} > 89\,\text{GeV}/c^2$ for left- and right-handed doubly charged Higgs bosons, respectively. Figure 1(right) shows these results. The recently added muon channel includes a tri-lepton event that passes all analysis requirements except for that of the reconstructed mass of the candidate H^{++}.

3. Large Extra Dimensions

Models of large extra dimensions propose a $4+n$ dimensional *bulk* of space-time where only gravitons can propagate in the n extra dimensions [4]. At the Tevatron, gravitons could be produced directly, in association with a gluon or quark, in $q\bar{q} \to gG$, $qg \to qG$, or $gg \to gG$. In each case, the signature is an energetic jet and missing transverse energy. This analysis requires jet $E_T > 150\,\text{GeV}$ and $\not{E}_T > 120\,\text{GeV}$.

The largest SM background is from $Z^0 \to \nu\nu$ +jets events, which give $\not{E}_T$ due to the neutrinos. There are smaller contributions from $W \to \ell\nu$ +jets, where there is $\not{E}_T$ from the neutrino or a lost lepton, and from QCD processes, where there is $\not{E}_T$ from the mis-measurement of jets. The total background expectation, for the $368\,\text{pb}^{-1}$ of CDF Run II data considered, is 265 ± 30 events[a]. The observation of 263 events is consistent with background and is used to place lower limits on the effective Planck scale, M_D, of the extra dimensions, and upper limits on the size, R, of the extra dimensions, assuming compactification on a torus. The two quantities are related to the Planck mass, M_{Planck}, through the expression $M^2_{\text{Planck}} \sim R^n M_D^{2+n}$. The 95% C.L. lower limits on M_D are given in Figure 2(left), with corresponding upper limits on R of $R < 0.36\,\text{mm}$, $R < 3.7 \times 10^{-6}\,\text{mm}$, $R < 1.1 \times 10^{-8}\,\text{mm}$, $R < 3.5 \times 10^{-10}\,\text{mm}$, and $R < 3.4 \times 10^{-11}\,\text{mm}$ for $n = 2, 3, 4, 5$, and 6.

4. Heavy Objects

A search for a heavy Z' object is carried out by looking for a peak in the di-electron mass, M_{ee}, and a distortion in the $\cos(\theta^*)$ distribution [5]. The dominant background is from Drell-Yan production and is estimated using Monte Carlo simulation. For the models tested, it is found that the data is more consistent with Z/γ^* plus backgrounds than with $Z/\gamma^*/Z'$ plus backgrounds, where this notation indicates that a Z' would interfere with the Z/γ^*, just as Z and γ^* interfere in the SM. In $448\,\text{pb}^{-1}$ of CDF Run II data, 120 events were observed for $M_{ee} > 200\,\text{GeV}/c^2$, compared to 115^{+16}_{-19} events expected from the SM. This is used to set a 95% C.L. lower limit on the mass of a Z' in the *sequential* model, which is often used for comparisons, of $M^{(\text{seq})}_{Z'} > 855\,\text{GeV}/c^2$. The M_{ee} and $\cos(\theta^*)$ distributions are shown in Figure 3.

Another broad approach, open to such models as heavy quarks, extra dimensions, and Supersymmetry, is to perform a signature based search for anomalous dilepton($e\mu$)+X events, where X can be large E_T jets, b-quark jets, a third lepton, large $\not{E}_T$, or large event H_T, where $H_T = E_T(e) + p_T(\mu) + E_T(\text{jets}) + \not{E}_T$. Here this strategy is applied to study a heavy quark model with down-type iso-singlet right-handed quarks [6]. The search looks

[a]Uncertainties include both statistical and systematic contributions.

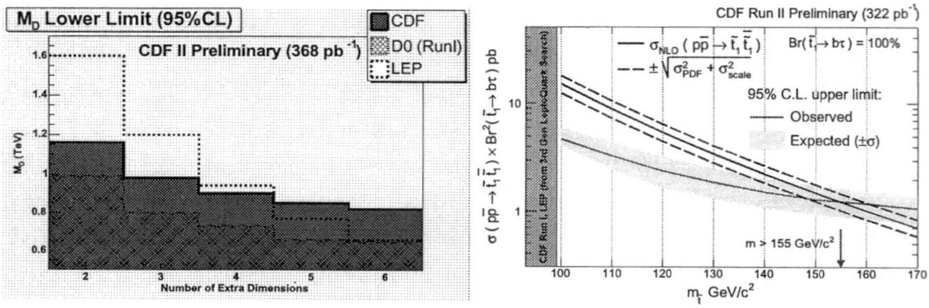

Figure 2. Left: Lower limits on the effective Planck scale, M_D, for extra dimensions. Right: The 95% C.L. limits for the R-parity violating Supersymmetric top search.

for an $e\mu$ pair plus two or more jets with $E_T > 50\,\text{GeV}$ and probes for any excess in the region $H_T > 400\,\text{GeV}$. With no events seen in $305\,\text{pb}^{-1}$ of CDF Run II data, compared to an expectation of 0.802 ± 0.440 events from the SM, a 90% C.L. limit is set at $4.49 \times \sigma_Q$, where $\sigma_Q = 0.290$ pb would be the cross section for a $300\,\text{GeV}/c^2$ heavy quark.

Finally, a search is made for heavy objects, which could be heavy quarks, Z', or Supersymmetric particles, for example, that decay to Z^0 bosons. High p_T Z^0 bosons are reconstructed in the $Z^0 \to ee$ and $Z^0 \to \mu\mu$ channels, with a requirement of $66 < M_{\ell\ell} < 116\,\text{GeV}/c^2$. In the future, W decays and the presence of additional objects such as a photon or b-quark jet will be considered. With $305\,\text{pb}^{-1}$ of CDF Run II data, the observed Z^0 p_T spectrum agrees with SM background predictions from Monte Carlo simulation. Limits are obtained on the differential cross section of extra Z^0 production as a function of p_T. For the $300\,\text{GeV}/c^2$ heavy quark model, the 95% C.L. upper limit on the cross section is $\sigma < 0.170 \pm 0.005$ pb.

5. Search for R-Parity Violating Supersymmetric Top

A search is carried out for pair produced Supersymmetric top (stop) quarks at the Tevatron. If R-parity is violated, each stop quark could decay to a tau lepton and b-quark. Our experimental signature is two jets, a lepton or muon from one tau decay, and a hadronic tau decay (τ_h). Backgrounds from Z^0+jets and QCD are reduced using a requirement of $p_T(\text{lepton}) + p_T(\tau_h) + \not{E}_T > 85\,\text{GeV}$. The W+jets background, where a jet fakes the hadronic tau, is reduced using the transverse mass of the lepton+$\not{E}_T$ system. In $322\,\text{pb}^{-1}$ of CDF Run II data, 2 events are seen, compared to $2.25^{+0.46}_{-0.22}$ events expected from SM sources. This gives a 95% C.L. lower limit on the stop mass of $m_{\text{stop}} > 155\,\text{GeV}/c^2$. The results are shown in Figure 2(right).

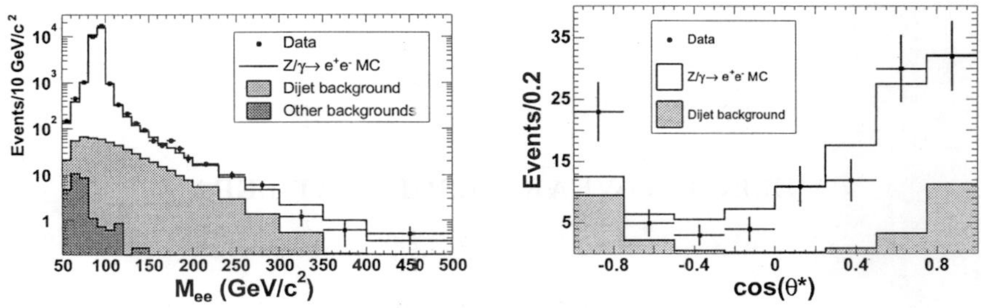

Figure 3. The M_{ee} (left) and $\cos(\theta^*)$ (right) distributions for a Z' search. The $\cos(\theta^*)$ distribution includes a requirement of $M_{ee} > 200\,\mathrm{GeV}/c^2$.

6. Conclusions

We have presented a portion of the new phenomena search results available from CDF. In the Higgs sector, the channel $h^0 \to WW^{(*)}$ is the most sensitive for higher mass SM Higgs, and limits have also been set for $H^{++}H^{--}$. In a search for large extra dimensions, limits have been set on the mass scale and radii. For heavy objects, limits have been set for Z', dilepton+X, and high p_T Z^0 production models. Finally, limits have been set for the production and decay of R-parity violating Supersymmetric top quarks. At CDF, both the data sample and potential for discovery continue to grow.

7. Acknowledgments

We thank our colleagues within the CDF collaboration, the Lake Louise conference organizers and participants, and the funding agencies for making this work possible.

References

1. F. Abe *et al.*, *Nucl. Instrum. Methods Phys. Res.* **A271**, 387 (1988); D. Amidei *et al.*, *Nucl. Instum. Methods Phys. Res.* **A350**, 73 (1994); F. Abe *et al.*, *Phys. Rev.* **D52**, 4784 (1995); P. Azzi *et al.*, *Nucl. Instrum. Methods Phys. Res.* **A360**, 137 (1995); The CDFII Detector Technical Design Report, Fermilab-Pub-96/390-E.
2. E. Arik, O. Cakir, S. A. Cetin, and S. Sultansoy, *Phys. Rev.* **D66**, 033003 (2002).
3. J. F. Gunion *et al.*, *Phys. Rev.* **D40**, 1546 (1989).
4. N. Arkani-Hamed, S. Dimopoulos, and G. Dvali, *Phys. Lett.* **B429** (1998).
5. Submitted to *Phys. Rev. Lett.*, hep-ex/0602045 (2006).
6. J. D. Bjorken, S. Pakvasa, and S. F. Tuan, *Phys. Rev.* **D66** 053008 (2002).

ELECTROWEAK RESULTS AT LEP2

PAOLO SPAGNOLO
Istituto Nazionale di Fisica Nucleare, Sezione di Pisa,
Largo B. Pontecorvo, 3, 56100 PISA, Italy

The review of the main analyses on the W Physics at LEP2 is presented. The measurement of the W decay branching ratios is a test of the lepton universality and the CKM matrix unitarity. The study of the WW cross section to prove the non-abelian structure of the $SU(2)_L \times U(1)_Y$ gauge simmetry. The direct reconstruction of WW invariant mass allows the precise measurement of the W mass.

1. Introduction

The LEP e^+e^-collider has been the most important experimental environment to study the electroweak interactions and precisely test the Standard Model. Since 1989 each of the four LEP experiments collected data corresponding to an integrated luminosity larger than 1000 pb^-1, of which about 200 pb^-1 at a centre-of-mass energy of the Z mass peak (from 1989 to 1995). After 1996 a second phase started, LEP2, with collected lumonosity of 750 pb^-1 per experiment centre-of-mass energy above 161 GeV, allowing to produce for the first time two real (on shell) massive bosons WW, ZZ, decaying in a four fermions final state. These new channels opened at LEP2, allow important measurements in W Physics, like the precise determination of the mass of the W and the decay W couplings and are also sensitive to the trilinear gauge couplings (TGC) predicted in the Standard Model [1]. This proceeding is a review of these important analyses. LEP ended collecting data in 2000 but only recently all the most important analyses in the W sector have been completed and the results of the four LEP experiments combined by the LEP Electroweak Working Group [2]. All the results here presented have been published by ALEPH [3], L3 [4], OPAL [5] and DELPHI [6] collaborations.

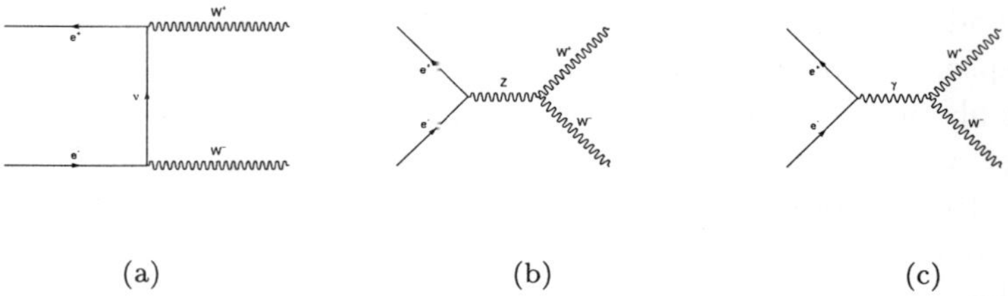

Figure 1. Diagrams of the WW production at LEP2

2. W couplings

Three diagrams are responsible for the WW production as shown in fig. 1: the *t-channel* with the neutrino exchange (fig. 1,a), the *s-channel* with a Z exchange (fig. 1,b) and the *s-channel* with a virtual photon exchange (fig. 1,c). All the possible 4-fermions processes from WW decays, shown in the tabel below together with the branching ratios and the typical efficiencies, are reconstructed at LEP experiments. Event selections are needed

Decay Mode	BR	averaged ε
qqqq	45 %	85 %
$\mu\nu$ qq	15 %	80 %
eν qq	15 %	80 %
$\tau\nu$ qq	15 %	60 %
$\ell\nu\ell\nu$	10 %	65 %

to disentangle WW events from the 2-fermion processes and to classify the events in three different channels (hadronic, leptonic and semileptonics). Decays to electrons, muons and taus can further be tagged. The typical signatures of the three topologies allow to separate WW events among each others and from background with good efficiencies and excellent purities. The cross sections can be readily computed from the number of events selected, the estimated background, the total luminosity and from the efficiency. By comparing the cross sections in the different final channels, it is possible to compute the branching ratios of the W in all the possible final states. The branching ratios of the leptonic and the hadronic channels measured by the four LEP experiments are shown in fig. 2. All the results are in good agreement with the Standard Model. From the mutual ratios of the leptonic branching ratios is it possible to test the W lepton universality at about 1% of precision level, obtaining $g(\mu)/g(e) = 0.997 \pm 0.010$, $g(\tau)/g(e) = 1.034 \pm 0.015$ and $g(\tau)/g(\mu) = 1.037 \pm 0.014$.

The electron and muon couplings are consistent, while the tau coupling is consitently larger:if the electron and muon couplings are assumed to be the same and combined, the tau result is 3σ higher. From the hadronic decay fraction it is possible to extract an hadronic coupling as well. The results is in impressive agreement with the leptonic coupling: $g(q)/g(\ell) = 1.000 \pm 0.006$. The W hadronic branching ratio can also be used to test the unitarity of the CKM matrix in the first two families. From the measured hadronic branching ratio one obtains $\Sigma|V_{ij}|^2 (i = u, c; j = d, s, b) = 2.000 \pm 0.026$ and eventually extract $|V_{cs}|$ from all the other known CKM parameters [7] $|V_{cs}| = 0.976 \pm 0.014$.

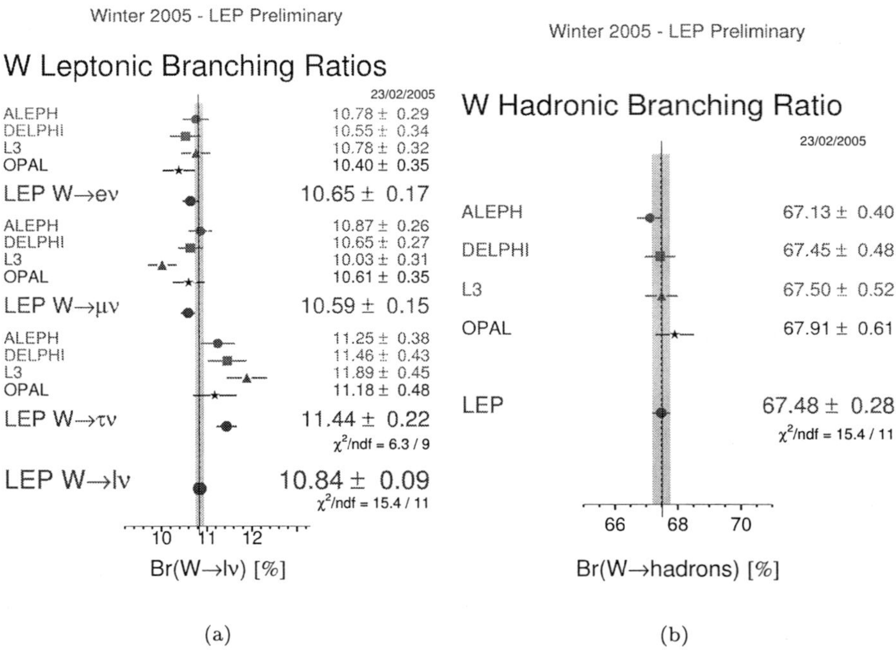

Figure 2. W decay branching ratios measured at LEP for the leptonic (a) and hadronic (b) channel

3. Triple gauge boson couplings

The triple gauge boson couplings (TGC) can be classified into charged couplings (cTGC) when the triple boson vertex is WWγ or WWZ and neutral couplings (nTGC) for the case of ZZZ, ZZγ or Z$\gamma\gamma$. The first are forseen in the Standard Model and are linked to the non-abelian structure of the $SU(2)_L \times U(1)_Y$ gauge simmetry. The neutral nTGC at the contrary, do not exist in the Standard Model at tree level. In order to measure the charged triple gauge couplings (cTGC) at LEP is necessary to have a vertex with two W and a neutral boson (Z,γ). Within the Standard Model, this vertex can be achieved through the production of two massive W with the $e^+e^- \to W^+W^-$ process available for the first time at LEP2 or, with lower sensitivity, through the single W production $e^+e^- \to W^\mp e^\pm \nu(\bar{\nu})$ and through the single photon production $e^+e^- \to \gamma\bar{\nu}_e\nu_e$.

The first proof of the existence of the charged TGC is in the measurement of the WW cross section. Both the *s-channel* diagrams in fig. 1 have a TGC vertex. All the three diagrams are needed to make the WW cross section converging to finite values at high energies. If the $SU(2)_L \times U(1)_Y$ was an abelian gauge simmetry, the TGC verteces would be forbidden and the WW cross section would diverge when increasing the centre-of-mass energy. The LEP2 data fit precisely the Standard Model predictions and confirm the presence of the TGCs and the non-abelian structure of the $SU(2)_L \times U(1)_Y$ gauge simmetry, as can be viewed in fig. 3.

The most general form for an effective charged TGC Lagrangian consistent with Lorentz invariance involves 14 complex couplings, 7 for the WWγ and 7 for the WWZ vertex. Most of these couplings are C- or P-violating while in the Standard Model C- and P-conservation are predicted in the TGCs sector. Assuming C and P conservation the 14 complex couplings are reduced to 6 real couplings: g_1^γ, g_1^Z, k_γ, k_Z, λ_γ and λ_Z. In the Standard model $g_1^\gamma = 1$, $g_1^Z = 1$, $k_\gamma = 1$, $k_Z = 1$, $\lambda_\gamma = 0$ and $\lambda_Z = 0$. The couplings can be related to physical properties of the gauge bosons, like the electric dipole, quadrupole and magnetic moment. For instance the W anomalous magnetic moment can be written as $\mu_W = \frac{e}{2M_W}(1 + k_\gamma + \lambda_\gamma)$. The requirement of local $SU(2)_L \times U(1)_Y$ gauge invariance introduce the further constraints $\Delta k_Z = -\Delta k_\gamma \tan^2\theta_W + \Delta g_1^Z$, $\lambda_\gamma = \lambda_Z$ and $g_1^\gamma = 1$ with Δ indicating the deviation from the Standard Model predictions and θ_W the electroweak mixing angle, leaving 3 independent real couplings [8]: g_1^Z, k_γ and λ_γ. These couplings have been experimentally tested with the $e^+e^- \to W^+W^-$ sample collected at LEP2. The angular distribution of the

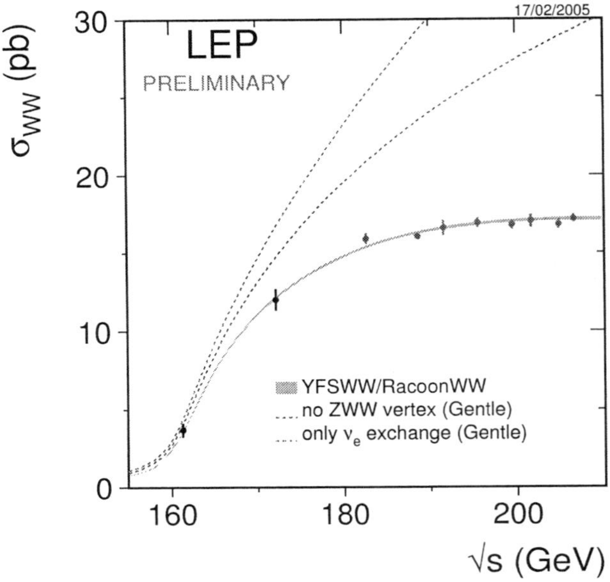

Figure 3. WW cross section measurement at LEP

$e^+e^- \to W^+W^-$ cross section are more sensitive to TGCs than the inclusive measurement. The results obtained on the three triple gauge couplings average from the combination of the LEP experiments are shown in fig. 4.

All the three cuplings are consistent with the Standard Model expectation, with a few percent precision.

ALEPH also performed a fit to all the 14 complex couplings, relaxing all the constraints on C- and P-conservation and $SU(2)_L \times U(1)_Y$ gauge invariance. Out of all the 28 real parameters, one at the time is allowed to vary and the other are fixed to the Standard model predictions. The results of this test are:

$$
\begin{aligned}
Re(g_1^\gamma) &= 1.123 \pm 0.091 \\
Re(g_1^Z) &= 1.066 \pm 0.073 \\
Re(k_\gamma) &= 1.071 \pm 0.062 \\
Re(k_Z) &= 1.065 \pm 0.061
\end{aligned}
\qquad (1)
$$

All the other 24 parameters are consistent with zero in perfect agreement with the Standard Model.

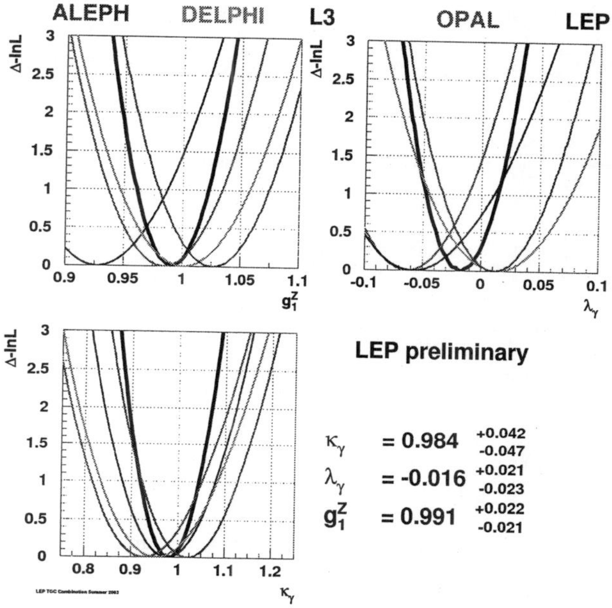

Figure 4. Fit results on the triple gauge couplings.

4. W mass and width

The LEP2 energies allow the direct measurement of the W mass from the event-by-event reconstruction of the W-pairs invariant mass. An alternative method measures the W mass from the threshold cross section neasurements of $e^+e^- \to W^+W^-$ events, yelding $m_W = 80.4 \pm 0.2$ GeV/c^2. All the possible semileptonic $qq\ell\nu$ and fully hadronic qqqq decays of the W-pairs are taken into account to directly measure the W mass with the neutrino energy estimated from the missing energy of the semileptonic events. The W mass is extracted from different fit methods to the invariant mass distribution. The resolution is improved by imposing the energy-momentum conservation constraints from the LEP2 energy and applying the equal mass constraint for the two Ws of each event. The statistical power of the direct measurement of the W mass is about 20 MeV, an order of magnitude more precise than the threshold method. However many systematic effects have to be taken into account. The fully hadronic channels also suffers for possible final state interconnection effects, like Bose-Einstein correlations and the Colour reconnection that increase the total systematic uncertainty and threfore the

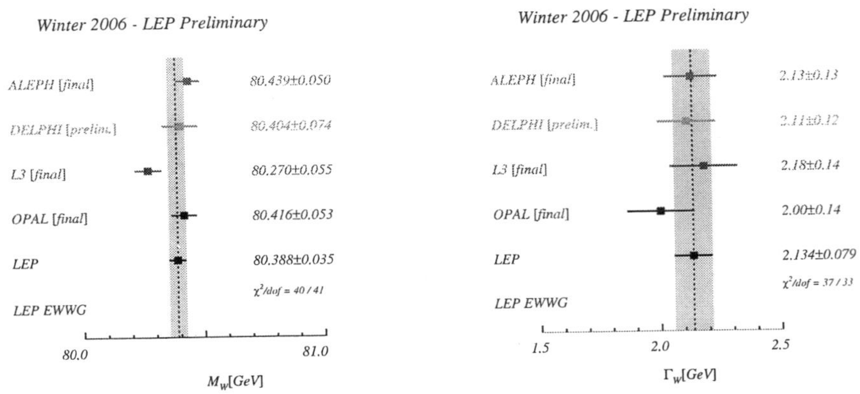

(a) (b)

Figure 5. W mass (a) and width (b) measured at LEP

total error. The measured W mass is $m_W(qq\ell\nu) = 80.385 \pm 0.038 \ \ GeV/c^2$ in the hadronic channel and $m_W(qqqq) = 80.391 \pm 0.061 \ \ GeV/c^2$ in the semileptonic decay. From the same fits of the mass measurements it is possible also to extract the width of the W shape. All the results of the four LEP experiments are showed in fig. 5 together with the LEP averages. The averaged measurements at LEP of the W mass and width are

$$m_W = 80.388 \pm 0.035 \ \ GeV/c^2 \tag{2}$$

$$\Gamma_W = 2.134 \pm 0.079 \ \ GeV/c^2. \tag{3}$$

In fig 6 (a) the LEP result is averaged with the Tevatron measurements. The inclusion of the current W mass measurement in the global electroweak fit yelds a constraint on the Higgs boson mass fitted in fig 6 (b). The 95% confidence level limits on the Higgs mass are:

$$114 < m_H < 219 \ \ GeV/c^2 \tag{4}$$

where the lower limit is the one excluded by LEP direct search.

5. Conclusions

Three important results in W Physics are obtained at LEP2. W couplings measurement allows a test of lepton universality and CKM matrix unitarity.

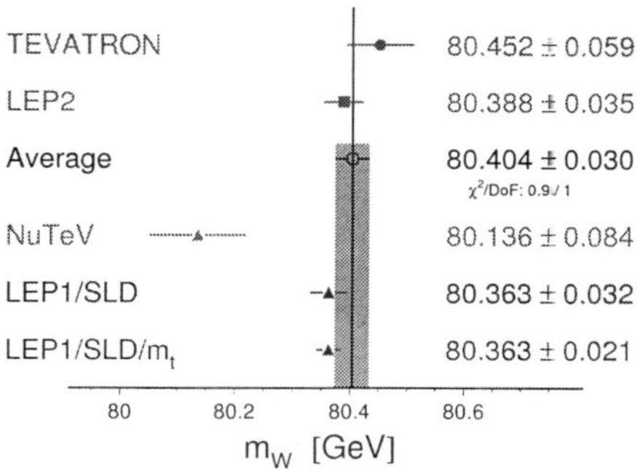

(a)

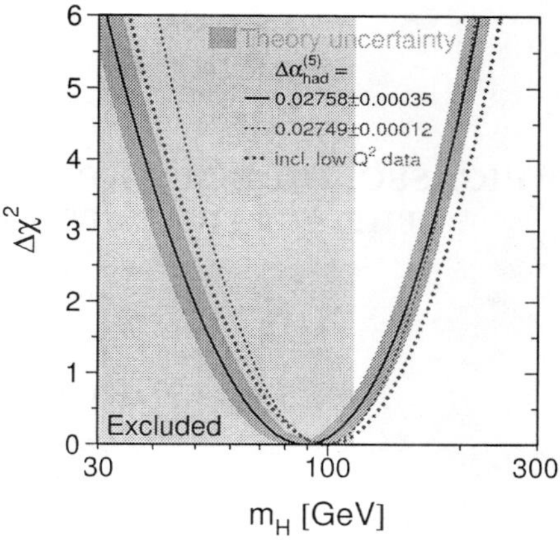

(b)

Figure 6. The LEP measured W mass is averaged with the Tevatron results (a) to costraint the Higgs mass (b) with the global electroweak fit

The trilinear couplings TGC are measured with a precision of few percent to prove for the first time the non-abelian stucture of the $SU(2)_L \times U(1)_Y$ gauge simmetry. The W mass is measured with a high precision, and the results is used to constraint the Higgs mass.

Acknowledgments

I would like to thank the LEP Electroweak Working Group for the combination results and the plots here presented.

References

1. G.Gounaris *et al*, in *Physics at LEP 2*, Report CERN 96-01 (1996), eds G.Altarelli, T. Sjostrand, F. Zwirner, Vol. 1, p.525.
2. The LEP-TGC combination group, LEPEWWG/TGC/2004-01, 2004.
3. ALEPH Collaboration, *Phys. Lett.* B **614**, 7 (2005).
4. L3 Collaboration, *Phys. Lett.* B **586**, 151 (2004).
5. OPAL Collaboration, *Eur. Phys.J.* C **33**, 463 (2004).
6. DELPHI Collaboration, *Measurement of charged TGC* DELPHI 2003-051 (July 2003) CONF-671.
7. Particle Data Group, S. Aidelman *et al*, *Phys. Lett.* B **592**, 1 (2004)
8. K.Gaemers and G.Gounaris*Z. Phys.* C **1**, 259 (1979).

PICASSO: DIRECT DARK MATTER DETECTION USING THE SUPERHEATED DROPLET TECHNIQUE

C. E. STOREY[1]
FOR THE PICASSO COLLABORATION
Queen's University
E-mail: connie@owl.phy.queensu.ca

K.CLARK[1], C.KRAUSS[1], A.NOBLE[1], F.AUBIN[2], M.AUGER[2], G.AZUELOS[2], M.BARNABÉ-HEIDER[2], M.DIMARCO[2], P.DOANE[2], M.-H.GENEST[2], R.GORNEA[2], R.GUÉNETTE[2], C.LEROY[2], L.LESSARD[2], J.-P.MARTIN[2], N.STARINSKY[2], U.WICHOSKI[2], V.ZACEK[2], E.BEHNKE[3], W.FEIGHERY[4], I.LEVINE[3], C.MUTHUSI[4], S.POSPISIL[5], J.SODOMKA[5], I.STEKL[5], S.KANAGALINGAM[6], R. NOULTY[6]

[1] *Department of Physics, Queen's University, Kingston, ON, Canada*
[2] *Départment de Physique, Université de Montéal, Montréal, QE, Canada*
[3] *Department of Physics & Astronomy, Indiana University, South Bend, South Bend, IN, USA*
[4] *Chemistry, Indiana University, South Bend, South Bend, IN, USA*
[5] *Czech Technical University, Prague, Czech Republic*
[6] *Bubble Technology Industries, Chalk River, ON, Canada*

The PICASSO experiment, located in the Creighton Mine in Sudbury, Ontario, Canada, has demonstrated the viability of the superheated droplet technique in direct dark matter detection. Experimental sensitivity is to spin dependent elastic collisions with ^{19}F as the active mass. Limits of $\sigma_p = 1.31$pb and $\sigma_n = 21.5$pb on the WIMP-proton and WIMP-neutron cross-sections respectively, at a 90% C.L. for a WIMP mass of 29 GeV/c^2 were determined in 2004[1] using a small scale setup with 20g of active mass. Details of the PICASSO experiment, along with the improvements incorporated into the upcoming PICASSO 32 phase will be outlined in these proceedings.

1. Dark Matter

Early indications that there is more mass in the universe than we can see were first introduced by Fritz Zwicky's work in 1933[2] with his studies of the Coma Cluster. Later, in the 1970s, Vera Rubin demonstrated that the mass of a galaxy calculated from its luminosity didn't agree with the mass required for the galaxy to maintain a constant velocity at increasing radii. To satisfy Newtonian physics, a dark matter halo was proposed to provide the missing mass[3]. These dark matter particles have been postulated to be massive, non-relativistic, abundant and weakly interacting. These particles have been termed Weakly Interacting Massive Particles (WIMPs).

All indications for the existence of dark matter are cosmological in origin. Studies of gravitational lensing, supernova redshifts, and galaxy clustering have all concluded that there is more contained in the universe than was previously thought. The Wilkinson Microwave Anisotropy Probe (WMAP) results[4] require the existence of dark matter, provided that the primordial fluctuations are adiabatic with a power law spectrum.

There are a variety of theories attempting to describe the universe and the particles contained within. One of these, Supersymmetry (SUSY), predicts that for every particle in the Standard Model, there exists a supersymmetric partner. Most of the SUSY models also predict the existence of a stable particle known as the LSP (Lightest Supersymmetric Particle) which, if neutral, is a strong WIMP candidate. The LSP is a combination of the SUSY partners of the Standard Model force carriers, and is commonly referred to as the neutralino. The goal of PICASSO is therefore to try to directly detect the neutralino, the preferred dark matter particle candidate.

2. PICASSO

PICASSO, (Project In CAnada to Search for Supersymmetric Objects), is a direct dark matter search. PICASSO uses superheated droplet detectors (SDD) with predominant sensitivity to the spin-dependent neutralino interaction on a C_4F_{10} target. An indepth understanding of detector response to various sources of background is essential, as are radioactively clean detectors, a low background environment, a large active mass, and proper modeling of detector response.

Neutralino interaction with the target medium depends on the nature of the neutralino and whether or not the target particle has an unpaired nucleon. PICASSO is a spin-dependent experiment since the chosen target nucleus ^{19}F has an unpaired proton. The interaction cross-section is therefore proportional to the square of the spin of the unpaired nucleon, and detector sensitivity is predominately in the WIMP-proton cross-section σ_p. The choice was based on a number of reasons, such as availability, safety, cost, and a low boiling point.

2.1. *Superheated Droplet Technique*

Similar to bubble chambers, PICASSO detectors use superheated fluid (SHF) as the active material. Any liquid held at a temperature above its boiling point is defined as being in a superheated state. This metastable state will undergo a violent and quick phase transition from liquid to vapour if the conditions for stability are no longer met. It is important to note that because of the SHF, SDD are therefore are only sensitive to radiation when operated in a specific range of pressures. At high pressure, any irradiated bubbles return to their fluid state.

Bubble chambers are filled entirely with SHF, so they can only be at operational temperature and pressure for the briefest of moments before the liquid has undergone the phase transition and therefore must be re-pressurized. In order to run the detectors for extended time periods, it is necessary to distribute the SHF such that the effects of one incident particle will remain localized and the remainder of the detector continues to be sensitive to radiation. Dispersal of the fluid in tiny droplet form uniformly throughout a gel matrix of equivalent density solves this issue and enables a high duty cycle.

2.2. *Creating and Understanding Signals*

The mechanics of bubble formation are founded in the Seitz thermal spike theory[5], and have been further applied to SDD by R. Apfel[6]. Based on Seitz's theory, bubble formation is governed by temperature, pressure, bubble size, surface tension and the energy deposited along the track of the incident particle. To initiate a phase change, an incident particle must create a nucleation site which will cause the remainder of the droplet to transition. To create this site, the particle needs to deposit a critical energy, E_c (Eq. 1) within a critical track length related to a critical radius, R_c (Eq. 2) by a constant a, $l_c = aR_c$.

Each of these factors are pressure dependent on $p_e = p_v(T) - p_0(T_0)$, as well as on the droplet surface tension, $\sigma(T)$. All factors are temperature dependent, where p_v is the vapour pressure, temperatures T and T_c are current and the critical values, and T_0, σ_0, and p_0 are the values taken at a reference point.

$$E_c = \frac{16\pi}{3} \frac{\sigma(T)^3}{(p_e)^2} \tag{1}$$

$$R_c = \frac{2\sigma(T)}{p_e} \tag{2}$$

$$\sigma(T) = \sigma_0 \frac{T_c - T}{T_c - T_0} \tag{3}$$

SDD have a high level of temperature dependence, as illustrated by the above arguments. SDD are threshold detectors; as temperature T increases, $p_e(T)$ increases while $\sigma(T)$ decreases. The combination causes $E_c(T)$ to decrease nonlinearly as T rises. Figure 1 illustrates the sensitivity of the detectors to different types of particles at different temperatures as a function of threshold energy. Low ionizing particles such as gammas cannot deposit

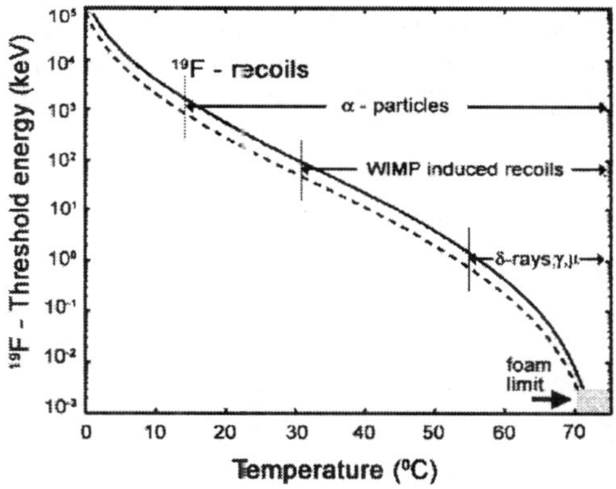

Figure 1. Energy thresholds as a function of temperature for ^{19}F recoils. Detector efficiencies of 50%(broken line) and 80% (continuous line), with sensitivity to alpha-induced recoils beginning at 15°C. WIMP-induced recoils are detected above 30°C. Sensitivity to minimally-ionizing particles such as γ-rays occurs around 55°C. At the foam limit, the detectors become unstable and begin to break down.

enough energy within the critical radius to cause bubble formation unless the detector is operating at a temperature high enough to have reduced the threshold energy. The PICASSO detectors are therefore operated over a temperature range which is almost totally insensitive to some of the most common backgrounds (gammas, betas, muons).

Projected WIMP count rates differ depending on the WIMP mass as illustrated in Figure 2. If the WIMP is a light particle, it will be more abundant, hence the higher count rates at increased temperature, where the detectors are more efficient.

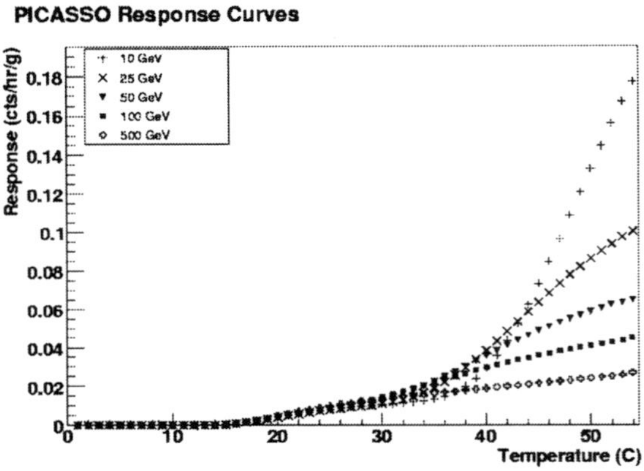

Figure 2. Detector response as a function of temperature, with a neutralino-fluorine cross-section of 2000pb. Projected count-rates are for combined alpha backgrounds and the neutralino signal, for neutralino masses of 10, 25, 50, 100 and 500 GeVc^{-2}

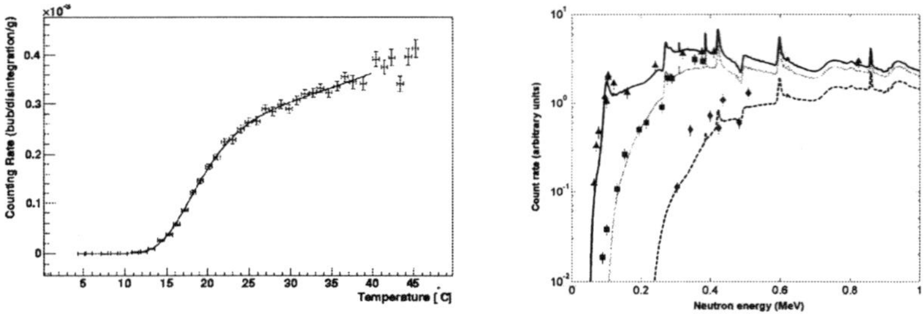

(a) Alpha Response

(b) Neutron Response

Figure 3. (a) Alpha response as a function of temperature ($^\circ C$) of a detector spiked with ^{241}Am. Data points with error bars are shown, with the fit drawn as the continuous line. Figure (b) Neutron response for temperatures of $10(^\circ C)$ (dashed line), $15(^\circ C)$ (dotted line), and $20(^\circ C)$ (plain line). The fits to the data (continuous lines) produce an exponential temperature dependence for threshold energy.

It is essential that all backgrounds and their effects are fully understood and accounted for. Calibrations performed at the Université de Montréal include spiking detectors with alpha sources, and irradiating the detectors with monoenergetic neutrons produced by a tandem accelerator from the

^{7}Li(n,p) reaction. Detector responses are measured over the temperature operating range of 18°C - 48°C, the results of which are shown in Figures 3a and 3b.

Extensive work has been done to understand the dominant alpha background. From the calibration data the alpha response is constructed and compared to the experimental data. This comparison indicates the quantity of counts which can be attributed to alpha sources.

The importance of neutron calibration is two-fold; not only is the fast neutron flux a background, and as such needs to be studied and understood, but neutralinos should behave like neutrons in the SDD. A controlled neutron source with known mass and energy has been used to predict what signals from neutralinos of various masses will be.

3. The Experiment

3.1. *The Location and Published Results*

Expected interactions with neutralinos are at such a low rate, approximately 1 event/kg/d, that there is no possibility of seeing the signals unless the background rate can be reduced to a level which does not suppress the neutralino signal. With this in mind, we have installed the PICASSO experiment in a corner of the Sudbury Neutrino Observatory (SNO) laboratory, located 2070m underground in the Creighton Mine near Subdury, Ontario. At this depth, the cosmic muon flux has been measured[7] to be less than 0.27 muons/m^2/day, the ambient thermal neutron flux from the rock is approximately 4.1x10^3 n/m^2/day, while the fast neutron flux which is the flux that the detectors are sensitive to, is around 4x10^3 n/m^2/day.

Six 1L, polypropylene detectors (Figure 6) were installed underground in 2004. Purification measures taken prior to installation included an adapted version of the HTiO ion exchange technique used by SNO on the gel compound, and a single distillation of the active liquid.

An effective exposure of 1.98±0.19 kgd was acquired, and analysis of the 3 detectors with the lowest backgrounds produced the results shown in Figure 4. The strong agreement between the alpha fit and the data points ruled out a significant WIMP signal, which would appear as a departure from the alpha fit. The limits that PICASSO was able to place on the neutron and proton cross-sections are shown in Figures 5a and 5b, respectively. Due to the choice of ^{19}F with an unpaired proton in the target, the experiment is far more sensitive to σ_p than to σ_n. Less stringent limits on σ_n were determined, and yet were competitive enough to rule out some parameter space not previously excluded.

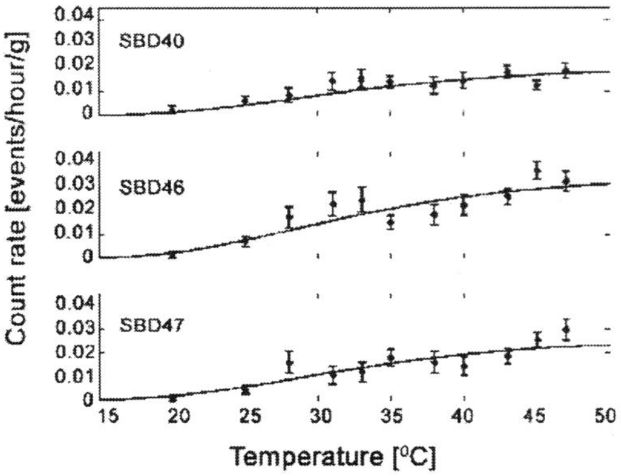

Figure 4. Published data from 3 detectors which were installed underground at SNO in 2004. Shown are the count rates as a function of temperature along with the alpha fit (continuous line) determined from calibrations.

3.2. *Improvements and PICASSO II*

With the viability of the PICASSO techniques now proven, far more competitive results can be obtained with a few experimental improvements. Internal alpha contamination reduction, larger detectors, improved temperature control, more piezoelectric sensors and a larger active mass fraction are many of the areas in which the PICASSO collaboration has worked hard at incorporating into the next phase, coined PICASSO II, which will have an active mass of 20kg.

PICASSO II has incorporated each of the aforementioned improvements, with installation to be completed by June 2006. After seven months of running, PICASSO II should have the sensitivity to probe the DAMA region. In 2000, the DAMA collaboration published[10] results of a positive signal, but in an area that has been excluded by spin-independent experiments (See Reference [11] for incompatible results from CDMS). The area has not yet been excluded by spin-dependent experiments.

Université de Montréal and Queen's University have both installed large cleanrooms to ensure cleanliness while handling detector ingredients and components. UdeM has also installed a large scale purification system and

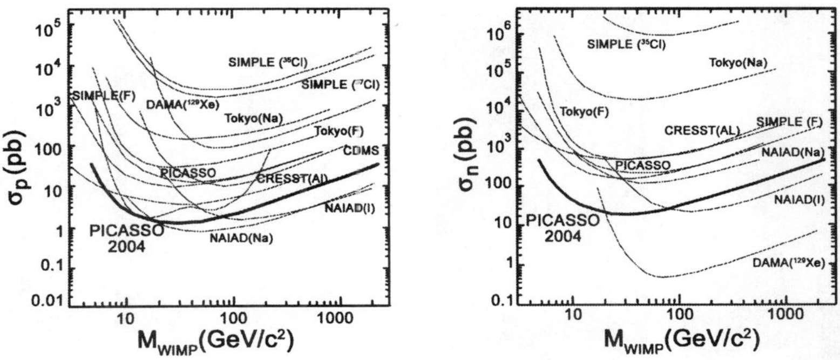

(a) WIMP-proton cross-section limits

(b) WIMP-neutron cross-section limits

Figure 5. Limits on the spin-dependent (a) WIMP-proton and (b) WIMP-neutron cross-sections, at a 90% C.L. as a function of WIMP mass, current at the time of publication (2005). Lines labeled as PICASSO and PICASSO 2004 refer to earlier published limits (2000) and those published in 2004. Plots were adopted from ref. [9], [10], and references therein.

together the two groups have come up with a better method of purifying the gel ingredients.

The individual detectors are now 4.5L in size, with each having 9 higher quality piezoelectric sensors mounted on the outside, allowing for improved data acquisition and for localization of the signals. The loading of the detectors has increased the active mass from 0.5% to 2.0%, and a new loading technique has increased bubbles sizes from 10-100μm up to 200μm. Due to their short range, the sensitivity becomes significantly reduced to alphas.

Long term stability is an important factor in the design and fabrication of the detectors. The detectors need to be recompressed periodically to convert the bubbles back to droplets, and to heal the gel which would be damaged by the presence of too many bubbles. Previously this was done pneumatically with N_2 gas. However, it became apparent over time that the gas was diffusing into the gel matrix, causing spurious background signals. To eliminate this effect, the detectors are now pressurized with a hydraulic system.

The temperature of the detectors from the published phase was con-

Figure 6. Photos of the 1L detectors installed underground in 2004 (left), the new detectors being installed in 2006 for PICASSO II (center), and the TPCS the new 4.5L detectors installed (right). Also visible in the center of the TPCS is a remote controlled calibration source.

trolled within 1.5°C by 2 Peltier junctions. In order to achieve the desired results, new temperature and pressure control systems (TPCS) were designed as depicted in Figure 6. The TPCS boxes consist of an aluminum box with resistors across the top and bottom plates for heating, and an internal fan to circulate the air to ensure an almost negligible temperature gradient within the boxes. Results from testing have temperature control accuracy within ±0.05°C. To monitor the temperatures, there is an array of sensors installed throughout the TPCS and on the detectors that are accurate to ±1/16°C.

The increase from 2 to 9 piezoelectric sensors will allow for an accurate localization algorithm. Localization has many benefits: identification of 'hot spots', determination of a fiducial volume, discrimination against signals that reconstruct outside of the detector. Several techniques are currently being studied, with forthcoming results as to the ideal method.

One localization approach is to use the correlation between 2 signals to determine the time difference between when each sensor 'hears' an event. The correlation approach has produced promising results, and it is expected that this will improve the analysis of the data collected from PICASSO II.

The expected results for PICASSO II are indicated in Figure 7. These limits have been generated with the accepted values of a local WIMP matter density of 0.3 GeV/c^2/cm^3, neutralino velocity dispersion of 230 km/s, an Earth-Halo relative velocity of 244 km/s, and an escape velocity of

600 km/s. They incorporate the increased active mass (2 kg), reduced backgrounds and larger bubble sizes. PICASSO II is projected to have a total exposure of 336 kgd, and the results should reach the lower limits on the cross-section that DAMA has placed.

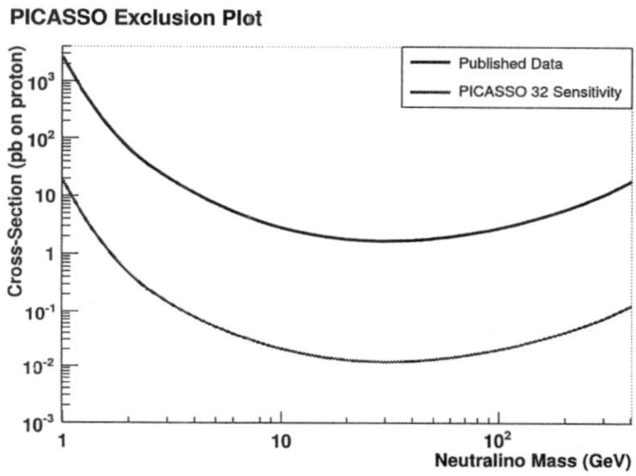

Figure 7. Current and projected limits on the WIMP-proton spin-dependent cross-section as a function of WIMP mass. PICASSO II has greatest sensitivity to a 29 GeV neutralino, with a 1.3pb

PICASSO II is only the second of five phases. Over the next few years, the PICASSO collaboration will be striving for lower backgrounds and increasing active mass to reach a final size of a 100 kg detector. This detector should be able to probe into the core of MSSM predictions, which would make PICASSO the world leader in sensitivity to spin-dependent dark matter interactions.

References

1. M. Barnabe-Heider, et al. *Phys Lett* **B624**, 186-194 (2005).
2. F. Zwicky, *Helv. Phys. Acta* **6**, 110 (1933).
3. G. Bertone, D. Hooper and J. Silk, *Phys Rept* **405**, 279-390 (2005).
4. D.N. Spergel, R. Bean, O. Doré, et al. *Wilkinson Microwave Anisotropy Probe (WMAP) ThreeYear Results: Implications for Cosmology* (Greenbelt, MD: NASA/GSFC) (2006).
5. F. Seitz, *Phys. Fluids* **1**, 2 (1958).
6. R. Apfel,*Nucl. Inst. and Meth.* **162**, 603 (1979).
7. SNO collaboration, private communication.
8. N. Boukhira et al., PICASSO Collaboration, *Astropart. Phys.* **14**, 227-237 (2000).
9. F. Guiliani and T.A. Girard, *Phys. Lett* **B588**, 151 (2004).

10. R. Bernabei et al., *Phys. Lett* **B480**, 23 (2000).
11. D.S. Akerib, J. Alvaro-Dean, M.S. Armel, et al., *Phys. Rev.* **D68**, 082002 (2003).

LHCB PARTICLE IDENTIFICATION AND PERFORMANCE.

P. SZCZYPKA*
Universtity of Bristol,
H.H. Wills Physics Laboratory,
Tyndall Avenue,
Bristol, BS8 1TL,
United Kingdom
E-mail: paul.szczypka@bristol.ac.uk

The LHCb particle identification system is presented. The system consists of two Ring-Imaging Cherenkov detectors which provide pion/kaon separation over the momentum range 1 to 100 GeV/c, the Calorimeter system, with which it is possible to identify neutral particles, and the Muon Detector.

1. Introduction

1.1. *The LHCb Experiment*

The Large Hadron Collider Beauty[1] (LHCb) experiment (Fig. 1) is a forward one-arm spectrometer dedicated to the study of CP violation and other rare phenomena in the decay of hadrons containing b-quarks at the Large Hadron Collider (LHC). The physics goals are to check the consistency of the Standard Model through precision measurements of the sides and angles of the Cabibbo-Kobayashi-Maskawa (CKM) triangle, and to search for new physics in decays that are rare, or forbidden, in the Standard Model.

Many of the decay modes of interest for CP-violation studies at LHCb have topologically similar backgrounds. It is therefore necessary to efficiently identify particles in order to maximize signal to background ratios. For the LHVb trigger, the Level 0 decision depends strongly on the identification of high p_T muons, electrons, photons and hadrons. Particle Iden-

*On behalf of the LHCb Collaboration.

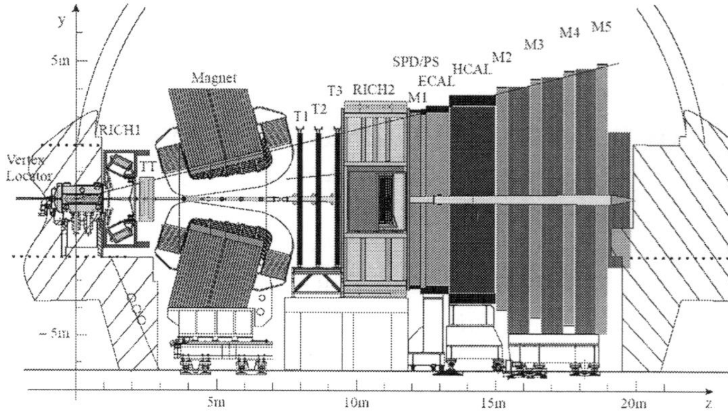

Figure 1. A schematic of the LHCb detector.

tification (PID) is also necessary in order to identify interesting events in the LHCb High Level Trigger (HLT). PID in the LHCb environment is achieved using data from a number of subdetector systems, namely the two Ring Imaging Cherenkov (RICH) detectors, the Calorimeter and the Muon systems.

2. LHCb Subdetectors

2.1. *The RICH Detectors*

Both LHCb RICH detectors[2] follow the same basic design. Cherenkov radiation is emitted by charged particles passing through the radiating material in the detector. The radiation is focused onto planes of photon detectors by sets of spherical mirrors which are tilted in order to allow positioning of the photon detectors outside the spectrometer acceptance. In addition to the spherical mirrors, secondary sets of plane mirrors are used to minimize the overall detector length. Both detectors use Hybrid Photon Detectors (HPDs) which are sensitive to single photons in the wavelength range 200nm $< \lambda <$ 600nm.

RICH 1 (Fig. 2) has an angular acceptance of 20 - 300 mrad and uses a combination of silicon aerogel ($n = 1.03$) and C_4F_{10} gas ($n = 1.0014$) radiators to identify low to intermediate momentum particles (up to $\sim 70GeV/c$). Higher momentum particles (up to $\sim 100GeV/c$) are detected by RICH 2 which has an acceptance of 15 - 120 mrad and uses a CF_4 gas

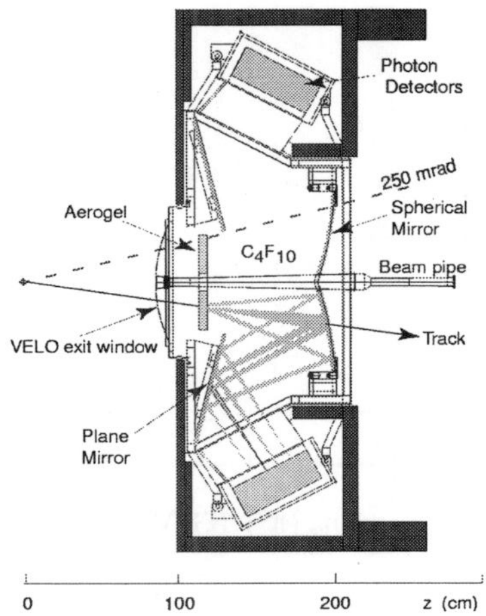

Figure 2. Schematic diagram of the RICH1 detector.

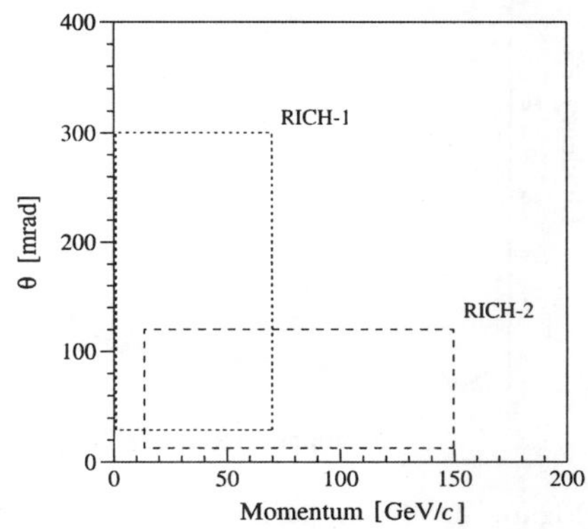

Figure 3. Track angle vs. momentum for all tracks from $B_0 \to \pi\pi$ events, showing the regions covered by the RICH detectors.

radiator ($n = 1.0005$). Together, the RICH detectors cover the phase space of interesting events almost completely (Fig. 3).

Particle identification using the RICH system is performed as follows. For each event, a set of mass hypotheses for each reconstructed track is calculated. Using these data, the probability distribution for finding photons in each pixel of the detector is determined and then compared to the observed hit distribution. A likelihood is determined from this comparison and then the track mass hypotheses are varied in order to maximise the likelihood. Efficient $\pi - K$ separation is achieved using this method, giving an average kaon identification efficiency of $\sim 88\%$ and an average pion misidentification probability of $\sim 3\%$ (Fig. 4).

2.2. *The Calorimeter system*

The LHCb Calorimeter System is composed of the Scintillator Pad/Preshower (SPD/PS), the Electromagnetic Calorimeter (ECAL) and the Hadronic Calorimeter (HCAL). The SPD/PS system is constructed from a lead converter plate sandwiched between two layers of scintillator pads. The ECAL is based on "Shashlik" technology and uses a sampling structure of 2 mm lead sheets interspersed with 4 mm thick scintillator

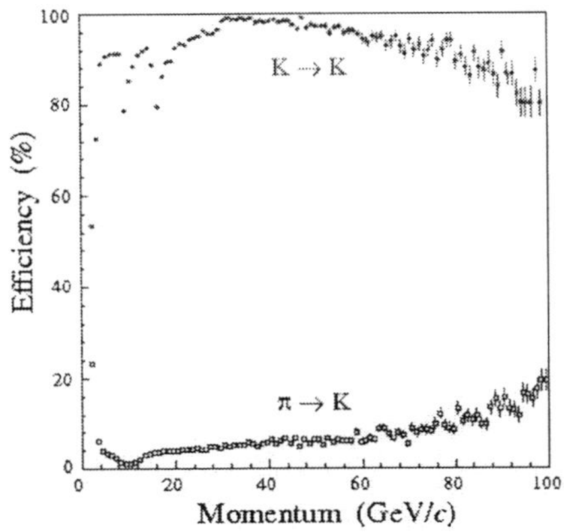

Figure 4. Kaon identification efficiency (top) and pion misidentification probability (bottom) using the RICH detectors.

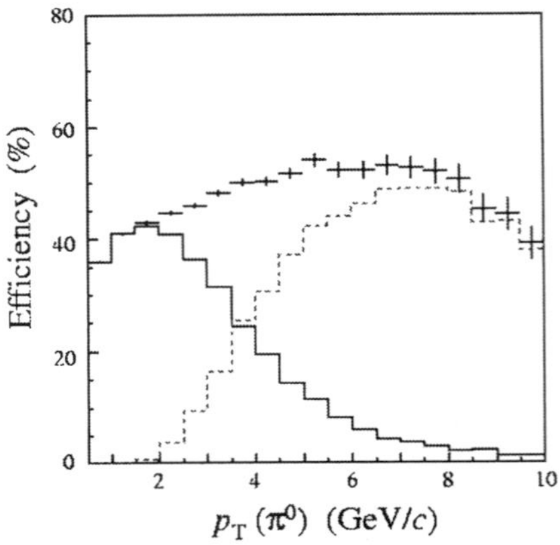

Figure 5. π^0 identification efficiency for both resolved pions (solid line), merged pions (dashed line) and the combination (error bars) using the calorimeter systems.

plates. The HCAL uses steel and scintillating tiles as absorber and active material respectively.

Information from the SPD/PS and ECAL can be used to reconstruct π^0s from $\pi^0 \to \gamma\gamma$ decays. The reconstructed photons may be separated into two catagories; resolved (where the two photon candidates are distinct) and merged (where the two photon candidates overlap forming a "merged" π^0 cluster). In the merged case, an algorithm disentangles the photon pair using directional information from the merged cluster and in the resolved case, photon candidates are paired to reconstruct the resolved π^0 mass. Using this method, overall π^0 identification can be achieved with ~ 53 % efficiency (Fig. 5). In $B^0 \to \pi^+\pi^-\pi^0$ decays, π^0 identification produces a B^0 mass resolution of $\sim 60MeV/c^2$.

2.3. *The Muon system*

The purpose of the LHCb Muon System is to provide fast triggering at Level-0 and for offline muon identification. Muon identification is important because muons are present in the final states of many CP-sensitive B decays, in particular the two "gold-plated" decays $B^0 \to J/\Psi(\mu\mu)K$ and $B_s^0 \to J/\Psi(\mu\mu)\phi$. The muon system consists of five stations (M1-M5) of rectangular construction, covering an acceptance of ± 300 mrad (horizontally) and ± 200 mrad (vertically).

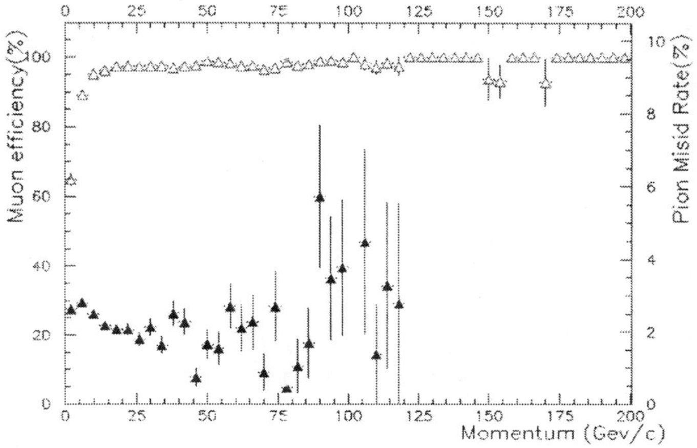

Figure 6. Muon identification efficiency (left axis) and pion misidentification probability (right axis).

A muon is selected by searching for muon station hits which are compatable with a reconstructed track extrapolation from the tracking stations. In order to reduce the particle misidentification rate, information from other subdetectors may be used e.g. π identification using the RICH detectors. A muon identification efficiency of 97.6 % ± 0.2 % with a pion misidentification probability of 2.50 % ± 0.04 % is obtained (Fig. 6). We expect to achieve a B^0 mass resolution of $\sim 9MeV/c^2$ in $B^0 \rightarrow J/\Psi(\mu\mu)K^0_S$ decays.

3. Conclusions

The particle identification provided by the RICH, Calorimeter and Muon systems is crucial to the LHCb physics programme. The LHCb RICH system will provide kaon/pion separation over the momentum range 1 to 100GeV/c, covering much of the required phase space. The Calorimeter system will achieve π^0 identification with approximately $\sim$ 53% efficiency. Excellent muon identification ($\sim$ 98% efficiency) will be also be achieved.

References

1. LHCb Collaboration. LHCb TDR 9 Reoptimised Detector-Design and Performance, September 2003. CERN-LHCC-2003-030.
2. LHCb Collaboration. LHCb TDR 3 RICH, September 2000. CERN-LHCC-2000-037.

B_c AT CDF*

W. WESTER†

Fermialb
MS 222
Batavia, IL 60510, USA
E-mail: wester@fnal.gov

We report CDF results on the B_c^- meson [1] in Run II. The B_c^- meson has been observed in semileptonic decays, $B_c^- \to J/\psi\ \ell^- \nu X$, where $\ell = e, \mu$ at a significance greater than 5σ in both channels. The $B_c^- \to J/\psi\ \ell^- \nu X$ observations have resulted in measurements of the relative production times branching ratio with respect to $B^- \to J/\psi\ K^-$ decays and a precise determination of the lifetime of the B_c^-: $\tau(B_c^-) = 0.474\ ^{+0.073}_{-0.066}$ (stat.) $\pm$ 0.033 (syst.) ps. Also, an observation of $B_c^- \to J/\psi\ \pi^-$ decays at a significance exceeding 6σ results in a precise determination of the mass of the B_c^-: $M(B_c^-) = 6275.2 \pm 4.3$ (stat.) $\pm$ 2.3 (syst.) MeV/c^2.

1. INTRODUCTION

1.1. *B Physics at Hadron Colliders*

The study of the B_c^- meson relies on the large production cross section at hadron colliders, triggerable low background decay modes, and the powerful capabilities of the modern multipurpose collider detectors.

1.2. *Tevatron in Run II*

The Tevatron at Fermilab is currently operating the Run II physics program where protons and anti-protons collide at energy $\sqrt{(s)} = 1.96$ TeV with over 1.5 fb^{-1} of integrated luminosity recorded. The CDF detector is instrumented with an inner silicon tracking system including Layer00 [2], which is mounted directly onto the beam pipe.

---

*This work is supported by the United States Department of Energy
†On behalf of the CDF Collaboration

1.3. B_c^- *properties*

The B_c^- meson is a special B meson composed of two distinct heavy quarks with different flavors. The presence of both such quarks impacts the production [3], decay [4], and mass properties [5,6] of the B_c^-. All of these theoretical ideas require testing through experimental measurements.

1.4. B_c^- *in Run I*

The observation by CDF in Run I [7] of a significant number of semileptonic candidates was hailed as the discovery of the last meson. The observation of 20.4 $^{+6.2}_{-5.5}$ signal events suggested that the study of B_c^- decays would be a fruitful enterprise in Run II.

2. Semileptonic B_c^- Decays

$B_c^- \to J/\psi\ \ell^- \nu X$ decays with $\ell = e$ or μ are not fully reconstructed due to the missing neutrino and possible missing particles, X. However, a $B_c^- \to J/\psi\ \ell^- \nu X$ signal can be identified over background and measurements of some of the B_c^- properties can be made. Understanding the background is a key component of the $B_c^- \to J/\psi\ \ell^- \nu X$ analyses with expected background contributions arising from $b\bar{b}$ events where one of the B mesons decays into a J/ψ and the other anti-B meson decays semileptonically. Backgrounds also include electrons and muon candidates that are "fake" in the sense that the lepton candidate comes from a hadronic track that happens to pass the lepton selection criteria. Fortunately, the study of the backgrounds can be performed both with Monte Carlo simulation and with the data itself. In particular, the larger J/ψ + *track* sample and the reference $B^- \to J/\psi\ K^-$ decay sample provide a means to help determine residual background.

3. $B_c^- \to J/\psi\ \ell^- \nu X$ in CDF

The CDF experiment has results of semileptonic B_c^- decays in both the final state μ [8] and e [9] channels In the $B_c^- \to J/\psi\ \mu^- X$ channel, CDF uses 0.36 fb^{-1} of integrated luminosity in which over 2.7 M J/ψ decays into dimuons are identified. Both the general J/ψ + *track* and $B^- \to J/\psi\ K^-$ samples are used to understand the sample composition. Fake muon backgrounds for B_c^- are estimated from the number of expected J/ψ + *track* combinations that have an invariant mass, M($J/\psi\mu$), in the 4-6 GeV/c^2 signal region where the track is mis-identified as a muon. The π, K, and p

composition as a function P_T is studied using dE/dx and time-of-flight particle identification capabilities of CDF. The fake rate can then be extracted using large samples of fully reconstructed $K_s^0 \to \pi^+ \pi^-$, $D^0 \to K^- \pi^+$ and $\Lambda^0 \to p\,\pi^-$ decays. The $b\bar{b}$ background studies use Monte Carlo simulation normalized to $B^- \to J/\psi\, K^-$ decays. Finally, the sidebands of the J/ψ are used to estimate the contribution arisings from fake-J/ψ events. Of the 106 events in the signal region, the three backgrounds are estimated to contribute approximately 16 (fake μ), 13 ($b\bar{b}$), and 19 (fake J/ψ) events resulting in a 5.3σ signal consisting of 60.0 ± 12.6 B_c^- candidates. A measurement is made of R, the production times branching ratio relative to $B^- \to J/\psi\, K^-$ decays with $P_T(B) > 4$ GeV/c and $|y| < 1$:
$R = 0.249 \pm 0.045$ (stat.) $\pm$ 0.069 (syst.) $^{+0.082}_{-0.033}$ (lifetime).

CDF has also studied semileptonic decays $B_c^- \to J/\psi\, e^- X$ where backgrounds are further complicated by the presence of conversion photons. The conversion background as a function of $\mathrm{M}(J/\psi e)$ is estimated by studing a sample of conversion photons and understanding the efficiency for identifying the electron track as a function of P_T. The conversion background contributes approximately 15 of the total 64 event background estimate. The remaining signal excess has a 5.9σ significance and contains $114.9 \pm 15.5 \pm 13.6$ B_c^- candidates. In this channel, a measurement of R is made for $P_T(B) > 4$ GeV/c and $|y| < 1$:
$R = 0.282 \pm 0.038$ (stat.) $\pm$ 0.035 (yield) $\pm$ 0.065 (acceptance).

The $B_c^- \to J/\psi\, e^- X$ sample is also used to measure the lifetime of the B_c^- meson. Figure 1 shows the pseudo-proper decay length distribution with superimposed signal and background contributions. With relaxed requirements on the pseudo-proper decay length compared with the R analysis, the fit finds a total of 238 signal events over 545 ± 55 background events in the $\mathrm{M}(J/\psi\mathrm{e})$ range between 4 and 6 GeV/c^2. From the fit, the B_c^- lifetime is measured to be: $c\tau(B_c^-) = 0.474\ ^{+0.073}_{-0.066}$ (stat.) $\pm$ 0.033 (syst.) ps.

4. $B_c^- \to J/\psi\, \pi^-$ in CDF

CDF reported initial evidence of $B_c^- \to J/\psi\, \pi^-$ decays with 0.36 fb^{-1} [10]. We now describe an independent analysis [11] using the full 0.8 fb^{-1} of collected and processed data available for analysis at the end of 2005. This search for $B_c^- \to J/\psi\, \pi^-$ uses a strategy that studies the effects of various selection criteria on the reference $B^- \to J/\psi\, K^-$ decay and the sideband background events below 5.5 GeV/c^2. The selection requirements include requiring the K candidate track to have an impact parameter that is significantly displaced from the primary vertex while pointing to a displaced J/ψ secondary vertex defined by the two muons. Only after the selection was approved internally by CDF was the K hypothesis changed to a π and the region of interest in M($J/\psi\pi$) examined. A small excess is observed in this analysis with 0.36 fb^{-1} of data and has become more signifiant with the full 0.8 fb^{-1} where a fit to a Gaussian signal and linear background gives 38.9 B_c^- signal events and 26.1 background events in the mass range between 6.24 and 6.30 GeV/c^2 - an observation with a significance greater than 6σ based upon simulations that include random fluctuations over a wide search window. Figure 2 shows the invariant mass distribution of $J/\psi\pi$ candidates in the range between 5.6 and 7.2 GeV/c^2. An unbinned likelihood fit is used to measure the mass of the B_c^-: M(B_c^-) = 6275.2 (stat.) $\pm$ 4.3 $\pm$ 2.3 (syst.) MeV/c^2.

5. Summary and Conclusions

Measurements of the B_c^- meson are happening at CDF and DØ using Run II data. The B_c^- is an unique system whose properties are currently best studied at the Tevatron collider where relatively large production rates, triggerable decays modes, and powerful detectors combine to provide the current best measurements of this meson. In particular, new precise measurements of the lifetime and mass of the B_c^- are challenging theoretical predictions.

References

1. Charge conjugate modes are implied.
2. C. S. Hill *et al.*, Nucl. Instrum. Methods **A530**, 1 (2004).
3. C. H. Chang and X. G. Wu, Eur. Phys. J. C. **38** (2004) 267.
4. V. V. Kislev, Phys. Atom. Nucl. **67**, 1559 (2004).
5. S. Godfrey, Phys. Rev. **D70**, 054017 (2004); E. J. Eichten and C. Quigg, Phys. Rev. **D49**, 5845 (1994); W. K. Kwong and J. L. Rosner, Phys Rev **D**44, 212 (1991).
6. I. F. Allison *et al.*, Phys. Rev. Lett. **94**, 172001 (2005).
7. F. Abe *et al.*, Phys. Rev. Lett. **81**, 2432 (1998) and F. Abe *et al.*, Phys. Rev. **D58**, 112004 (1998).
8. CDF Collaboration, CDF note 7649 (public).
9. CDF Collaboration, CDF note 7926 (public).
10. CDF Collaboration, hep-ex/0505076.
11. CDF Collaboration, CDF note 8004 (public).

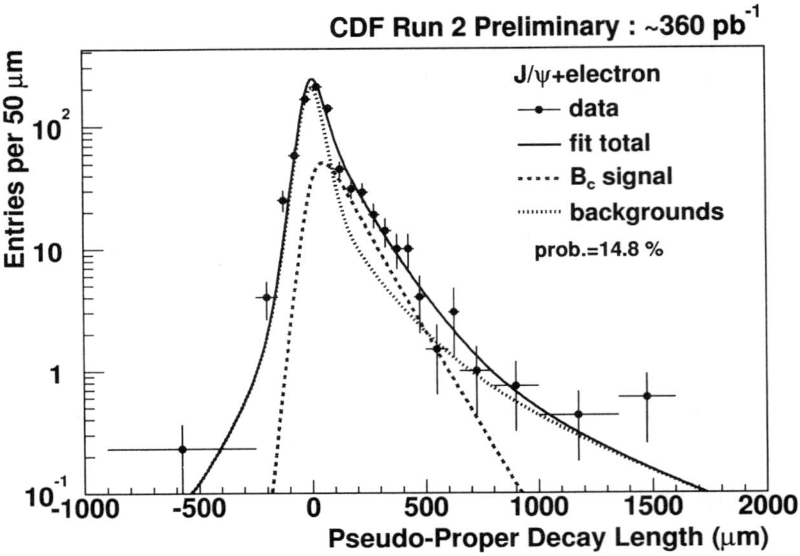

Figure 1. The pseudo proper lifetime distribution of $B_c^- \to J/\psi\ e^- X$ candidates in the CDF Run II data showing the contributions of background and B_c^- signal.

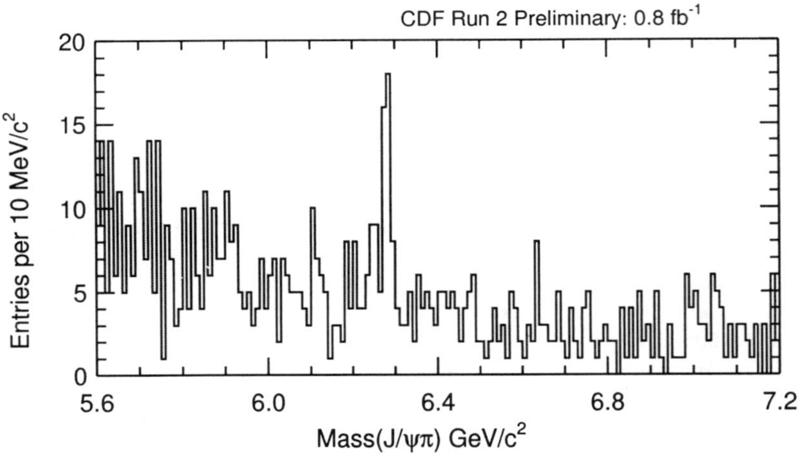

Figure 2. Invariant mass distribution of $J/\psi\ \pi^-$ in 0.8 fb^{-1} of CDF data.

TRANSVERSITY MEASUREMENTS AT HERMES

B. ZIHLMANN

(ON BEHALF OF THE HERMES COLLABORATION)

University of Gent
Dept. Subatomic Physics
Proeftuinstraat 86
B-9000 Gent Belgium
E-mail: zihlmann@hermes.desy.de

Azimuthal single spin asymmetries (SSA) in semi inclusive deep inelastic lepton scattering (SIDIS) provide a tool to access transversity, the distribution of transversely polarized quarks in a transversely polarized nucleon. Using a transversely polarized hydrogen target SSA are measured for positively and negatively charged pions. For both particle types the extracted Collins moments are significantly different from zero in the kinematic region covered by the HERMES detector.

1. Introduction

The deep inelastic scattering (DIS) cross section is proportional to the product of the leptonic $L^{\mu\nu}$ and hadronic $W^{\mu\nu}$ tensors. While $L^{\mu\nu}$ can be calculated in quantum electrodynamics, the hadronic Tensor $W^{\mu\nu}$, which is represented by the handbag diagram (see fig.1), can be separated into a hard and a soft part. The hard part, describing the absorption of the virtual photon by the quark, can be calculated. The soft part, which cannot be calculated from first principles, contains the structure of the nucleon. One

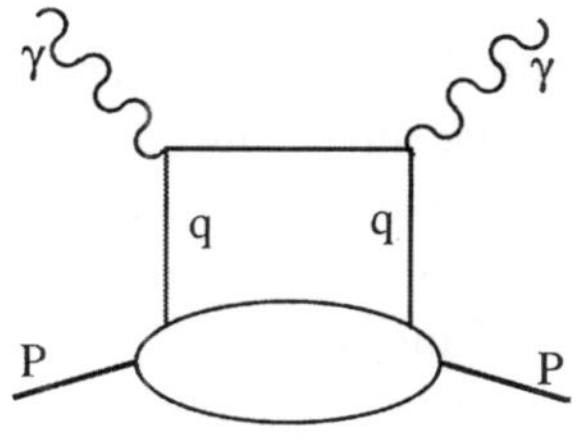

Figure 1. Handbag diagram for inclusive DIS.

approach to quantify the soft part is to describe it in terms of helicity amplitudes. These amplitudes are defined with respect to the helicity of the nucleon and the quark that absorbs the virtual photon. Since the total helicity of the nucleon-quark system and parity have to be conserved only three of the possible 16 amplitudes are independent and non zero [1]. These are

$$A_{++,++};\ A_{+-,+-};\ A_{+-,-+}\ ,$$

of which the first two conserve the individual helicities while the last flips the helicity of both the quark and the nucleon. These amplitudes or combinations thereof can be related to the quark distribution functions in the nucleon. The sum of the first two helicity conserving amplitudes is related to the quark longitudinal momentum distribution $q(x)$. The difference of these two amplitudes is related to the quark longitudinal helicity distribution, traditionally denoted as $\Delta q(x)$. It represents the difference of the number density of quarks polarized longitudinally and parallel to the spin of the parent nucleon minus the number density of quarks polarized opposite to the longitudinal polarization vector of the parent nucleon.

The helicity flip amplitude is related to transversity $h_1(x)$. Switching from a helicity basis to a transverse polarization basis the helicity flip amplitude can be rewritten as the difference of two amplitudes in this new basis where the quark and nucleon polarization vectors are either parallel or anti-parallel and the initial polarization vectors are conserved:

$$A_{+-,-+} \sim A_{\uparrow\uparrow,\downarrow\downarrow} - A_{\uparrow\downarrow,\uparrow\downarrow}.$$

This relates then to the difference of distributions of transversely polarized quarks in a transversely polarized nucleon where the quarks are polarized either parallel or anti-parallel to the spin of the parent nucleon.

Since the DIS process conserves helicity the helicity flip amplitude cannot be accessed by inclusive DIS measurements. An additional process is needed to flip the helicity of the struck quark. The Collins fragmentation function is such a chiral-odd function that describes the fragmentation of a transversely polarized struck quark into a hadron. Thus, semi inclusive deep inelastic scattering processes need to be studied with a transversely polarized hydrogen target. In such experiments one measures the convolution of a quark distribution function (DF) with a fragmentation function (FF), where the DF will be the transversity distribution function and the FF will be the Collins function.

2. SSA and Transversity

Figure 2 illustrates the definition of the azimuthal angles ϕ and ϕ_S, and the kinematics of detecting the scattered lepton and a hadron in the final state. Asymmetries $A_{UT}^{\pi^\pm}(\phi, \phi_S)$ are calculated as a function of ϕ, the angle between the hadron production plane and the lepton scattering plane, and ϕ_S, the angle between the target polarization vector and the lepton scattering plane. The subscript UT denotes unpolarized beam and transversely polarized target while $\pi^\pm$ refers to either positively or negatively charged pions detected in the final state together with the scattered lepton. Charged pions are identified by means of a dual-radiator Ring-imaging Čerenkov detector. Separation of hadrons and leptons is achieved with a transition radiation detector, a pre-shower detector and an electro-magnetic calorimeter. The $A_{UT}^{\pi^\pm}(\phi, \phi_S)$ asymmetries are differences between the yields of SIDIS events with opposite target spin states where the scattered lepton and a pion are detected in the final state. The transverse target polarization is being flipped every 90 seconds by 180°. The characteristic signature for the

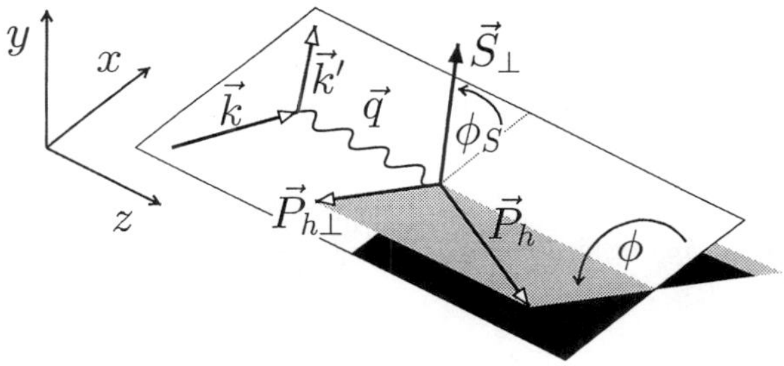

Figure 2. Kinematics of semi inclusive DIS on a transversely polarized target

Collins mechanism is a $\sin(\phi+\phi_S)$ modulation of the asymmetries [2]. Other physics processes can also generate azimuthal asymmetries. One example is the Sivers mechanism which can cause a $\sin(\phi - \phi_S)$ modulation. It is proportional to the convolution of a quark transverse momentum distribution function (Sivers function) and the known unpolarized fragmentation function. The Sivers function describes the correlation of the transverse momentum of quarks with the transverse polarization of the nucleon [3]. This process can cause a non-zero modulation in the asymmetry due to final state interactions of the struck quark with the rest of the nucleon. This effect does not vanish by choosing an appropriate gauge link. In Drell-Yan processes the Sivers mechanism is related to initial state interaction of the two scattering quarks. As a consequence the Sivers function has the opposite sign in Drell-Yan as compared to SIDIS. The asymmetry calculated from the SIDIS yields is related to the Sivers and Collins mechanism as

$$\begin{aligned} A_{UT}(\phi,\phi_S) &= \frac{1}{S_\perp}\frac{N^\uparrow(\phi,\phi_S)-N^\downarrow(\phi,\phi_S)}{N^\uparrow(\phi,\phi_S)+N^\downarrow(\phi,\phi_S)} \\ &\propto \sin(\phi+\phi_S)\frac{\sum_q e_q^2\cdot I[h_1^q(x,p_T^2)\cdot H_1^{\perp q}(z,(zk_T)^2)]}{\sum_q e_q^2\cdot q(x)\cdot D_1^q(z)} \\ &\quad +\sin(\phi-\phi_S)\frac{\sum_q e_q^2\cdot I[f_{1T}^{\perp q}(x,p_T^2)\cdot D_1^q(z,(zk_T)^2))]}{\sum_q e_q^2\cdot q(x)\cdot D_1^q(z)} \\ &\quad +\dots \quad . \end{aligned} \tag{1}$$

Here, $S_\perp$ is the transverse component of the hydrogen target spin. $H_1^{\perp q}$ is the Collins FF, h_1^q the transversity distribution, f_{1T}^q the Sivers DF and D_1^q the unpolarized FF for a given quark flavor q. Since the intrinsic momenta p_T and k_T of the quark before and after the hard scattering process do appear explicitly in the convolution integral (indicated by $I[...]$) and not just in the FF and DF themselves, it is necessary to assume some distribution of these momenta in order to factorize the FF and DF. Here we assume a Gaussian ansatz for the distribution of the transverse momenta reducing the asymmetry to

$$\begin{aligned} A_{UT}(\phi,\phi_S) \propto \sin(\phi+\phi_S)&\frac{\sum_q e_q^2\cdot h_1^q(x)\cdot H_1^{(1/2)\perp q}(z)}{\sum_q e_q^2\cdot q(x)D_1^q(z)} \\ +\sin(\phi-\phi_S)&\frac{\sum_q e_q^2\cdot f_{1T}^{(1/2)\perp q}(x)\cdot D_1^q(z))}{\sum_q e_q^2\cdot q(x)D_1^q(z)} \\ +\dots& \quad . \end{aligned} \tag{2}$$

where the superscript (1/2) indicates that the convolution integral contains the factor $\frac{|p_T|}{2M}$ or $\frac{|k_T|}{2M}$ for the DF and FF, respectively. Hence, in order to extract the Sivers and Collins contributions, the asymmetry has to be fitted with a function that contains the amplitudes of the $\sin(\phi+\phi_S)$ and $\sin(\phi-\phi_S)$ modulations as free parameters[4]. The $\sin(\phi+\phi_S)$ and $\sin(\phi-\phi_S)$ amplitudes are extracted from the asymmetries simultaneously by fitting the calculated asymmetries $A_{UT}(\phi,\phi_S)$ on a grid of 12 by 12 bins in ϕ and ϕ_S.

The resulting amplitudes for the Collins and Sivers moments for the π^+ and π^- mesons in the final state are shown in figure 3 and 4 as a function of the Bjorken scaling variable x, the fractional energy of the hadron z and the transverse momentum of the hadron $P_{h\perp}$. A significant positive signal

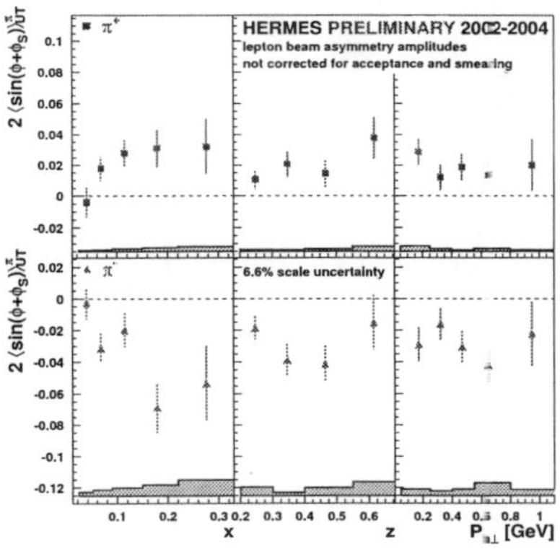

Figure 3. Amplitudes for the Collins mechanism on the π^+ and π^-

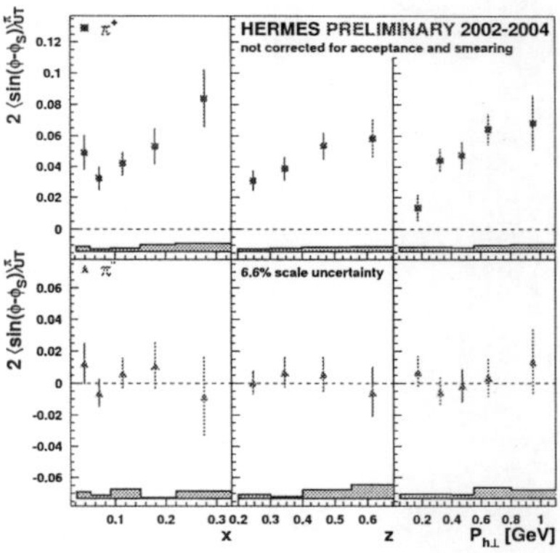

Figure 4. Amplitude for the Sivers mechanism on the π^+ and π^-

is observed for the π^+ Collins azimuthal moment, and a negative signal of similar magnitude for the π^- moment. Assuming u-quark dominance the large negative signal for the π^- moment on the hydrogen target is rather surprising, suggesting a substantial disfavored Collins function with opposite sign to that of the favored function.

A large positive signal is observed for the π^+ Sivers moment. This is the first observation of transverse momenta of quarks in the nucleon, and therefore provides the first evidence in leptoproduction for the T-odd Sivers parton distribution. The results for π^- are consistent with zero.

References

1. V. Barone, and P. G. Ratcliffe, *Transverse Spin Physics*, World Scinetific, ISBN 981-238-101-5.
2. J. C. Collins, *Nucl. Phys. B,* **396**, 161 (1993)
3. D. W. Sivers, *Phys. Rev. D,* **41**, 83 (1990)
4. A. Airapetian et al., *Phys. Rev. Lett.* **94**, 012002 (2005)

LIST OF PARTICIPANTS

Terry Awes	Oak Ridge National Laboratory
Birger Back	Argonne National Laboratory
Stephanie Beauceron	DAPNIA/SPP, CEA Saclay
Matthew Beckingham	DESY, Hamburg
Selemon Bekele	The University of Kansas
Jaroslav Bielcik	Yale University/BNL
Cristina Biino	INFN Torino
David Boersma	University of Wisconsin
Kieran Boyle	Stony Brook University
Marko Bracko	University of Maribor
Pawel Bruckman de Renstrom	University of Oxford and H. Niewodniczanski Institute PAN
Topher Cawlfield	University of Illinois
Ming-Chuan Chang	Tohoku University, Japan
Benoit Clement	Institut de Recherches Subatomique Strasbourg, France
Andrzej Czarnecki	University of Alberta
Georgios Daskalakis	Imperial College
Jerome de Favereau	Universite Catholique de Louvain
James Degenhardt	University of Michigan
Pierre-Antoine Delsart	Universite de Montreal
Volker Drollinger	University of Padova
Ivan Fedorko	Fermilab
Henry Frisch	Enrico Fermi Institute
Bostjon Golob	University of Ljubljana
Ulrich Harbach	J.W. Goethe Universitat Frankfurt
John Harris	Yale University
Carsten Hast	Stanford Linear Accelerator Center
Nicholas Hastings	The University of Tokyo
Matthias Hoek	University of Giessen
Akimasa Ishikawa	The University of Tokyo
Rod Johnson	Naval Postgraduate School
Benjamin Kahle	DESY/ZEUS
Hyejoo Kang	Stanford University
David Kaplan	The Johns Hopkins University
Mahmud Khan	University of Michigan
Faqir Khanna	University of Alberta
H.-Juergen Kluge	GSI, Darmstadt, Germany

Benjamin Koch	ICTP Frankfurt
Tomas Kopf	Silesian University in Opava
Andrey Korytov	University of Florida
Dan Krop	Indiana University
Paul Kutter	University of Wisconsin at Madison
Sabine Lammers	Columbia University
Olivier Leroy	CPPM-IN2P3-CNRS, Marseille, FR
Loren Linden Levy	University of Illinois
Jenny List	DESY-FLC
Michael Melich	Naval Postgraduate School
Roger Moore	University of Alberta
Thomas Moore	University of Massachusetts, Amherst
Ben Morgan	University of Warwick
Kunihiro Nagano	KEK/IPNS
Jacopo Nardulli	NIKEF
Matteo Negrini	Ferrara University/INFN
Nicola Neri	INFN Pisa
Carsten Noeding	University of Freiburg
Scott Oser	University of British Columbia
Arantza Oyanguren	LAL – Orsay
Alexey Pak	University of Alberta
Enrique Palencia	Instituto de Fisica de Cantabria
James Pivarski	CLEO, Cornell University
Maxim Pospelov	University of Victoria
Jean-Michel Poutissou	TRIUMF
Paul Prideaux	The University of Liverpool/DESY
Chiara Rovelli	Milano Bicocca University
Gray Rybka	MIT
Stefan Schael	RWTH Aachen
Anna Sfyria	University of Geneva
Aron Soha	University of California at Davis
Paolo Spagnolo	INFN Pisa and CERN
Connie Storey	Queen's University
Paul Szczypka	University of Bristol
Osamu Tajima	KEK
Benoit Viaud	Universite de Montreal
Kai Voss	University of Victoria
William Wester	Fermilab
Benedikt Zihlmann	University of Gent, Belgium